# Sol-Gel Technology Applied to Materials Science: Synthesis, Characterization and Applications

# Sol-Gel Technology Applied to Materials Science: Synthesis, Characterization and Applications

Editor

**Aleksej Zarkov**

Basel • Beijing • Wuhan • Barcelona • Belgrade • Novi Sad • Cluj • Manchester

*Editor*
Aleksej Zarkov
Institute of Chemistry
Vilnius University
Vilnius
Lithuania

*Editorial Office*
MDPI
St. Alban-Anlage 66
4052 Basel, Switzerland

This is a reprint of articles from the Special Issue published online in the open access journal *Materials* (ISSN 1996-1944) (available at: www.mdpi.com/journal/materials/special_issues/Sol_Gel_Materials_Science).

For citation purposes, cite each article independently as indicated on the article page online and as indicated below:

Lastname, A.A.; Lastname, B.B. Article Title. *Journal Name* **Year**, *Volume Number*, Page Range.

**ISBN 978-3-7258-0160-2 (Hbk)**
**ISBN 978-3-7258-0159-6 (PDF)**
doi.org/10.3390/books978-3-7258-0159-6

© 2024 by the authors. Articles in this book are Open Access and distributed under the Creative Commons Attribution (CC BY) license. The book as a whole is distributed by MDPI under the terms and conditions of the Creative Commons Attribution-NonCommercial-NoDerivs (CC BY-NC-ND) license.

# Contents

**About the Editor** .................................................................. vii

**Aleksej Zarkov**
Sol–Gel Technology Applied to Materials Science: Synthesis, Characterization and Applications
Reprinted from: *Materials* **2024**, *17*, 462, doi:10.3390/ma17020462 ...................... 1

**Anna Lucia Pellegrino, Emil Milan, Adolfo Speghini and Graziella Malandrino**
Fabrication of Europium-Doped $CaF_2$ Films via Sol-Gel Synthesis as Down-Shifting Layers for Solar Cell Applications
Reprinted from: *Materials* **2023**, *16*, 6889, doi:10.3390/ma16216889 ...................... 4

**Nawel Ghezali, Álvaro Díaz Verde and María José Illán Gómez**
Screening $Ba_{0.9}A_{0.1}MnO_3$ and $Ba_{0.9}A_{0.1}Mn_{0.7}Cu_{0.3}O_3$ (A = Mg, Ca, Sr, Ce, La) Sol-Gel Synthesised Perovskites as GPF Catalysts
Reprinted from: *Materials* **2023**, *16*, 6899, doi:10.3390/ma16216899 ...................... 16

**Maria Covei, Cristina Bogatu, Silvioara Gheorghita, Anca Duta, Hermine Stroescu, Madalina Nicolescu, et al.**
Influence of the Deposition Parameters on the Properties of $TiO_2$ Thin Films on Spherical Substrates
Reprinted from: *Materials* **2023**, *16*, 4899, doi:10.3390/ma16144899 ...................... 38

**Guillermo Monrós, Mario Llusar and José A. Badenes**
High NIR Reflectance and Photocatalytic Ceramic Pigments Based on M-Doped Clinobisvanite $BiVO_4$ (M = Ca, Cr) from Gels
Reprinted from: *Materials* **2023**, *16*, 3722, doi:10.3390/ma16103722 ...................... 56

**Katarzyna Wojtasik, Magdalena Zięba, Cuma Tyszkiewicz, Wojciech Pakieła, Grażyna Żak, Olgierd Jeremiasz, et al.**
Zinc Oxide Films Fabricated via Sol-Gel Method and Dip-Coating Technique–Effect of Sol Aging on Optical Properties, Morphology and Photocatalytic Activity
Reprinted from: *Materials* **2023**, *16*, 1898, doi:10.3390/ma16051898 ...................... 77

**Dovydas Karoblis, Kestutis Mazeika, Rimantas Raudonis, Aleksej Zarkov and Aivaras Kareiva**
Sol-Gel Synthesis and Characterization of Yttrium-Doped $MgFe_2O_4$ Spinel
Reprinted from: *Materials* **2022**, *15*, 7547, doi:10.3390/ma15217547 ...................... 96

**Ixchel Alejandra Mejia-Estrella, Alejandro Pérez Larios, Belkis Sulbarán-Rangel and Carlos Alberto Guzmán González**
Reaching Visible Light Photocatalysts with Pt Nanoparticles Supported in $TiO_2$-$CeO_2$
Reprinted from: *Materials* **2022**, *15*, 6784, doi:10.3390/ma15196784 ...................... 106

**Nicolas Crespo-Monteiro, Arnaud Valour, Victor Vallejo-Otero, Marie Traynar, Stéphanie Reynaud, Emilie Gamet and Yves Jourlin**
Versatile Zirconium Oxide ($ZrO_2$) Sol-Gel Development for the Micro-Structuring of Various Substrates (Nature and Shape) by Optical and Nano-Imprint Lithography
Reprinted from: *Materials* **2022**, *15*, 5596, doi:10.3390/ma15165596 ...................... 118

**Beatriz Toirac, Amaya Garcia-Casas, Miguel A. Monclús, John J. Aguilera-Correa, Jaime Esteban and Antonia Jiménez-Morales**
Influence of Addition of Antibiotics on Chemical and Surface Properties of Sol-Gel Coatings
Reprinted from: *Materials* **2022**, *15*, 4752, doi:10.3390/ma15144752 ...................... 128

Timofey M. Karnaukhov, Grigory B. Veselov, Svetlana V. Cherepanova and Aleksey A. Vedyagin
Sol-Gel Synthesis and Characterization of the Cu-Mg-O System for Chemical Looping Application
Reprinted from: *Materials* **2022**, *15*, 2021, doi:10.3390/ma15062021 . . . . . . . . . . . . . . . . . **148**

# About the Editor

**Aleksej Zarkov**

As of 2023, Prof. Aleksej Zarkov has held a Full Professor position at the Institute of Chemistry of Vilnius University. He obtained a PhD in 2016 at Vilnius University where he conducted research on the development of synthetic approaches for the preparation of oxide-based solid electrolytes. His research interests encompass the synthesis and characterization of functional inorganic materials in the form of bulk materials, nanoparticles, and thin films. Recently, A. Zarkov has focused on the synthesis and investigation of calcium phosphates for medical, optical, and environmental applications. He is a co-author of 99 articles and has participated at numerous conference announcements and presentations. During his scientific career, A. Zarkov has held long-term internships at various prestigious scientific institutions, including Osaka University (Osaka, Japan), Georgetown University (Washington D.C., USA), University of Cologne (Cologne, Germany), University of Aveiro (Aveiro, Portugal), and others. For his scientific achievements, A. Zarkov has been presented with several awards and scholarships from Vilnius University Rector and the Lithuanian Academy of Sciences.

*Editorial*

# Sol–Gel Technology Applied to Materials Science: Synthesis, Characterization and Applications

Aleksej Zarkov

Institute of Chemistry, Vilnius University, Naugarduko 24, LT-03225 Vilnius, Lithuania; aleksej.zarkov@chf.vu.lt

Citation: Zarkov, A. Sol–Gel Technology Applied to Materials Science: Synthesis, Characterization and Applications. *Materials* 2024, 17, 462. https://doi.org/10.3390/ma17020462

Received: 27 December 2023
Accepted: 17 January 2024
Published: 18 January 2024

**Copyright:** © 2024 by the author. Licensee MDPI, Basel, Switzerland. This article is an open access article distributed under the terms and conditions of the Creative Commons Attribution (CC BY) license (https://creativecommons.org/licenses/by/4.0/).

The rapid advances in technologies around the globe necessitate the development of new materials, nanostructures, and multicomponent composites with specific chemical and physical properties that can meet the requirements of modern technologies. Using appropriate synthetic approaches is crucial for the preparation of inorganic materials with designed microstructure and properties. Among the different technologies currently available, the sol–gel method is very well known for its versatility, simplicity, and time- and cost-efficiency. The mixing of starting materials on an atomic level provides high homogeneity and stoichiometry of the products, facilitating the fabrication of high-quality materials at low temperatures. The versatility of the Sol–Gel method allows for the development of materials for a wide range of applications in electronics, optoelectronics, catalysis, biomedicine, and many other areas. The scope of this Special Issue of Materials, entitled "Sol-Gel Technology Applied to Materials Science: Synthesis, Characterization and Applications", is focused on, but not limited to, the preparation, characterization, and application of functional inorganic materials, as well as hybrid materials, which are important in the field of catalysis, electronics, optics, biomedicine, etc.

Due to the uniqueness of the sol–gel approach, it can be used for the preparation of both powders and thin films. Several contributions in this Special Issue investigate powdered sol–gel-derived materials. Ghezali et al. [1] investigated compositionally complex $BaMnO_3$-based perovskite-type materials. Two series of materials with the chemical formulas $Ba_{0.9}A_{0.1}MnO_3$ (BM-A) and $Ba_{0.9}A_{0.1}Mn_{0.7}Cu_{0.3}O_3$ (BMC-A) (A = Mg, Ca, Sr, Ce, and La) were synthesized and applied for soot oxidation in simulated gasoline direct injection engine exhaust conditions. The soot conversion data reveal that $Ba_{0.9}La_{0.1}Mn_{0.7}Cu_{0.3}O_3$ (BMC-La) is the most active catalyst in an inert (100% He) reaction atmosphere, and $Ba_{0.9}Ce_{0.1}MnO_3$ (BM-Ce) is the best catalyst if a low amount of $O_2$ (1% $O_2$ in He) is present.

Karnaukhov et al. [2] used the sol–gel technique to prepare the two-component oxide system Cu–Mg–O, where MgO acts as the oxide matrix, and CuO is an active chemical looping component. The reduction behavior of the Cu–Mg–O system was examined in nine consecutive reduction/oxidation cycles. The main characteristics of the oxide system underwent noticeable changes during the first reduction/oxidation cycle. Starting from the third cycle, the system stabilized, providing the uptake of similar hydrogen amounts within the same temperature range.

The design, synthesis, and tuning of the properties of magnetic materials are among the important tasks of materials science [3]. Karoblis et al. [4] investigated the influence of processing conditions on the phase purity and the structural and magnetic properties of sol-gel-derived yttrium-doped $MgFe_2O_4$. The authors found that the iron content in the initial reaction mixture significantly contributed to the phase purity of the final products substituted with aliovalent $Y^{3+}$ ions. According to the Mössbauer spectroscopy studies, with the increase in the dopant amount in solid solutions, the amount of iron in the tetrahedral position increased. The coercivity as well as saturation magnetization decreased with the increase in the yttrium content.

Mejia-Estrella et al. [5] aimed to obtain a viable photocatalytic material able to oxidize organic pollutants under the visible light spectrum. For this purpose, the authors fabricated different Pt-TiO$_2$ and Pt-TiO$_2$-CeO$_2$ composite materials employing sol–gel and impregnation methods. The obtained materials were tested for their photocatalytic oxidation activity on a herbicide 2,4-dichlorophenoxyacetic acid, frequently used in agriculture. The activity of the materials reached a removal efficiency of 98% of the initial concentration of the pollutant in 6 h.

Monros et al. [6] applied the sol–gel technique for the preparation of Ca- and Cr-doped BiVO$_4$-based compounds. This multifunctional host material has raised considerable interest in the scientific community as a wideband semiconductor with photocatalytic activity, a material with high NIR reflectance for camouflage and cool pigments, and a photoanode for photoelectrochemical seawater splitting. Ca- or Cr-doped BiVO$_4$ pigments have been prepared for paints and glazes, with colors ranging from turquoise to black depending on the processing route.

The deposition of thin films employing the sol–gel approach allows for the fabrication of coatings on both flat and shaped surfaces. Covei et al. [7] studied the influence of the deposition parameters on the properties of TiO$_2$ thin films on spherical substrates. The complex influence of the substrate morphology, the sol dilution with ethanol, and the number of layers on the structure, morphology, chemical composition, and photocatalytic performance of TiO$_2$ thin films were investigated. As a result, photocatalytically active TiO$_2$ thin films with high surface area were deposited on spherical beads (2 mm diameter), ensuring easy recovery from wastewater.

Wojtasik et al. [8] fabricated photocatalytically active ZnO coatings on soda lime glass substrates using the sol–gel method and the dip-coating technique. Their study aimed to determine the effect of the duration of the sol aging process on the properties of the fabricated ZnO films. The photocatalytic properties of ZnO layers were studied by observing the degradation of methylene blue dye in an aqueous solution under UV illumination. The strongest photocatalytic activity was observed for the coatings produced from a sol system that was aged over 30 days. These layers also had the highest porosity (37.1%) and the largest water contact angle (68.53°).

Crespo-Monteiro et al. [9] investigated a sol–gel procedure that allows for the direct micro–nanostructuring of ZrO$_2$ layers without physical or chemical etching processes, using optical or nanoimprint lithography. The synthesis of ZrO$_2$ and the micro–nanostructuring process were presented by masking, colloidal lithography, and nanoimprint lithography on glass and plastic substrates as well as on plane and curved substrates.

The sol–gel approach also allows for the easy modification and functionalization of the final products. Toirac et al. [10] investigated biodegradable sol–gel coatings as a promising method for the controlled release of antibiotics for the local prevention of infection in joint prostheses. Sols were prepared from a mixture of MAPTMS and TMOS silanes, tris(trimethylsilyl)phosphite, and eight different individually loaded antimicrobials. The results revealed that the coatings had a microscale roughness attributed to the accumulation of antibiotics and organophosphites in the surface protrusions; no evidence was found for the existence of chemical bonds between antibiotics and the siloxane network.

Although the sol–gel method is mostly used for the synthesis of oxide materials, the synthesis of halides is also possible. Pellegrino et al. [11] successfully applied the sol–gel process for the fabrication of Eu-doped CaF$_2$ thin films as down-shifting layers for solar cell applications. The Ca(hfa)$_2$·diglyme·H$_2$O and Eu(hfa)$_3$·diglyme were used as precursors and mixed in various proportions in a water–ethanol solution. The optimization of deposition process parameters in terms of annealing temperature, substrate nature, and doping ion percentage was performed. The down-shifting properties were validated by taking luminescence measurements under UV excitation.

**Funding:** This research received no external funding.

**Conflicts of Interest:** The author declares no conflicts of interest.

## References

1. Ghezali, N.; Díaz Verde, Á.; Illán Gómez, M.J. Screening $Ba_{0.9}A_{0.1}MnO_3$ and $Ba_{0.9}A_{0.1}Mn_{0.7}Cu_{0.3}O_3$ (A = Mg, Ca, Sr, Ce, La) Sol-Gel Synthesised Perovskites as GPF Catalysts. *Materials* **2023**, *16*, 6899. [CrossRef] [PubMed]
2. Karnaukhov, T.M.; Veselov, G.B.; Cherepanova, S.V.; Vedyagin, A.A. Sol-Gel Synthesis and Characterization of the Cu-Mg-O System for Chemical Looping Application. *Materials* **2022**, *15*, 2021. [CrossRef] [PubMed]
3. Karoblis, D.; Stewart, O.C., Jr.; Glaser, P.; El Jamal, S.E.; Kizalaite, A.; Murauskas, T.; Zarkov, A.; Kareiva, A.; Stoll, S.L. Low-Temperature Synthesis of Magnetic Pyrochlores ($R_2Mn_2O_7$, R = Y, Ho–Lu) at Ambient Pressure and Potential for High-Entropy Oxide Synthesis. *Inorg. Chem.* **2023**, *62*, 10635–10644. [CrossRef] [PubMed]
4. Karoblis, D.; Mazeika, K.; Raudonis, R.; Zarkov, A.; Kareiva, A. Sol-Gel Synthesis and Characterization of Yttrium-Doped $MgFe_2O_4$ Spinel. *Materials* **2022**, *15*, 7547. [PubMed]
5. Mejia-Estrella, I.A.; Pérez Larios, A.; Sulbarán-Rangel, B.; Guzmán González, C.A. Reaching Visible Light Photocatalysts with Pt Nanoparticles Supported in $TiO_2$-$CeO_2$. *Materials* **2022**, *15*, 6784. [PubMed]
6. Monrós, G.; Llusar, M.; Badenes, J.A. High NIR Reflectance and Photocatalytic Ceramic Pigments Based on M-Doped Clinobisvanite $BiVO_4$ (M = Ca, Cr) from Gels. *Materials* **2023**, *16*, 3722. [CrossRef] [PubMed]
7. Covei, M.; Bogatu, C.; Gheorghita, S.; Duta, A.; Stroescu, H.; Nicolescu, M.; Calderon-Moreno, J.M.; Atkinson, I.; Bratan, V.; Gartner, M. Influence of the Deposition Parameters on the Properties of $TiO_2$ Thin Films on Spherical Substrates. *Materials* **2023**, *16*, 4899. [CrossRef] [PubMed]
8. Wojtasik, K.; Zięba, M.; Tyszkiewicz, C.; Pakieła, W.; Żak, G.; Jeremiasz, O.; Gondek, E.; Drabczyk, K.; Karasiński, P. Zinc Oxide Films Fabricated via Sol-Gel Method and Dip-Coating Technique—Effect of Sol Aging on Optical Properties, Morphology and Photocatalytic Activity. *Materials* **2023**, *16*, 1898. [CrossRef] [PubMed]
9. Crespo-Monteiro, N.; Valour, A.; Vallejo-Otero, V.; Traynar, M.; Reynaud, S.; Gamet, E.; Jourlin, Y. Versatile Zirconium Oxide ($ZrO_2$) Sol-Gel Development for the Micro-Structuring of Various Substrates (Nature and Shape) by Optical and Nano-Imprint Lithography. *Materials* **2022**, *15*, 5596.
10. Toirac, B.; Garcia-Casas, A.; Monclús, M.A.; Aguilera-Correa, J.J.; Esteban, J.; Jiménez-Morales, A. Influence of Addition of Antibiotics on Chemical and Surface Properties of Sol-Gel Coatings. *Materials* **2022**, *15*, 4752. [PubMed]
11. Pellegrino, A.L.; Milan, E.; Speghini, A.; Malandrino, G. Fabrication of Europium-Doped $CaF_2$ Films via Sol-Gel Synthesis as Down-Shifting Layers for Solar Cell Applications. *Materials* **2023**, *16*, 6889. [CrossRef]

**Disclaimer/Publisher's Note:** The statements, opinions and data contained in all publications are solely those of the individual author(s) and contributor(s) and not of MDPI and/or the editor(s). MDPI and/or the editor(s) disclaim responsibility for any injury to people or property resulting from any ideas, methods, instructions or products referred to in the content.

Article

# Fabrication of Europium-Doped CaF$_2$ Films via Sol-Gel Synthesis as Down-Shifting Layers for Solar Cell Applications

Anna Lucia Pellegrino [1], Emil Milan [2], Adolfo Speghini [2] and Graziella Malandrino [1,*]

[1] Dipartimento di Scienze Chimiche, Università di Catania and INSTM UdR Catania, Viale A. Doria 6, I-95125 Catania, Italy; annalucia.pellegrino@unict.it
[2] Nanomaterials Research Group, Dipartimento di Biotecnologie, Università di Verona and INSTM UdR Verona, Strada le Grazie 15, I-37134 Verona, Italy; emil.milan@univr.it (E.M.); adolfo.speghini@univr.it (A.S.)
* Correspondence: gmalandrino@unict.it

**Abstract:** In the present work, an in-depth study on the sol-gel process for the fabrication of Eu-doped CaF$_2$ materials in the form of thin films has been addressed for the production of down-shifting layers. Fine-tuning of the operative parameters, such as the annealing temperature, substrate nature and doping ion percentage, has been finalized in order to obtain Eu(III)-doped CaF$_2$ thin films via a reproducible and selective solution process for down-shifting applications. An accurate balance of such parameters allows for obtaining films with high uniformity in terms of both their structural and compositional features. The starting point of the synthesis is the use of a mixture of Ca(hfa)$_2$•diglyme•H$_2$O and Eu(hfa)$_3$•diglyme adducts, with a suited ratio to produce 5%, 10% and 15% Eu-doped CaF$_2$ films, in a water/ethanol solution. A full investigation of the structural, morphological and compositional features of the films, inspected using X-ray diffraction analysis (XRD), field emission scanning electron microscopy (FE-SEM) and energy dispersive X-ray analysis (EDX), respectively, has stated a correlation between the annealing temperature and the structural characteristics and morphology of the CaF$_2$ thin films. Interestingly, crystalline CaF$_2$ films are obtained at quite low temperatures of 350–400 °C. The down-shifting properties, validated by taking luminescence measurements under UV excitation, have allowed us to correlate the local environment in terms of the degree of symmetry around the europium ions with the relative doping ion percentages.

**Keywords:** metalorganic complexes; morphological control; luminescence; energy conversion

## 1. Introduction

Recently, great effort has been devoted to the development of innovative materials with high impact in an energetic scenario. In particular, the possibility of engineering new devices with increasing efficiency of energy production represents the main challenge in the photovoltaic field [1–3]. Among a plethora of inorganic functional materials, fluoride-based systems are one of the most promising classes in this field due to their excellent chemical stability and optical properties [4,5], which allow them to extend their uses in a wide range of microelectronic, photonic and nanomedicine applications [6–8]. Particularly, the doping of fluoride inorganic systems with luminescent species, such as trivalent lanthanide ions (Ln$^{3+}$), represents one of the most intriguing strategies to obtain more efficient photovoltaic devices through energy conversion processes [9,10]. This class of materials combines the excellent optical properties of the fluoride matrices, such as low phonon energy and high optical transparency [11,12], with the energy conversion properties of the doping ions both as down- and up-conversion systems. In fact, appropriate combinations of luminescent lanthanide ions, such as Yb$^{3+}$/Er$^{3+}$ or Yb$^{3+}$/Tm$^{3+}$ for up-conversion (UC) and Eu$^{3+}$ for down-conversion/down-shifting (DC and DS) systems, are able to collect the radiation outside of the absorption range of the active photovoltaic material and shift its energy to

a more suitable optical region, with the aim of enhancing the efficiency of solar devices. Specifically, in UC processes, photons in the near-infrared region are converted into higher-energy photons, i.e., in the ultraviolet (UV) or in the visible regions; in DC processes, one high-energy photon is converted into two or more photons at lower energy and in the DS phenomena, high-energy photons, usually in the UV region, are absorbed and converted into lower-energy photons [13]. Among the several DS species, europium (III) is one of the most promising down-shifters, due to its high emission efficiency and long lifetimes [14–16], thus allowing the exploitation of the UV range of the solar spectrum. Furthermore, the inorganic host component also plays a crucial role in the engineering of efficient energy conversion systems. The most extensively investigated inorganic matrices are represented by binary and multicomponent fluorides of the type $CaF_2$, $BaF_2$ and $SrF_2$ or $NaYF_4$, $NaGdF_4$ and $LiYF_4$ [17–22]. Among them, calcium fluoride has been regarded as one of the most efficient matrices for lanthanide ions in energy conversion systems, due to its optical properties [23,24] and chemical stability. However, calcium fluoride has been reported in the literature in the form of lanthanide-doped nanoparticles for applications in nanomedicine [25], while in the form of thin films, it has been much less investigated. Nevertheless, the bottleneck for its massive application in optics, microelectronics and the PV field is related to the possibility of fabrication process development, which allows for the growth of nanostructured doped fluoride systems in the form of thin film with a wide range of structural, morphological and compositional control. Up to now, the most explored synthetic routes for pure and doped $CaF_2$ are related to vapor deposition growth, such as electron beam evaporation, molecular beam epitaxy, chemical vapor deposition and solution routes [26–29].

In this scenario, the present sol-gel approach represents the first report on a cheap and industrially appealing solution process for the synthesis of compact and homogenous Eu-doped $CaF_2$ thin film with high structural and compositional tunable properties. In particular, we report an in-depth study on a combined sol-gel/spin-coating approach to the fabrication of a down-shifting layer based on an Eu-doped $CaF_2$ thin film starting from a mixture of $Ca(hfa)_2 \bullet diglyme \bullet H_2O$ and $Eu(hfa)_3 \bullet diglyme$ complex in a water/ethanol solution. The annealing temperature, substrate nature and doping europium percentage are the key parameters to fine-tune both the morphological and luminescence features of the as-synthesized $CaF_2$-based films. A complete investigation of the structural, morphological and compositional characterization of the Eu-doped $CaF_2$ thin film has been executed using X-ray diffraction analysis (XRD), field emission scanning electron microscopy (FE-SEM) and energy dispersive X-ray analysis (EDX), respectively. Finally, the luminescent properties as down-shifting layers as a function of europium ion concentration have been investigated using luminescence measurements.

## 2. Materials and Methods

### 2.1. $CaF_2$: Eu Synthesis

The $Ca(hfa)_2 \bullet diglyme \bullet H_2O$ and $Eu(hfa)_3 \bullet diglyme$ precursors were synthesized as previously reported in [14,30].

The sol–gel reaction was conducted in a water/ethanol solution with trifluoroacetic acid as a catalyst for the hydrolysis reactions, starting from a mixture of Ca and Eu precursors. The adducts were mixed in molar ratio values of 0.95:0.05, 0.9:0.1 and 0.85:0.15 mmol for the preparation of 5%, 10% and 15% of Eu-doping $CaF_2$ films, respectively. The reaction and the different molar ratios used are reported in the following:

0.95 $Ca(hfa)_2 \bullet diglyme \bullet H_2O$:0.05 $Eu(hfa)_3 \bullet diglyme$:87 $C_2H_5OH$:3 $H_2O$:0.8 $CF_3COOH$ for $CaF_2$:Eu (5%).

0.90 $Ca(hfa)_2 \bullet diglyme \bullet H_2O$:0.10 $Eu(hfa)_3 \bullet diglyme$:87 $C_2H_5OH$:3 $H_2O$:0.8 $CF_3COOH$ for $CaF_2$:Eu (10%).

0.85 $Ca(hfa)_2 \bullet diglyme \bullet H_2O$:0.15 $Eu(hfa)_3 \bullet diglyme$:87 $C_2H_5OH$:3 $H_2O$:0.8 $CF_3COOH$ for $CaF_2$:Eu (15%).

The different solutions were submitted to hydrolysis and aging reactions under stirring at 60 °C for 20 h. The spin-coating process was carried out on Si (100) and glass substrates about 1 cm × 2 cm, cut from microscope glass slides (Forlab, Carlo Erba, Milan, Italy) of 0.8 mm thickness. The films were deposited via spin-coating using a multistep approach with a four-time deposition. After each spin, a 10 min annealing step at 350 °C or 400 °C in air was carried out. Finally, the films were annealed at 350 °C or 400 °C in air for 1 h. The spin-coating was carried out with a ramping rate of 1000 revolutions per minute (rpm), a spinning rate of 3000 rpm and a time of 60 s using a Spincoater SPIN150 (SPS Europe, Putten, The Netherlands) system.

## 2.2. Characterization

The film structures were analyzed using XRD in grazing incidence mode (0.5°) and using a SmartLab Rigaku diffractometer (Rigaku, Tokyo, Japan) operating at 45 kV and 200 mA, equipped with a rotating anode of Cu $K_\alpha$ radiation. The film morphologies were analyzed via FE-SEM using a ZEISS SUPRA 55 VP field emission microscope (ZEISS, Jena, Germany). For the FE-SEM characterization, the films deposited onto Si were analyzed as prepared, while a very thin Au layer was sputtered onto the films and deposited onto a non-conducting substrate such as glass. Energy-dispersive X-ray (EDX) analysis allowed the atomic compositional analysis of the samples. An Oxford INCA windowless detector (Oxford Instruments, Abingdon, UK), having a resolution of 127 eV as the full-width half maximum (FWHM) of the Mn $K_\alpha$, was used. For measuring the luminescence spectra, the samples were excited using an LED flashlight and a band-pass filter centered at 390 nm. The emission spectra were measured using a 4× microscopy objective at a 90° geometry and an edge filter at 532 nm (Semrock 532 nm RazorEdge® ultrasteep, IDEX Health & Science, LLC, Rochester, NY, USA) to reject the exciting radiation. The emission spectra were dispersed using a single monochromator (Andor Shamrock 500i, 300 lines/mm grating, Andor technology, Belfast, Northern Ireland) with an optical resolution of 1.3 nm and measured with a Peltier-cooled (−90 °C) CCD camera (Andor, iDus, Andor technology, Belfast, Northern Ireland).

## 3. Results and Discussion

An optimized procedure of sol-gel reaction and spin-coating deposition was applied for the fabrication of Eu-doped $CaF_2$ thin films. Particularly, the operative parameters of the process, such as the annealing temperature, nature of the substrate and europium concentration were finely tuned in order to obtain full control of the structural, morphological and luminescent properties of the down-shifting layers. It is worth noting that the present work represents, to the best of our knowledge, the first report on the fabrication of down-shifting $CaF_2$ thin films using a sol-gel process. Similar synthetic strategies have been reported in the last years for the fabrication of pure binary fluoride $CaF_2$ and the up-converting systems of $CaF_2$: $Yb^{3+}$, $Er^{3+}$, $Tm^{3+}$ [28] and multicomponent fluoride $NaYF_4$: $Yb^{3+}$, $Er^{3+}$ and $NaYF_4$: $Yb^{3+}$, $Tm^{3+}$ [20]. The approach herein represents an easy synthetic method for producing Eu-doped $CaF_2$ films with inexpensive equipment and high crystallinity. Furthermore, the use of $Ca(hfa)_2$•diglyme•$H_2O$ and $Eu(hfa)_3$•diglyme adducts, as a starting mixture, provides a unique source for the Ca, Eu and F components.

### 3.1. Structural Investigation

Structural characterization of the prepared films at different doping ion percentages was conducted using X-ray diffraction (XRD) analysis and has been reported in Figure 1. The patterns show the formation of pure and nanostructured $CaF_2$ as confirmed by the presence of characteristic peaks at 2θ = 28.29, 47.09 and 55.81° corresponding to reflections of the (111), (220) and (311) lattice planes, respectively, in accordance with the ICDD card n.35-0816 of the $CaF_2$ phase. As a reference, also an undoped $CaF_2$ sample obtained at 400 °C on a silicon substrate has been reported, confirming a perfect match with the ICDD card of the pure $CaF_2$ phase. The small peak around 51° arises from the machine when measurements are carried out in grazing incidence. Furthermore, an in-depth investigation was conducted with the use of graphite as an internal standard for each sample in order to evaluate the peak position and the potential shift of the CaF$_2$ signals. A comparison of the three patterns reported as a function of the Eu-doping concentrations displays some differences in terms of the peak positions with respect to those reported in the diffraction card (black lines in Figure 1). In particular, a magnification of the 220 reflection, shown in the inset of Figure 1, displays a perfect match of the peak arising from the CaF$_2$: Eu (5%) sample in comparison with the position reported in the ICDD database for the CaF$_2$ phase, and a slight shift toward higher angles for CaF$_2$: Eu (10%) and CaF$_2$: Eu (15%).

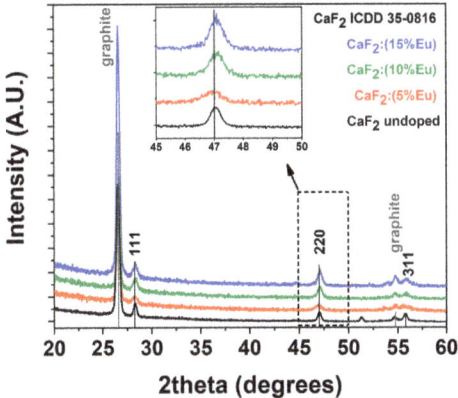

**Figure 1.** XRD patterns of undoped CaF$_2$ (gray line), CaF$_2$: Eu (5%) (red line), CaF$_2$: Eu (10%) (green line) and CaF$_2$: Eu (15%) (blue line) thin films on Si (100) substrate at the annealing temperature of 400 °C. Graphite is used as an internal standard.

This behavior can be rationalized considering the smaller ionic radius of the Eu-doping ion, which is assumed to be in a substitutional position with respect to Ca$^{2+}$. In fact, Eu$^{3+}$ in an eight coordination has an ionic radius of 1.066 Å, which is slightly smaller than the Ca$^{2+}$ ionic radius of 1.12 Å for the same coordination [31]. This effect results in a slight contraction of the lattice structure at a higher doping ion concentration and thus in a shift toward the higher angles of the diffraction peaks. Notably, considering the need for charge balance due to the different charges, 2+ for calcium and 3+ for europium, both the formation of interstitial fluoride ions or clusters can be possible [17,32,33].

### 3.2. Morphological Characterization

The different conditions of the synthetic process affect the morphology of the films both in terms of the uniformity of the substrate coverage and in the general aspect of the layers. In particular, a full overview of the different morphologies obtained for the Eu-doped CaF$_2$ films has been reported for the Si substrates in Figure 2 and for the glass substrates in Figure 3.

The FE-SEM images of the CaF$_2$: Eu (5%) films on the Si (100) substrate are reported in Figure 2a,d for the two annealing temperatures 350 °C and 400 °C, respectively. The images display, for both samples, similar morphologies with the formation of discontinuous coverage being independent from the annealing temperature. The growth is quite porous and the presence of regions not covered could be attributed to two aspects: (i) the density of the incipient gel used in the spin-coating deposition for this Ca: Eu ratio and (ii) the poor wettability of the silicon substrate. Therefore, under these operative parameters, the annealing temperature slightly affects the morphology of the films. The observed morphology may be compared to the sponge-like or coral-like shape observed for the CaF$_2$ nanostructures prepared using pulsed electron beam evaporation and thermally annealed at 200 °C [26]. On the other hand, the morphology is completely different from that observed in the MBE-grown CaF$_2$ films, where nanocrystals are derived from epitaxial growth [26], or from the liquid phase epitaxy grown layers, where the CaF$_2$ films were deposited onto a CaF$_2$ single-crystal substrate [29].

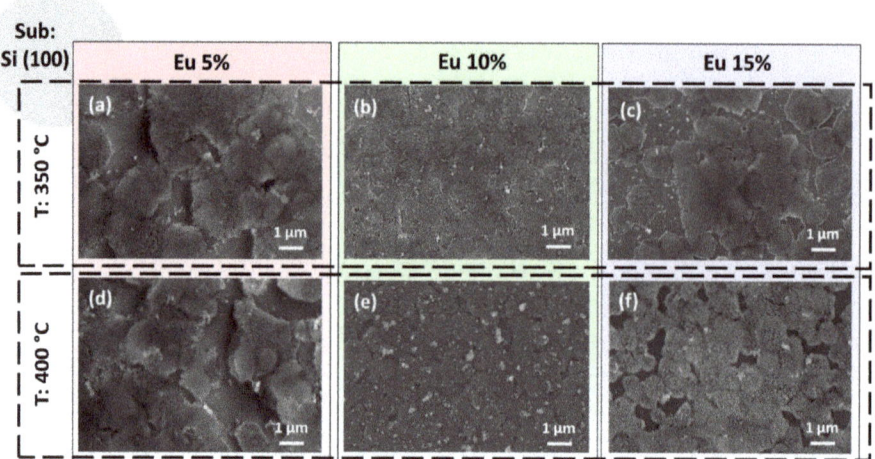

**Figure 2.** FE-SEM images of the CaF$_2$ films deposited onto Si (100) substrate at (**a**) 350 °C and 5% Eu doping; (**b**) 350 °C and 10% Eu doping; (**c**) 350 °C and 15% Eu doping; (**d**) 400 °C and 5% Eu doping; (**e**) 400 °C and 10% Eu doping and (**f**) 400 °C and 15% Eu doping.

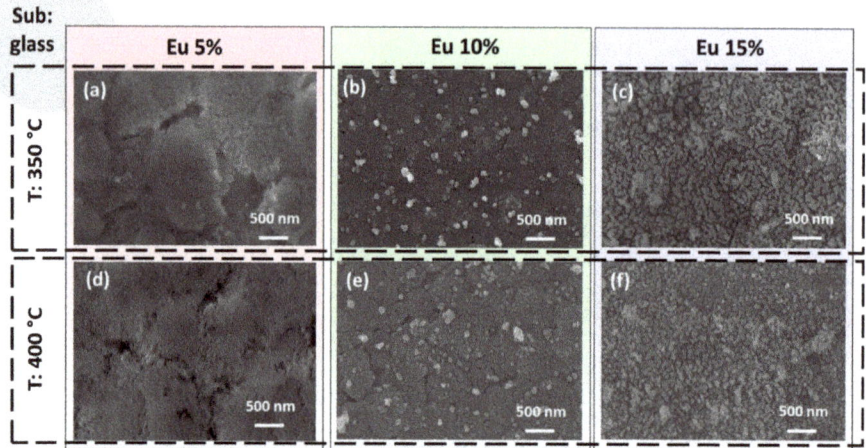

**Figure 3.** FE-SEM images of the CaF$_2$ films deposited onto glass substrate at (**a**) 350 °C and 5% Eu doping; (**b**) 350 °C and 10% Eu doping; (**c**) 350 °C and 15% Eu doping; (**d**) 400 °C and 5% Eu doping; (**e**) 400 °C and 10% Eu doping and (**f**) 400 °C and 15% Eu doping.

In Figure 2b,e are shown the CaF$_2$: Eu (10%) films on Si (100) at 350 °C and 400 °C annealing temperatures, respectively. Under these process parameters, the films appear more compact and homogenous, even if some outgrowths are visible and are more evident in Figure 2e for the sample obtained at 400 °C. This feature has been already observed for pure CaF$_2$ film growth on Si [28] and could arise during the annealing step process. Finally, the samples with the highest percentage of Eu-doping ions, i.e., $CaF_2$: Eu (15%), are reported in Figure 2c,f for the two annealing temperatures. They display partial coverage of the silicon substrates and the presence of some outgrowths similar to those found in Figure 2b,e, but aggregated with each other. However, in these last two samples, the coverage of the Si substrate, even if it appears somewhat similar to what we have observed for the $CaF_2$: Eu (5%) films (Figure 2a,d), presents a more compact and flat surface, especially for the sample in Figure 2f. This last aspect could be associated with the higher treatment temperature, i.e., 400 °C, used in these last films.

The morphologies obtained on glass substrates are presented in the overview in Figure 3 as a function of the annealing temperatures and Eu-doping percentage. The $CaF_2$: Eu (5%) films on glass obtained at 350 °C and 400 °C shown in Figure 3a,d present both a very flat and smooth morphology, similar to the one observed for the same film composition on the Si substrate (see Figure 2a,d). However, for these samples, the coverage seems to be more compact, with only a few voids for the sample at 350 °C in Figure 2a. This tendency is likely due to the better wettability of the glass substrate compared to silicon, which seems the main aspect in determining the morphology features of the samples in this case. The $CaF_2$: Eu (10%) films on glass annealed at 350 °C and 400 °C, respectively, are reported in Figure 2b,e. For both these systems, we observe a similar morphology, regardless of the annealing temperature, with a uniform and compact layer characterized by the presence of some outgrowths being uniformly distributed on the surfaces. These features are very similar to what we found for the analogous films obtained on silicon (see Figure 2b,e). Finally, the $CaF_2$: Eu (15%) films on glass are displayed in Figure 3c,f for 350 °C and 400 °C annealing treatments. For these samples, we observe a different morphology characterized by homogeneous coating with small grains in the order of tens of nanometers barely visible. This morphology suggests grain coalescence phenomena in the coating formation during the annealing treatment, as already observed in similar conditions for the up-converting $CaF_2$: $Yb^{3+}$, $Er^{3+}$ and $Tm^{3+}$ layers [28]. Notably, at higher temperatures (see Figure 3f), the films appear more compact and uniform. Therefore, the different morphologies observed using FE-SEM analyses can be ascribed to many factors: (i) the different wettability of the substrates, which is the main reason for the films obtained on silicon; (ii) the density of the sol used, which is directly correlated with the starting mixture of the Ca and Eu adducts in the gel; (iii) a slight variation in the spin-coating step; (iv) annealing temperature of the final treatments, as confirmed by the trend of the best uniformity being observed at a higher temperature. In each case, the balance of all these aspects results in a plethora of different morphologies.

Then, the thickness of the as-prepared films was evaluated using FE-SEM analysis in cross-sectional mode. The images of the $CaF_2$: Eu (10%) samples on silicon at 350 °C and 400 °C, for which plain-view analyses reveal more uniform and compact films, have been reported in Figure 4.

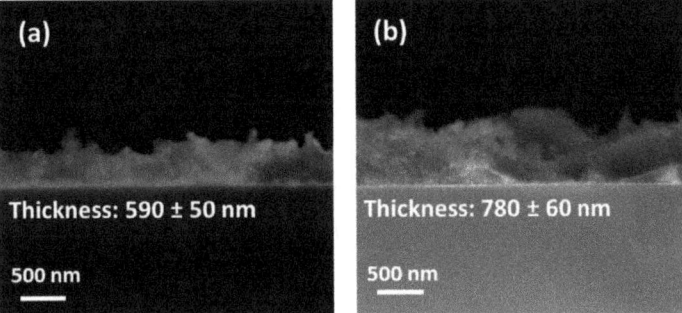

**Figure 4.** FE-SEM images in cross-section of $CaF_2$: Eu (10%) films deposited onto Si (100) at (**a**) 350 °C and (**b**) 400 °C.

At a lower annealing temperature, i.e., 350 °C, the cross-sectional image in Figure 4a confirms the formation of a compact layer with a thickness of about 590 ± 50 nm. The $CaF_2$: Eu (10%) sample obtained at the higher temperature of 400 °C in Figure 4b displays a similar compact and uniform film of 780 ± 60 nm.

The increment in the thickness value at a higher annealing temperature can be explained considering that already during the deposition step, which occurs via a multistep procedure consisting of four-time spin-coating depositions alternated with fast annealing in the air for 10 min, the crystallization of the layers takes place. The deposition process hence is more efficient at higher annealing temperatures, allowing the formation of slightly thicker films.

## 3.3. Compositional Characterization

In order to ensure the down-shifting properties of the Eu-doped $CaF_2$ films, a preliminary study on the sample composition has been conducted using energy dispersive X-ray (EDX) analysis. In Figure 5a, the EDX spectrum of the $CaF_2$: Eu (10%) film deposited onto Si (100) at 400 °C is reported as a case study. The spectrum shows the $K_\alpha$ peak of fluorine, the $K_\alpha$ peak of the silicon substrate and the $K_\alpha$ and $K_\beta$ peaks of calcium, at values of 0.67, 1.74, 3.70 and 4.02 keV, respectively. In addition, the presence of europium is confirmed by the signals at 1.14 and 5.81 keV, related to the M and L lines, respectively. Notably, the absence of C and O elements in the spectrum points to the good reactivity and clean decomposition of the Ca and Eu precursors during the sol-gel process. The quantitative analysis shown in the inset in Figure 5a confirms the correct stoichiometry of the Eu concentration of the film at about 10%, which is coincident with the value set in the starting mixture, i.e., 0.90:0.10 of $Ca(hfa)_2 \bullet diglyme \bullet H_2O$: $Eu(hfa)_3 \bullet diglyme$.

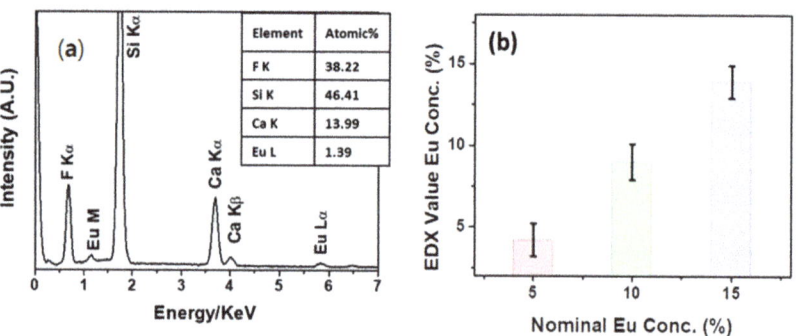

**Figure 5.** (a) EDX spectrum of the $CaF_2$: Eu (10%) film deposited onto Si (100) at 400 °C and (b) relationship between nominal Eu concentrations in the Eu-doped $CaF_2$ films deposited onto Si (100) at 400 °C and the values extrapolated from EDX quantitative analyses.

Additionally, quantitative analysis was conducted for all the Eu-doped $CaF_2$ samples deposited onto silicon and is reported in the graph in Figure 5b. Several different regions for each kind of samples have been analyzed in order to ensure both a more representative value of Eu concentration and a good homogeneity of the layers. The graph displays the average value of the Eu concentration recorded using EDX quantitative analysis versus the nominal concentration used in the starting mixture of each Eu-doped $CaF_2$ sample. Concentrations of 4.2 ± 1%, 9.0 ± 1.1% and 13.9 ± 1.0% have been obtained for the $CaF_2$: Eu (5%), $CaF_2$: Eu (10%) and $CaF_2$: Eu (15%) films, respectively, confirming a good match between the nominal and experimental Eu-doping concentrations.

Finally, the homogeneity of the systems has been evaluated using EDX map analysis over a large area. In particular, for the $CaF_2$: Eu (10%) onto Si at 400 °C sample, the maps of the F, Ca and Eu elements are reported in Figure 6. Notably, the doping ion distribution is very uniform over the analyzed area (about 20 × 28 µm), confirming the suitability of the synthetic process for the fabrication of down-converting system with good luminescence performance.

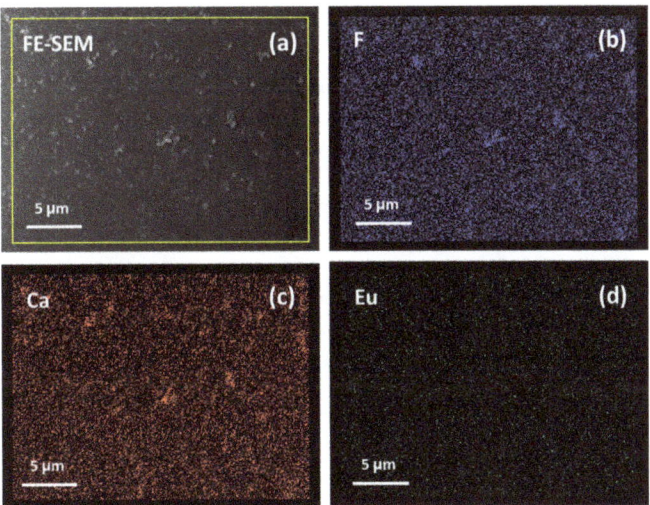

**Figure 6.** FE-SEM images (**a**) and EDX map analysis of F (**b**), Ca (**c**) and Eu (**d**) elements for the CaF$_2$: Eu (10%) film deposited onto Si (100) at 400 °C.

### 3.4. Luminescence Properties

Radiation centred around 390 nm was chosen as the excitation source because at this wavelength, there is a strong absorption by Eu$^{3+}$ ions due to the $^7F_0 \rightarrow ^5L_6$ transition. It is worth noting that the oscillator strength of this transition is among the strongest ones for Eu$^{3+}$ ions in the CaF$_2$ host in the UV-visible range [32]. The room temperature emission spectra of the Eu$^{3+}$-doped CaF$_2$ films deposited onto the glass substrate at 400 °C are shown in Figure 7. The observed emission bands are due to transitions of the Eu$^{3+}$ ions, ranging from the $^5D_0$ excited level to the $^7F_J$ multiplet (J = 0, 1, 2, 3, 4). In particular, the bands assigned to the $^5D_0 \rightarrow ^7F_1$ (in the 580–600 nm range) and $^5D_0 \rightarrow ^7F_2$ transitions (in the 605–630 nm) dominate the emission spectrum for all the samples. The emission spectra are compatible with those observed for similar samples, e.g., CaF$_2$: Eu$^{3+}$ nanoparticles prepared using a hydrothermal technique [34] and using a sol-gel technique [35], Eu$^{3+}$-doped CaF$_2$ thin films prepared using electrochemical processing [36] and thin films of Eu$^{3+}$-doped CaF$_2$ deposited onto an aluminum layer using a vacuum deposition approach [37]. It has to be noted that the $^5D_0 \rightarrow ^7F_1$ transition is a magnetic dipole one, not dependent on the local environment around the lanthanide ion [38]. On the other hand, the $^5D_0 \rightarrow ^7F_2$ transition is an electric dipole, allowed only for lanthanide sites without inversion symmetry. This property makes the latter transition highly sensitive to probing the local environment of the Eu$^{3+}$ ions. The ratio between the intensities of the bands due to the two abovementioned transitions is useful to get insights about the degree of asymmetry in the local environment of the lanthanide ion [39]. The asymmetry ratio $R$ is defined as:

$$R = \frac{I(^5D_0 \rightarrow ^7F_2)}{I(^5D_0 \rightarrow ^7F_1)} \quad (1)$$

where $I$ is the integrated area of the $^5D_0 \rightarrow ^7F_J$ (J = 1, 2) transitions. The $R$ values increase on decreasing the degree of symmetry around the lanthanide ion. The calculated $R$ values for the samples are 3.12, 2.70 and 1.92 for Eu$^{3+}$ concentrations of 5%, 10% and 15%, respectively. A decrease in the $R$ values is observed on increasing the lanthanide concentration, indicating that on average, the symmetry of the lanthanide ions accordingly increases. It is well known that lanthanide ions can be accommodated in the $CaF_2$ structure in several sites with different symmetries [40–42]. In the present case, an increased occupation of more symmetric sites on increasing the lanthanide concentration is compatible with the observed spectra, inducing an overall increase in the symmetry around the lanthanide ions and therefore a decrease in the $R$ value. For instance, on increasing the lanthanide concentration, a higher number of $Eu^{3+}$ ions could substitute the regular $Ca^{2+}$ lattice sites and fewer $Eu^{3+}$ ions could occupy the particle surface or defect sites [43].

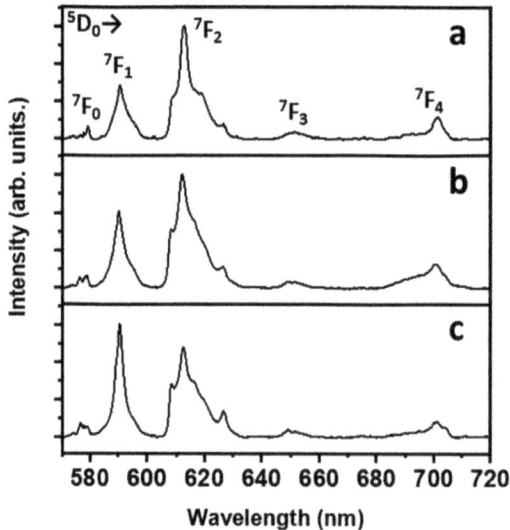

**Figure 7.** Room temperature emission spectra with transition assignments of the $CaF_2$ thin films deposited onto glass substrate at 400 °C at different $Eu^{3+}$ ion concentrations: (**a**) 5%; (**b**) 10%; (**c**) 15%. All transitions originate from the $^5D_0$ level. The end level is shown in the picture close to the corresponding emission band.

Any possibility of the presence of $Eu^{2+}$ can be excluded since no peaks related to the $Eu^{2+}$ species are observed in the spectrum. This outcome may be compared to the findings reported in the study by Secu et al. [44], who synthesized Eu(II)-doped $CaF_2$ films using sol-gel and annealing techniques under a reduced atmosphere, while the present thermal treatments are carried out in air.

## 4. Conclusions

In conclusion, an in-depth study is herein described for the fabrication of europium-doped $CaF_2$ systems in the form of thin films using a combined sol-gel/spin-coating deposition process. To the best of our knowledge, the present work is the first report on the production of $Eu^{3+}$-doped $CaF_2$-based thin films using a sol-gel approach. Specifically, through fine-tuning the operative parameters, such as the annealing temperature, substrate nature and doping ion concentration, we have managed to optimize the formation of homogeneous films of $CaF_2$ on both silicon and glass substrates. XRD characterization assessed that crystalline films are obtained with a thermal treatment of 350–400 °C on both substrates. EDX microanalyses confirmed the film purity since neither C nor O is observed in the spectrum, pointing to a clean decomposition of the precursor, and the europium percentage in the film is parallel with the nominal amount in the initial sol mixture. Actually, the sol-gel process combined with a spin-coating deposition represents an appealing method at an industrial level due to the low-cost equipment involved and the relatively low temperature of the annealing treatment. In alternative to spin-coating, spraying could be applied for a potential scaling-up of the process. Finally, the luminescence properties of the samples have been investigated by taking spectroscopy measurements under UV excitation. The down-shifting features have confirmed the functional properties of the material and thus its potential use in solar cell devices.

**Author Contributions:** Conceptualization, A.L.P.; validation and investigation, A.L.P. and E.M.; writing—original draft preparation, A.L.P.; writing—review and editing, G.M. and A.S.; funding acquisition, G.M. and A.S.; supervision, G.M. and A.S. All authors have read and agreed to the published version of the manuscript.

**Funding:** This research has been funded by the European Union (NextGeneration EU), through the MUR-PNRR project SAMOTHRACE (ECS00000022).

**Institutional Review Board Statement:** Not applicable.

**Informed Consent Statement:** Not applicable.

**Data Availability Statement:** Data are available from the authors upon request.

**Acknowledgments:** A.L.P. and G.M. acknowledge the Bio-Nanotech Research and Innovation Tower (BRIT) laboratory of the University of Catania (grant no. PONa3_00136, financed by the MIUR) for the SmartLab diffractometer facility. A.L.P. thanks the Ministero dell'Università e della Ricerca within the PON "Ricerca e Innovazione" 2014–2020 Azioni IV.4 program. A.S and E.M acknowledge the University of Verona for the funding of research projects in the Joint Research 2022 framework. The Technological Platform Center of the University of Verona is gratefully acknowledged for the use of its instrumental facilities.

**Conflicts of Interest:** The authors declare no conflict of interest.

# References

1. Fagiolari, L.; Sampo, M.; Lamberti, A.; Amici, J.; Francia, C.; Bodoardo, S.; Bella, F. Integrated energy conversion and storage devices: Interfacing solar cells, batteries and supercapacitors. *Energy Storage Mater.* **2022**, *51*, 400–434. [CrossRef]
2. Chen, J.; Xu, K.; Xie, W.; Zheng, L.; Tian, Y.; Zhang, J.; Chen, J.; Liu, T.; Xu, H.; Cheng, K.; et al. Enhancing perovskite solar cells efficiency through cesium fluoride mediated surface lead iodide modulation. *J. Colloid Interface Sci.* **2023**, *652*, 1726–1733. [CrossRef] [PubMed]
3. Wei, Y.; Zhao, Y.; Liu, C.; Wang, Z.; Jiang, F.; Liu, Y.; Zhao, Q.; Yu, D.; Hong, M. Constructing All-Inorganic Perovskite/Fluoride Nanocomposites for Efficient and Ultra-Stable Perovskite Solar Cells. *Adv. Funct. Mater.* **2021**, *31*, 2106386. [CrossRef]
4. Gan, F. Optical properties of fluoride glasses: A review. *J. Non-Cryst. Solids* **1995**, *184*, 9–20. [CrossRef]
5. Sharma, R.K.; Mudring, A.-V.; Ghosh, P. Recent trends in binary and ternary rare-earth fluoride nanophosphors: How structural and physical properties influence optical behavior. *J. Lumin.* **2017**, *189*, 44–63. [CrossRef]
6. Wen, C.; Lanza, M. Calcium fluoride as high-k dielectric for 2D electronics. *Appl. Phys. Rev.* **2021**, *8*, 021307. [CrossRef]
7. Jiang, C.; Brik, M.G.; Li, L.; Li, L.; Peng, J.; Wu, J.; Molokeev, M.S.; Wong, K.-L.; Peng, M. The electronic and optical properties of a narrow-band red-emitting nanophosphor $K_2NaGaF_6$:$Mn^{4+}$ for warm white light-emitting diodes. *J. Mater. Chem. C* **2018**, *6*, 3016–3025. [CrossRef]
8. Ansari, A.A.; Parchur, A.K.; Thorat, N.D.; Chen, G. New advances in pre-clinical diagnostic imaging perspectives of functionalized upconversion nanoparticle-based nanomedicine. *Coord. Chem. Rev.* **2021**, *440*, 213971. [CrossRef]
9. Mehrdel, B.; Nikbakht, A.; Aziz, A.A.; Jameel, M.S.; Dheyab, M.A.; Khaniabadi, P.M. Upconversion lanthanide nanomaterials: Basics introduction, synthesis approaches, mechanism and application in photodetector and photovoltaic devices. *Nanotechnology* **2022**, *33*, 082001. [CrossRef]
10. Goldschmidt, J.C.; Fischer, S. Upconversion for Photovoltaics—A Review of Materials, Devices and Concepts for Performance Enhancement. *Adv. Optical Mater.* **2015**, *3*, 510–535. [CrossRef]
11. Wells, J.-P.R.; Reeves, R.J. Up-conversion fluorescence of $Eu^{3+}$ doped alkaline earth fluoride crystals. *J. Lumin.* **1995**, *66*, 219–223. [CrossRef]
12. Naccache, R.; Yu, Q.; Capobianco, J.A. The Fluoride Host: Nucleation, Growth, and Upconversion of Lanthanide-Doped Nanoparticles. *Adv. Optical Mater.* **2015**, *3*, 482–509. [CrossRef]
13. Bünzli, J.-C.G.; Chauvin, A.-S. Lanthanides in Solar Energy Conversion. In *Handbook on the Physics and Chemistry of Rare Earths*; Elsevier: Amsterdam, The Netherlands, 2014; Volume 44, pp. 169–281.
14. Fagnani, F.; Colombo, A.; Malandrino, G.; Dragonetti, C.; Pellegrino, A.L. Luminescent 1,10-Phenanthroline β-Diketonate Europium Complexes with Large Second-Order Nonlinear Optical Properties. *Molecules* **2022**, *27*, 6990. [CrossRef] [PubMed]
15. Bünzli, J.-C.G.; Eliseeva, S.V. Intriguing aspects of lanthanide luminescence. *Chem. Sci.* **2013**, *4*, 1939–1949. [CrossRef]

16. Yang, D.; Liang, H.; Liu, Y.; Hou, M.; Kan, L.; Yang, Y.; Zang, Z. A large-area luminescent downshifting layer containing an $Eu^{3+}$ complex for crystalline silicon solar cells. *Dalton Trans.* **2020**, *49*, 4725–4731. [CrossRef]
17. Pellegrino, A.L.; Cortelletti, P.; Pedroni, M.; Speghini, A.; Malandrino, G. Nanostructured $CaF_2$:$Ln^{3+}$ ($Ln^{3+}$ = $Yb^{3+}/Er^{3+}$, $Yb^{3+}/Tm^{3+}$) thin films: MOCVD fabrication and their upconversion properties. *Adv. Mater. Interfaces* **2017**, *4*, 1700245. [CrossRef]
18. Lo Presti, F.; Pellegrino, A.L.; Milan, E.; Radicchi, E.; Speghini, A.; Malandrino, G. $Eu^{3+}$ activated $BaF_2$ nanostructured thin films: Fabrication and a combined experimental and computational study of the energy conversion process. *J. Mater. Chem. C* **2023**, *11*, 12195–12205. [CrossRef]
19. Andrade, A.B.; Ferreira, N.S.; Valerio, M.E.G. Particle size effects on structural and optical properties of $BaF_2$ nanoparticles. *RSC Adv.* **2017**, *7*, 26839–268485. [CrossRef]
20. Pellegrino, A.L.; Lucchini, G.; Speghini, A.; Malandrino, G. Energy conversion systems: Molecular architecture engineering of metal precursors and their applications to vapor phase and solution routes. *J. Mater. Res.* **2020**, *35*, 2950–2966. [CrossRef]
21. Vetrone, F.; Naccache, R.; Mahalingam, V.; Morgan, C.G.; Capobianco, J.A. The Active-Core/Active-Shell Approach: A Strategy to Enhance the Upconversion Luminescence in Lanthanide-Doped Nanoparticles. *Adv. Funct. Mater.* **2009**, *19*, 2924–2929. [CrossRef]
22. Cheng, T.; Marin, R.; Skripka, A.; Vetrone, F. Small and Bright Lithium-Based Upconverting Nanoparticles. *J. Am. Chem. Soc.* **2018**, *140*, 12890–12899. [CrossRef] [PubMed]
23. Malitson, I.H. A Redetermination of Some Optical Properties of Calcium Fluoride. *Appl. Opt.* **1963**, *2*, 1103. [CrossRef]
24. Normani, S.; Braud, A.; Soulard, R.; Doualan, J.L.; Benayad, A.; Menard, V.; Brasse, G.; Moncorge, R.; Goossens, J.P.; Camy, P. Site selective analysis of $Nd^{3+}$–$Lu^{3+}$ codoped $CaF_2$ laser crystals. *Cryst. Eng. Comm.* **2016**, *18*, 9016–9025. [CrossRef]
25. Cantarelli, I.X.; Pedroni, M.; Piccinelli, F.; Marzola, P.; Boschi, F.; Conti, G.; Sbarbati, A.; Bernardi, P.; Mosconi, E.; Perbellini, L.; et al. Multifunctional nanoprobes based on upconverting lanthanide doped $CaF_2$: Towards biocompatible materials for biomedical imaging. *Biomater. Sci.* **2014**, *2*, 1158–1171. [CrossRef] [PubMed]
26. Sokovnin, S.Y.; Il'ves, V.G.; Zuev, M.G.; Uimin, M.A. Physical properties of fluorides barium and calcium nanopowders produced by the pulsed electron beam evaporation method. *J. Phys. Conf. Ser.* **2018**, *1115*, 032092. [CrossRef]
27. Sokolov, N.S.; Suturin, S.M. MBE growth of calcium and cadmium fluoride nanostructures on silicon. *Appl. Surf. Sci.* **2001**, *175*, 619–628. [CrossRef]
28. Pellegrino, A.L.; La Manna, S.; Bartasyte, A.; Cortelletti, P.; Lucchini, G.; Speghini, A.; Malandrino, G. Fabrication of doped calcium fluoride thin films for energy upconversion in photovoltaics: A comparison of MOCVD and sol-gel approaches. *J. Mater. Chem. C* **2020**, *8*, 3865–3877. [CrossRef]
29. Brasse, G.; Loiko, P.; Grygiel, C.; Leprince, P.; Benayad, A.; Lemarie, F.; Doualan, J.-L.; Braud, A.; Camy, P. Liquid Phase Epitaxy growth of $Tm^{3+}$-doped $CaF_2$ thin-films based on LiF solvent. *J. Alloys Compd.* **2019**, *803*, 442–449. [CrossRef]
30. Kuzmina, N.P.; Tsymbarenko, D.M.; Korsakov, I.E.; Starikova, Z.A.; Lysenko, K.A.; Boytsova, O.V.; Mironov, A.V.; Malkerova, I.P.; Alikhanyan, A.S. Mixed ligand complexes of AEE hexafluoroacetylacetonates with diglyme: Synthesis, crystal structure and thermal behavior. *Polyhedron* **2008**, *27*, 2811–2818. [CrossRef]
31. Shannon, D. Revised effective ionic radii and systematic studies of interatomic distances in halides and chalcogenides. *Acta Crystallogr. Sect. A* **1976**, *32*, 751. [CrossRef]
32. Laval, J.P.; Mikou, A.; Frit, B.; Roult, G. Short-range order in heavily lanthanide(3+) doped calcium fluoride fluorites: A powder neutron diffraction study. *Solid State Ion.* **1988**, *28*, 1300–1304. [CrossRef]
33. Wang, F.; Fan, X.; Pi, D.; Wang, M. Synthesis and luminescence behavior of $Eu^{3+}$-doped $CaF_2$ nanoparticles. *Solid State Commun.* **2005**, *133*, 775–779. [CrossRef]
34. Cortelletti, P.; Pedroni, M.; Boschi, F.; Pin, S.; Ghigna, P.; Canton, P.; Vetrone, F.; Speghini, A. Luminescence of $Eu^{3+}$ Activated $CaF_2$ and $SrF_2$ Nanoparticles: Effect of the Particle Size and Codoping with Alkaline Ions. *Cryst. Growth Des.* **2018**, *18*, 686–694. [CrossRef]
35. Ritter, B.; Haida, P.; Fink, F.; Krahl, T.; Gawlitza, K.; Rurack, K.; Scholz, G.; Kemnitz, E. Novel and easy access to highly luminescent Eu and Tb doped ultra-small $CaF_2$, $SrF_2$ and $BaF_2$ nanoparticles–structure and luminescence. *Dalton Trans.* **2017**, *46*, 2925–2936. [CrossRef] [PubMed]
36. Bustamante, A.; Barranco, J.; Calisto, M.E.; López-Cruz, E.; Aguilar-Zarate, P. Alkaline earth fluoride and Eu doped thin films obtained by electrochemical processing. *J. Solid State Electrochem.* **2023**, *27*, 2115–2125. [CrossRef]
37. Zikmund, T.; Bulìr, J.; Novotný, M.; Jiříček, P.; Houdková, J.; Lančok, J. Electric and magnetic dipole emission of $Eu^{3+}$: Effect of proximity to a thin aluminum film. *J. Lumin.* **2022**, *246*, 118778. [CrossRef]
38. Bondzior, B.; Dereń, P.J. The role of hypersensitive transition in $Eu^{3+}$ optical probe for site symmetry determination in BaScBO-SrScBO solid-solution phosphor. *J. Lumin.* **2018**, *201*, 298–302. [CrossRef]
39. Srivastava, A.M.; Brik, M.G.; Beers, W.W.; Cohen, W. The influence of nd0 transition metal cations on the $Eu^{3+}$ asymmetry ratio $R = \frac{I(^5D_0 \to ^7F_2)}{I(^5D_0 \to ^7F_1)}$ and crystal field splitting of $^7F_1$ manifold in pyrochlore and zircon compounds. *Opt. Mater.* **2021**, *114*, 110931. [CrossRef]
40. Racu, A.V.; Ristić, Z.; Ćirić, A.; Đorđević, D.; Bușe, G.; Poienar, M.; Gutmann, M.J.; Ivashko, O.; Ștef, M.; Vizman, D.; et al. Analysis of site symmetries of $Er^{3+}$ doped $CaF_2$ and $BaF_2$ crystals by high resolution photoluminescence spectroscopy. *Opt. Mater.* **2023**, *136*, 113337. [CrossRef]
41. Hu, L.; Reid, M.F.; Duan, C.-K.; Xia, S.; Yin, M. Extraction of crystal-field parameters for lanthanide ions from quantum-chemical calculations. *J. Phys. Condens. Matter* **2011**, *23*, 045501. [CrossRef]

42. Czaja, M.; Bodył-Gajowska, S.; Lisiecki, R.; Meijerink, A.; Mazurak, Z. The luminescence properties of rare-earth ions in natural fluorite. *Phys. Chem. Miner.* **2012**, *39*, 639–648. [CrossRef]
43. Petit, V.; Camy, P.; Doualan, J.-L.; Portier, X.; Moncorgé, R. Spectroscopy of $Yb^{3+}:CaF_2$: From isolated centers to clusters. *Phys. Rev. B* **2008**, *78*, 085131. [CrossRef]
44. Secu, C.; Rostas, A.-M.; Secu, M. Europium (II)-Doped $CaF_2$ Nanocrystals in Sol-Gel Derived Glass-Ceramic: Luminescence and EPR Spectroscopy Investigations. *Nanomaterials* **2022**, *12*, 3016. [CrossRef] [PubMed]

**Disclaimer/Publisher's Note:** The statements, opinions and data contained in all publications are solely those of the individual author(s) and contributor(s) and not of MDPI and/or the editor(s). MDPI and/or the editor(s) disclaim responsibility for any injury to people or property resulting from any ideas, methods, instructions or products referred to in the content.

Article

# Screening $Ba_{0.9}A_{0.1}MnO_3$ and $Ba_{0.9}A_{0.1}Mn_{0.7}Cu_{0.3}O_3$ (A = Mg, Ca, Sr, Ce, La) Sol-Gel Synthesised Perovskites as GPF Catalysts

Nawel Ghezali, Álvaro Díaz Verde and María José Illán Gómez *

MCMA Group, Inorganic Chemistry Department and Institute of Materials of the University of Alicante (IUMA), Faculty of Sciences, University of Alicante, 03690 Alicante, Spain; ghezalinawel34@gmail.com (N.G.); alvaro.diaz@ua.es (Á.D.V.)
* Correspondence: illan@ua.es

**Abstract:** $Ba_{0.9}A_{0.1}MnO_3$ (BM-A) and $Ba_{0.9}A_{0.1}Mn_{0.7}Cu_{0.3}O_3$ (BMC-A) (A = Mg, Ca, Sr, Ce, La) perovskite-type mixed oxides were synthesised, characterised, and used for soot oxidation in simulated Gasoline Direct Injection (GDI) engine exhaust conditions. The samples have been obtained by the sol-gel method in an aqueous medium and deeply characterised. The characterization results indicate that the partial substitution of Ba by A metal in $BaMnO_3$ (BM) and $BaMn_{0.7}Cu_{0.3}O_3$ (BMC) perovskites: (i) favours the hexagonal structure of perovskite; (ii) improves the reducibility and the oxygen desorption during Temperature-Programmed Desorption ($O_2$-TPD) tests and, consequently, the oxygen mobility; (iii) mantains the amount of oxygen vacancies and of Mn(IV) and Mn(III) oxidation states, being Mn(IV) the main one; and (iv) for $Ba_{0.9}A_{0.1}Mn_{0.7}Cu_{0.3}O_3$ (BMC-A) series, copper is partially incorporated into the structure. The soot conversion data reveal that $Ba_{0.9}La_{0.1}Mn_{0.7}Cu_{0.3}O_3$ (BMC-La) is the most active catalyst in an inert (100% He) reaction atmosphere, as it presents the highest amount of copper on the surface, and that $Ba_{0.9}Ce_{0.1}MnO_3$ (BM-Ce) is the best one if a low amount of $O_2$ (1% $O_2$ in He) is present, as it combines the highest emission of oxygen with the good redox properties of Ce(IV)/Ce(III) and Mn(IV)/Mn(III) pairs.

**Keywords:** perovskite; cerium; lanthanum; calcium; magnesium; strontium; copper; soot oxidation; GDI engines

**Citation:** Ghezali, N.; Díaz Verde, Á.; Illán Gómez, M.J. Screening $Ba_{0.9}A_{0.1}MnO_3$ and $Ba_{0.9}A_{0.1}Mn_{0.7}Cu_{0.3}O_3$ (A = Mg, Ca, Sr, Ce, La) Sol-Gel Synthesised Perovskites as GPF Catalysts. *Materials* **2023**, *16*, 6899. https://doi.org/10.3390/ma16216899

Academic Editor: Aleksej Zarkov

Received: 28 September 2023
Revised: 20 October 2023
Accepted: 25 October 2023
Published: 27 October 2023

**Copyright:** © 2023 by the authors. Licensee MDPI, Basel, Switzerland. This article is an open access article distributed under the terms and conditions of the Creative Commons Attribution (CC BY) license (https://creativecommons.org/licenses/by/4.0/).

## 1. Introduction

Nowadays, the accumulation in the atmosphere of pollutants generated by automobile engines is one of the main environmental problems due to their negative effects. During the combustion process, gasoline and diesel engines emit harmful compounds such as carbon monoxide, nitrogen oxides, and particulate matter (PM) that seriously affect the air quality [1]. These primary pollutants react with other atmospheric compounds to form secondary pollutants, such as, among others, ground-level ozone, which stands out for being harmful to plants and causing respiratory issues in humans. GDI engines present high fuel efficiency and low $CO_2$ emissions but generate a high amount of PM, especially with a diameter lower than 10 μm, which is the most dangerous one as it can deeply penetrate into the organism (lungs, bloodstream, etc.) [2]. As a result, the European Commission (by the European Green Deal's Zero Pollution Action Plan [3]) established the 2030 target, which consists of lowering the number of particulate matter with a diameter of 2.5 μm (PM2.5) by at least 44% of 2005 levels. To deal with this problem, gasoline particulate filters (GPFs) are being used in GDI engine vehicles, which must undergo routine regeneration to avoid soot accumulation in the inner channel. On the other hand, three-way catalysts (TWCs) have been employed to remove the gaseous pollutants from gasoline engines since the 1980s. Thus, both TWCs and GPFs are essential after-treatment devices for GDI engines in order to meet the Euro 6 limits [4]. In this context, catalysts for soot removal at low temperatures and at low oxygen (or zero) partial pressures are highly demanded for GPF devices [5]. Until now, noble metal-based catalysts were the most commonly employed formulations for GDI soot removal, especially those composed by Pt. However, noble metals are scarce and expensive [6–8], making it mandatory to look for alternative and more economically accessible formulations. Recently, ceria-based mixed oxides have been proposed as promising catalysts due to their oxygen storage properties and the versatility of Ce to modify its oxidation state [9].

In this context, perovskite-type mixed oxides ($ABO_3$) are also considered potential catalytic formulations for soot oxidation, as they exhibit intriguing and adjustable physicochemical properties that could be improved by using different strategies, such as, among others, the modification of the composition by the partial substitution of A and/or B cations [10–13]. In general, transition metals (Fe, Co, Mn, Cr, Cu, V, etc.) are usually located in the B position, while lanthanide and/or alkaline earth metals (Sr, Ca, Ba, etc.) typically occupy the A position [10,14]. Because of the doping with A and/or B cations with different sizes and oxidation states, the redox properties, the generation of oxygen vacancies and the oxygen mobility of perovskites can be greatly improved [10,15]. Additionally, the doping could also modify the electronic structure and, consequently, the semiconductor properties, making these solids good candidates for several catalytic and electrocatalytic applications, such as the $CO_2$ reduction reaction ($CO_2RR$) [16–18], the oxygen reduction reaction (ORR) [19], the oxygen evolution reaction (OER) [20], and some photocatalytic processes [19,21,22]. Thus, in previous studies, the authors determined that the partial substitution of iron by copper in a $BaFe_{1-x}Cu_xO_3$ catalyst series (x = 0, 0.1, 0.3, and 0.4) modifies the catalytic performance for soot oxidation under GDI and diesel engine exhaust conditions [23,24]. These perovskites catalyzed the oxidation of soot in both exhaust conditions that are in the presence of $NO_2$ (diesel engines) and the 1% $O_2$ ("fuel cut" stage of GDI engine exhaust). $BaFeO_3$ perovskite was the most active catalyst, whose performance was mainly related to the lattice oxygen mobility, which decreased with copper content.

On the other hand, it is well established that $AMnO_3$ perovskites, due to their redox properties related to the electronic configuration of Mn(III) and Mn(IV) [25–27], are active catalysts for oxidation reactions. Moreover, Mn(III) presents the Jahn–Teller effect, which is a distortion that provokes some structural defects that generate active sites for oxidation reactions [28–30]. Thus, the presence of these two oxidation states, especially an enriched Mn(IV) surface, increases the oxygen mobility and, consequently, the catalytic activity for the oxidation of soot [31]. In fact, in a previous work, the authors used $BaMnO_3$ and $BaMn_{1-x}Cu_xO_3$ mixed oxides with a perovskite-like structure (obtained by employing various synthesis methods that allow particular chemical and physical properties) as feasible catalysts for GPF systems [31]. The results obtained for soot removal in simulated GDI engine exhaust conditions (i.e., low percentage of oxygen) reveal that, on the one hand, the presence of oxygen vacancies is required to adsorb and activate oxygen, and, on the other hand, a labile Mn(IV)/Mn(III) redox pair is needed to dissociate the adsorbed oxygen. Thus, the coexistence of both properties allows the transport of the activated oxygen towards the soot.

Considering these conclusions, the aim of this work is the synthesis by the sol-gel method of two series of barium manganese perovskite-type mixed oxides in which 10% of barium has been replaced (i.e., $Ba_{0.9}A_{0.1}MnO_3$ and $Ba_{0.9}A_{0.1}Mn_{0.7}Cu_{0.3}O_3$, where A = Ca, Sr, Mg, Ce, or La). These samples will be tested as catalysts for GPFs to be used for soot removal in simulated GDI engine exhaust conditions.

## 2. Materials and Methods

### 2.1. Synthesis of Catalyst

The sol-gel method adapted to aqueous medium [32–34] was used for the synthesis of the two series ($Ba_{0.9}A_{0.1}MnO_3$ and $Ba_{0.9}A_{0.1}Mn_{0.7}Cu_{0.3}O_3$) of samples. The metal precursors used are the following: barium acetate ($Ba(CH_3COO)_2$, Sigma-Aldrich, St. Louis, MO, USA 99.0% purity), calcium nitrate tetrahydrate ($Ca(NO_3)_2 \cdot 4H_2O$, Sigma-Aldrich, 99.0% purity), lanthanum(III) nitrate hydrate ($La(NO_3)_3 \cdot H_2O$, Sigma-Aldrich,

99.0% purity), magnesium nitrate hexahydrate (Mg(NO$_3$)$_2$*6H$_2$O, Sigma-Aldrich, 99.0% purity), cerium(III) nitrate hexahydrate (Ce(NO$_3$)$_3$*6H$_2$O, Sigma-Aldrich, 99.0% purity), strontium nitrate (Sr(NO$_3$)$_2$, Sigma-Aldrich, 99.0% purity), copper(II) nitrate trihydrate (Cu(NO$_3$)$_2$*3H$_2$O, Panreac, Castellar del Vallès, Spain, 99.0% purity), and manganese(II) nitrate tetrahydrate (Mn(NO$_3$)$_2$*4H$_2$O, Sigma-Aldrich, 99.0% purity). Additionally, citric acid (C$_6$H$_8$O$_7$, Sigma-Aldrich, 99.0% purity) has been employed as a complexing agent (using a citric acid/Ba ratio of 2), and EDTA (Sigma-Aldrich, 98.5% purity) has also been added as a chelating agent (EDTA/Ba = 2) for the synthesis of the BaMnO$_3$ reference sample to avoid the precipitation of metal precursors. To obtain the gel, citric acid was dissolved in 40 mL of distilled water at 60 °C, and then the metal precursors, in the same order in which the metals appear in the perovskite formulae, are added. In the case of the BaMnO$_3$ reference sample, EDTA is incorporated into the dissolution before the metal precursors, and, finally, citric acid is included. After that, the solution was stirred at 65 °C for 5 h. Throughout the process, the pH was maintained at 8.5 by adding an ammonia solution (Panreac, 30.0 wt%). Then, the gel was dried at 90 °C for 48 h, and the resulting powder was calcined at 850 °C for 6 h.

## 2.2. Characterization

For sample characterization, the following techniques were employed.

The elemental composition was obtained by Inductively Coupled Plasma Optical Emission Spectroscopy (ICP-OES) on a Perkin-Elmer device model Optimal 4300 DV (Waltham, MA, USA). For each experiment, 10 mg of catalyst was dissolved in a mixture of 5 mL of aqua regia and 10 mL of distilled water.

The textural properties were obtained by N$_2$ adsorption (at $-196$ °C) in an Autosorb-6B device (Quantachrome, Anton Paar Austria GmbH, Graz, Austria). Before the adsorption experiments, degassification at 250 °C for 4 h was carried out.

X-ray Diffraction (XRD) was used for determining the crystalline structure, using the XRD patterns recorded (in a Bruker D8-Advance device, Billerica, MA, USA) between 20° and 80° 2θ angles (step rate of 0.4°/min) and using Cu K$_\alpha$ (0.15418 nm) radiation.

Surface chemistry composition was obtained by X-ray photoelectron spectroscopy (XPS) in a K-Alpha photoelectron spectrometer device (Thermo-Scientific, Waltham, MA, USA) with an Al K$_\alpha$ (1486.7 eV) radiation source. To obtain XPS spectra, the pressure of the analysis chamber was held at $5 \times 10^{-10}$ mbar. The binding energy (BE) and kinetic energy (KE) scales were adjusted by setting the C 1 s transition to 284.6 eV, and the BE and KE values were determined with the peak-fit software of the spectrometer (Thermo Avantage v5.9929).

Temperature-Programmed Reduction with H$_2$ (H$_2$-TPR) in a Pulse Chemisorb 2705 (from Micromeritics, Norcross, GA, USA) provided by a Thermal Conductivity Detector (TCD) was used to estimate the reducibility of samples. For the tests, 30 mg of sample, heated at 10 °C/min from 25 °C to 1000 °C in a 5% H$_2$/Ar atmosphere (40 mL/min), was used. A CuO reference sample was employed for the quantification of H$_2$ consumption.

Oxygen Temperature-Programmed Desorption (O$_2$-TPD) experiments were performed in a Thermal Gravimetric Mass Spectrometry (TG-MS) device (Q-600-TA and Thermostar from Balzers Instruments (Balzers, Liechtenstein), respectively). Sixteen milligrams of sample (heated at 10 °C/min from room temperature to 950 °C under a 100 mL/min helium atmosphere) was used. Before the experiments, samples were preheated to 150 °C for 1 h for moisture removal. The 18, 28, 32, and 44 $m/z$ signals were recorded for the quantification of H$_2$O, CO, O$_2$, and CO$_2$ evolved. The amount of oxygen was calculated based on a CuO reference sample.

## 2.3. Activity Tests

The soot oxidation tests (under simulated GDI engine exhaust conditions) were developed on the TG-MS device employed for O$_2$-TPD. Thus, 16 mg of a catalyst and soot mixture (soot:catalyst ratio of 1:8, using Printex-U as model soot in loose contact mode) was preheated at 150 °C (1 h) in a 1% O$_2$/He mixture (100 mL/min); subsequently, the temperature was increased until 900 °C at 10 °C/min (soot-TPR). Two different reactant mixtures were employed: (i) 1% $O_2$/He, which reproduces "fuel cuts" GDI engine exhaust conditions, and (ii) 100% He, which represents regular stoichiometric GDI engine operations [23,24,31].

The soot conversion and the selectivity to $CO_2$ percentages were calculated by these equations:

$$\text{Soot conversion (\%)} = \frac{\Sigma_0^t CO_2 + CO}{\Sigma_0^{final}(CO_2 + CO)} \times 100 \quad (1)$$

$$\text{Selectivity to } CO_2 \text{ (\%)} = \frac{CO_2 \text{ total}}{(CO_2 + CO)_{total}} \cdot 100 \quad (2)$$

where $\Sigma_0^t(CO_2 + CO)$ is the amount of $CO_2$ and CO generated at time $t$, while $\Sigma_0^{final}(CO_2 + CO)$ is the total amount of CO + $CO_2$ evolved during the experiment, coming from the oxidation of the total amount of soot.

## 3. Results and Discussion

### 3.1. Characterization

#### 3.1.1. $Ba_{0.9}A_{0.1}MnO_3$ Series

The nomenclature for the mixed oxides is shown in Table 1, along with the specific surface area values (obtained by applying the Brunauer-Emmett-Teller (BET) equation to $N_2$ adsorption data), the A metal content (determined by ICP-OES), and some XRD data.

**Table 1.** Nomenclature, XRD data, metal content and BET surface area of BM-A catalysts.

| Nomenclature | Molecular Formula | Specific Surface Area (m²/g) | A (wt%) | Intensity (a.u) [a] | Average Crystal Size (nm) | Cell Parameters (Å) [b] | |
|---|---|---|---|---|---|---|---|
| | | | | | | a | c |
| BM | $BaMnO_3$ | 3 | - | 1154 | 46 | 5.7 | 4.9 |
| BM-Ca | $Ba_{0.9}Ca_{0.1}MnO_3$ | 11 | 1.6 | 1913 | 25 | 5.7 | 4.8 |
| BM-La | $Ba_{0.9}La_{0.1}MnO_3$ | 7 | 1.1 | 1562 | 28 | 5.7 | 4.8 |
| BM-Mg | $Ba_{0.9}Mg_{0.1}MnO_3$ | 7 | 4.2 | 2382 | 28 | 5.7 | 4.8 |
| BM-Sr | $Ba_{0.9}Sr_{0.1}MnO_3$ | 5 | 4.8 | 2382 | 18 | 5.6 | 4.9 |
| BM-Ce | $Ba_{0.9}Ce_{0.1}MnO_3$ | 10 | 1.3 | 1913 | 22 | 5.5 | 5.0 |

[a] Corresponding to the main XRD hexagonal perovskite peak, [b] Calculated using the main XRD hexagonal perovskite peak.

The low surface areas presented in Table 1 suggest, as expected for perovskite-type mixed oxides [10], that all samples present low specific surface areas and extremely small porosity development that, according to K. Akinlolu [35], could be a consequence of the calcination conditions used in the synthesis (850 °C). Note that the calcination temperature used for synthesis was the minimum one needed to obtain the perovskite phase [36] because, if higher temperatures were used, larger crystal sizes would be obtained, thus promoting a decrease in the number of active sites [37]. On the other hand, ICP-OES data confirm that the mixed oxides contain almost all the amount of A metal supplied during the synthesis.

Figure 1a shows the XRD patterns of samples, including the raw $BaMnO_3$ (BM), obtained also by the sol-gel method. The hexagonal 2H-$BaMnO_3$ perovskite structure (PDF number: 026-0168, denoted by the ICDD, the International Centre of Diffraction Data) is the main crystalline phase for all samples [32]. This structure is formed by chains of face-sharing $MnO_6$ units rather than by corner-sharing $MnO_6$ units [32,36,38]. Note that the partial substitution of the A cation does not cause a significant change to the $BaMnO_3$ hexagonal perovskite structure, as the position of the main peak (shown in Figure 1b) and the values of the $a$ and $c$ cell parameters (included in Table 1) are similar to those shown by

BM. However, the intensity of the main XRD peak of the hexagonal perovskite structure is higher for all BM-A samples than for BM perovskite, suggesting that the addition of A metal promotes crystal growth during the calcination step. On the other hand, $Ba_3Mn_2O_8$ and $MnO_2$ (PDF numbers: 073-0997 and 024-0735, respectively) are detected as minority phases. The Williamson-Hall method [39] was used to determine the average crystallite size shown in Table 1, and a decrease is detected after the partial substitution of Ba by A metal, presenting BM-Sr the lowest value.

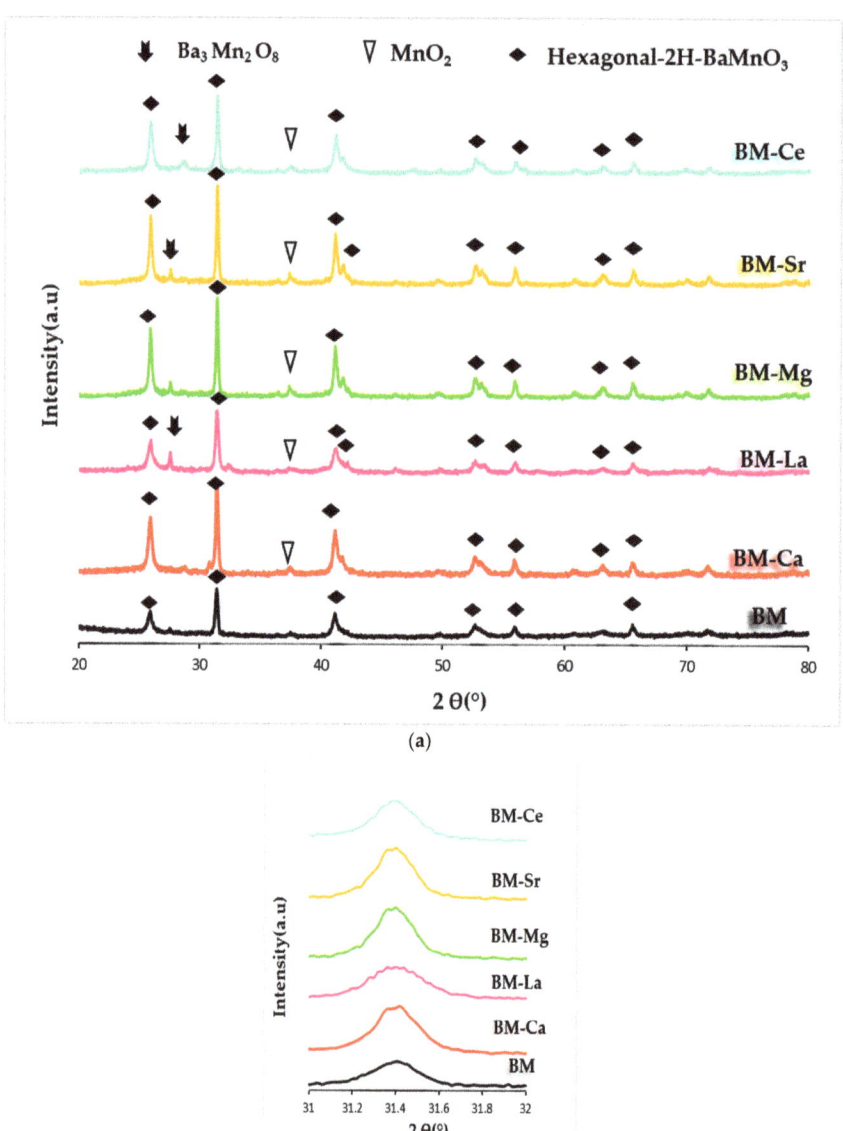

**Figure 1.** (a) XRD profiles of BM-A and BM catalysts, and (b) magnification corresponding to the hexagonal 2H-$BaMnO_3$ diffraction peak.

XPS gives information about the species present on the surface and Figure 2 shows the O 1s (a) and Mn $2p_{3/2}$ (b) XPS spectra for BM and BM-A samples.

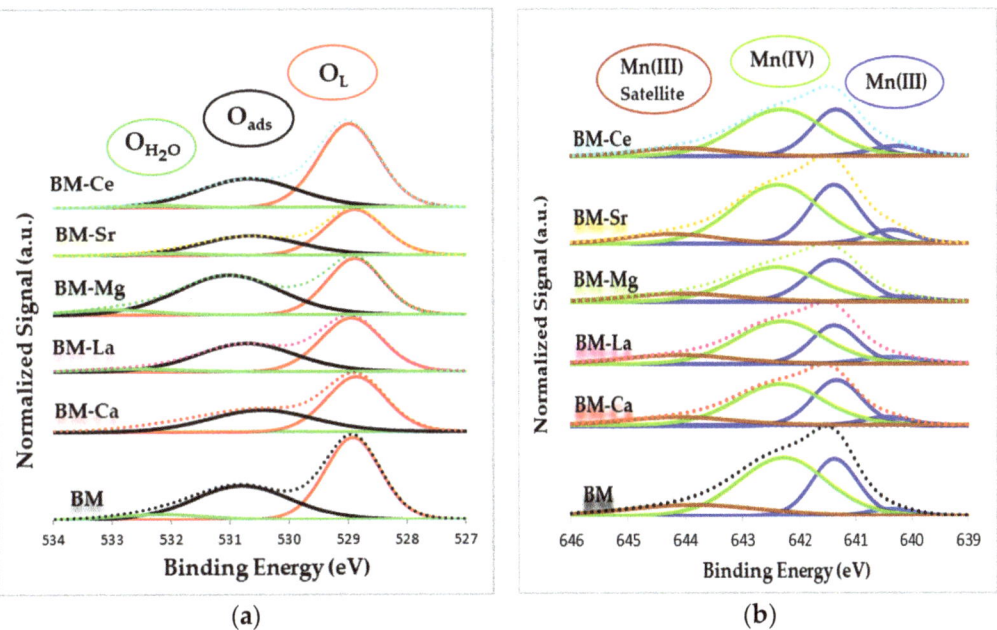

**Figure 2.** XPS spectra of BM-A and BM catalysts in the (a) O 1s and (b) Mn $2p_{3/2}$ core levels regions.

In the O 1s XPS spectra, featured in Figure 2a, three different contributions are observed [40,41]: (i) at low binding energies, a sharp peak centered around 529 eV, which is attributed to lattice oxygen (the so-called "$O_L$"), (ii) at intermediate binding energies (around 531 eV), a wide peak which corresponds to adsorbed oxygen species (named "$O_{ads}$"), such as surface carbonate ($CO_3^{2-}$), hydroxyl groups ($OH^-$) and peroxide ($O_2^{2-}$) or superoxide ($O_2^-$) ions, and (iii) a peak at higher binding energies (532 eV) associated to chemisorbed water (labelled as "$O_{H2O}$"). From these XPS spectra, the binding energies for the maximum of the $O_L$ and $O_{ads}$ deconvoluted bands have been obtained, which are included in Table 2, and these data indicate that the partial Ba replacement does not significantly affect the binding energy of the deconvoluted peak maxima. The $O_L/(Ba + Mn + A)$ ratio, calculated using the area under the $O_L$ peak and the areas under the signals associated with cations (Table 2), would reveal that surface oxygen vacancies exist if the $O_L/(Ba + Mn + A)$ XPS ratio is lower than the nominal one (1.5, calculated based on the chemical formulae of the perovskite) [32,40,42]. So, because all the XPS ratios are lower than 1.5, oxygen vacancies exist on the surface of all samples, and the partial replacement of Ba does not substantially change these values. The generation of oxygen surface vacancies [32,36,40], which is required to achieve the positive charge imbalance caused by Mn(III) (see next paragraph), is relevant in the soot oxidation reaction [42] because it allows the creation of the reactive oxygen species.

In the Mn $2p_{3/2}$ XPS spectra shown in Figure 2b, the following contributions are detected: (i) Mn(III), located at around 641 eV; (ii) Mn(IV), found at approximately 642 eV; and (iii) the Mn(III) satellite peaks at ca 644 eV [32,42–45]. These spectra reveal that Mn(III) and Mn(IV) coexist on the surface. This fact, previously observed [32,36,45,46] for manganese-based perovskites, indicates that the Mn(II) precursor is being oxidized to Mn(III)/Mn(IV) to achieve electroneutrality on the surface. Note that on the surface of an ideal $BaMnO_3$ perovskite (considering the charges of barium ($Ba^{2+}$) and oxygen ($O^{2-}$)

ions), Mn(IV) species must be present to preserve the electroneutrality. However, as oxygen vacancies exist on the surface (as already demonstrated by O 1s transition data), Mn(III) should be present to compensate the negative charge deficiency. The Mn(IV)/Mn(III) ratio, calculated with the area of the deconvoluted XPS peaks and included in Table 2, features that Mn(IV) is the predominant oxidation state for all catalysts, as all the Mn(IV)/Mn(III) values are above 1. However, a decrease in the Mn(IV) amount is observed after partial substitution of Ba by A metal, being more significant for BM-Mg. As indicated in the introduction, the presence of both oxidation states is relevant, and it could be expected that, as it was observed in other studies focused on La-Mn and Ba-Mn perovskite-type catalysts [32,38,42,45,46], a surface enriched with Mn(IV) allows improved oxygen mobility, thus enhancing the catalytic activity.

Table 2. XPS data of BM and BM-A catalysts.

| Catalyst | B.E Max Mn(III) (eV) | B.E Max Mn(VI) (eV) | B.E Max $O_L$ (eV) | B.E Max $O_{ads}$ (eV) | $\frac{O_L}{(Ba + Mn + A)}$ (Nominal = 1.5) | $\frac{Mn(IV)}{Mn(III)}$ |
|---|---|---|---|---|---|---|
| BM | 641.4 | 642.3 | 528.9 | 530.8 | 1.0 | 1.7 |
| BM-Ca | 641.3 | 642.3 | 528.9 | 530.5 | 1.0 | 1.6 |
| BM-La | 641.4 | 642.3 | 529.0 | 530.7 | 1.0 | 1.6 |
| BM-Mg | 641.4 | 642.3 | 528.9 | 531.0 | 1.0 | 1.1 |
| BM-Sr | 641.4 | 642.4 | 528.9 | 530.7 | 1.1 | 1.4 |
| BM-Ce | 641.4 | 642.4 | 529.0 | 530.7 | 1.0 | 1.4 |

Temperature-Programmed Reduction tests using hydrogen as a reducing agent ($H_2$-TPR) allow estimating the reducibility and redox properties of samples. The $H_2$ consumption profiles for perovskites are determined by the oxidation state and redox properties of the B-site metal [30,45], and the number and sequence of these peaks are strongly dependent on their identity [10,38]. On the other hand, according to the literature, the metal at the A-site typically shows a unique oxidation state and is hardly reduced [10,47]. The $H_2$-TPR profile for $MnO_2$, used as a reference, presents two overlapping peaks at around 400 and 500 °C that correspond to the reduction of $MnO_2/Mn_2O_3$ to $Mn_3O_4$ and of $Mn_3O_4$ to MnO, respectively [48]. Usually [30,32,42,45,47], a multiple-step reduction was observed for the manganese-based samples, showing: (i) an intense peak centered between 400–600 °C, corresponding to the Mn(IV) and Mn(III) reduction to Mn(II); (ii) a small peak from 700 °C to 800 °C due to the oxygen species reduction; and (iii) a third peak with a maximum between 900 °C and 1000 °C, corresponding to the Mn(III) to Mn(II) reduction in the bulk [47]. The $H_2$-TPR profiles of BM-A samples (Figure 3a) reveal that only for BM-La and BM-Mg, the higher reduction peak appears at lower temperatures than in BM, indicating an improvement in the redox properties. Based on the hydrogen consumption profiles shown in Figure 3a, the experimental hydrogen consumption per gram of sample has been calculated over the temperature range between 450 °C and 600 °C, and these data have been compared in Figure 3b with the theoretical hydrogen consumption calculated assuming a complete reduction of manganese, being as pure Mn(III) (blue line) or pure Mn(IV) (red line). Note that, for BM-A samples, the experimental $H_2$ consumptions are placed between both Mn(IV) and Mn(III) nominal values, being closer to Mn(III) for BM-Ca and to Mn(IV) for BM-Mg, as well as for raw BM. Thus, it seems that, as observed on the surface of catalysts (see XPS results), Mn(III)/Mn(IV) oxidation states are present in the bulk of all samples, and the predominant oxidation state is determined by the A metal.

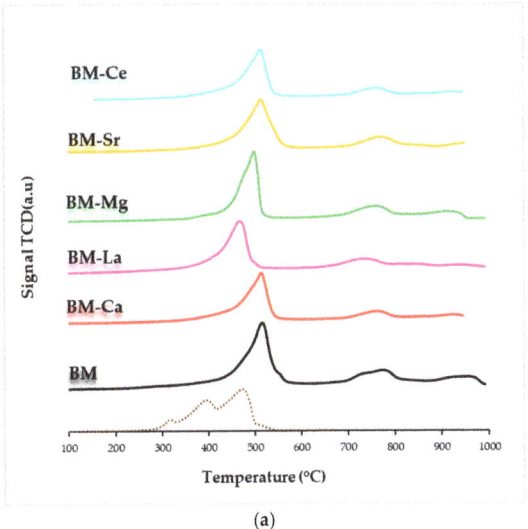
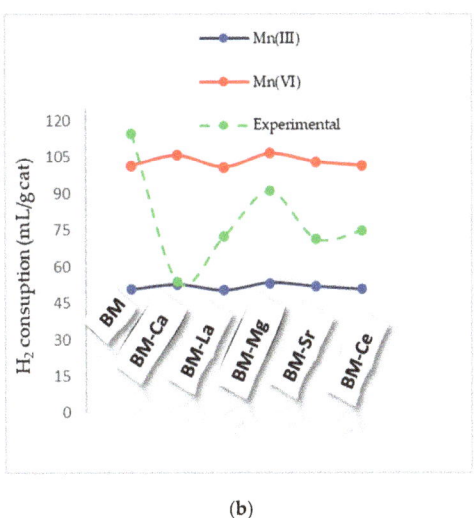

**Figure 3.** $H_2$-TPR profiles of BM-A, BM catalysts, and $MnO_2$, used as a reference (**a**) and $H_2$ consumption (mL/g of catalyst) (**b**).

$O_2$-TPD tests provide information about the labile oxygen species of the catalysts and about their oxygen mobility [10,36,45], as shown in the profiles displayed in Figure 4a. For perovskites, three oxygen desorption regions are usually detected: (i) the peak located between 150 °C and 350 °C, for the oxygen evolving from the adsorbed on the surface vacancies (called α-$O_2$); (ii) the peak appering between 500 °C and 700 °C, corresponding to the oxygen located on surface lattice defects, such as dislocation or grain frontiers (denoted as α'-$O_2$); and (iii) the peak over 700 °C, generated by the desorption of lattice oxygen (designed as β-$O_2$) due to the reduction of manganese in the perovskite lattice and related to the oxygen mobility through the bulk [10,32,42,45]. Thus, BM-Ca, BM-Ce, and BM-Mg samples evolve oxygen above 700 °C, coming from the perovskite lattice (β-oxygen), which depends on the partial reduction of Mn(IV) to Mn(III) [32,42,46,49], and for BM-Ce, also on the Ce(IV)/Ce(III) redox pair [50]. However, BM-La and BM-Sr do not follow the described trend, as BM-La shows a certain oxygen desorption at a temperature below 700 °C, which corresponds to α'-$O_2$, and BM-Sr presents a desorption profile similar to the raw perovskite (BM) without clearly defined peaks. Figure 4b features the amount of β-$O_2$ evolved, calculated with the area of the peak between 700 °C and 950 °C (except for BM-La) and CuO as a reference for the quantification. It is observed that, due to the partial replacement of Ba with A metals, the amount of $O_2$ released increases, mainly for BM-Ce due to the contribution of the Ce(IV)/Ce(III) redox pair. This finding reveals an increase in the mobility of β-$O_2$ due to the presence of A metals, showing BM-Ce the highest value.

### 3.1.2. $Ba_{0.9}A_{0.1}Mn_{0.7}Cu_{0.3}O_3$ Series

The nomenclature of catalysts, the BET surface area, the XRD data, and the A and Cu metal contents (obtained by ICP-OES) are reported in Table 3. The ICP-OES data confirm that almost all samples contain the amount of metals (Cu and A) supplied during the synthesis. Additionally, a low surface area is shown for BMC, and the addition of A metal does not significantly affect this parameter. This observation remains consistent with what was previously mentioned for the BM-A series of catalysts.

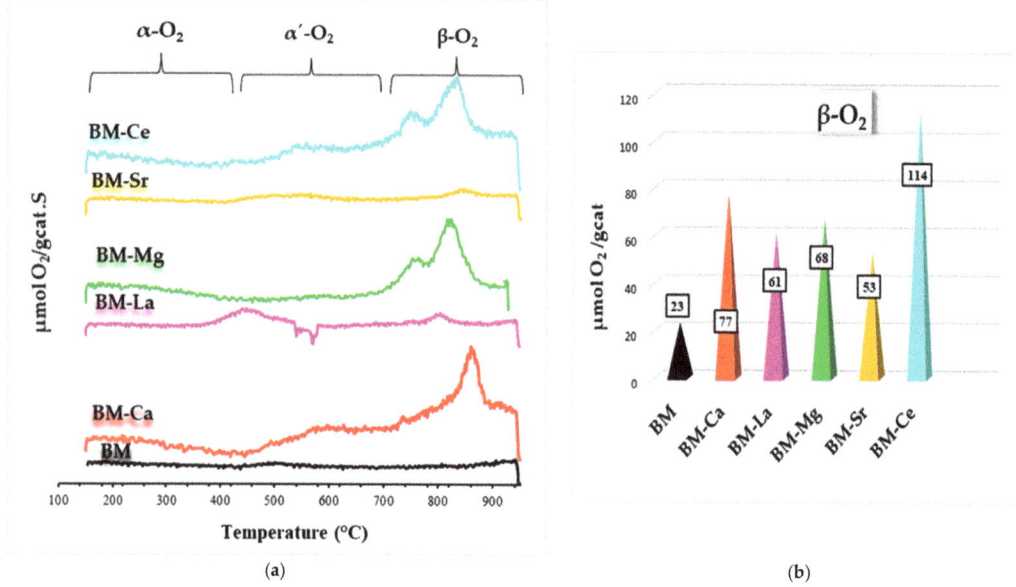

**Figure 4.** $O_2$-TPD profiles (a) and $\beta$-$O_2$ amount ($\mu$mol/g cat) (b) of BM and BM-A catalysts.

**Table 3.** Nomenclature, XRD data, A and Cu metal contents, and BET surface area of BMC-A catalysts.

| Nomenclature | Molecular Formula | A (wt%) | Cu (wt%) | BET (m²/g) | 2θ (°) [a] | Intensity (a.u) [a] | Average Crystal (nm) | Cell Parameters (Å) [b] | |
|---|---|---|---|---|---|---|---|---|---|
| | | | | | | | | a | c |
| BMC | $BaMn_{0.7}Cu_{0.3}O_3$ | - | 8.0 | 3 | 30.9 | 2448 | 30.7 | 5.8 | 4.3 |
| BMC-Ca | $Ba_{0.9}Ca_{0.1}Mn_{0.7}Cu_{0.3}O_3$ | 2.0 | 9.8 | 7 | 31.0 | 1154 | 29.3 | 5.7 | 4.3 |
| BMC-La | $Ba_{0.9}La_{0.1}Mn_{0.7}Cu_{0.3}O_3$ | 5.4 | 9.8 | 7 | 30.9 | 2064 | 18.6 | 5.8 | 4.2 |
| BMC-Mg | $Ba_{0.9}Mg_{0.1}Mn_{0.7}Cu_{0.3}O_3$ | 1.0 | 9.6 | 3 | 30.9 | 1246 | 25.9 | 5.8 | 4.3 |
| BMC-Sr | $Ba_{0.9}Sr_{0.1}Mn_{0.7}Cu_{0.3}O_3$ | 3.9 | 9.1 | 9 | 31.1 | 1913 | 25.0 | 5.8 | 4.3 |
| BMC-Ce | $Ba_{0.9}Ce_{0.1}Mn_{0.7}Cu_{0.3}O_3$ | 2.1 | 9.2 | 6 | 30.9 | 1441 | 22.4 | 5.6 | 4.3 |

[a] Corresponding to the main peak of $BaMnO_3$ polytype structure, [b] Calculated using the main diffraction peak of $BaMnO_3$ polytype structure.

Figure 5 features the XRD patterns of all samples. The diffraction peaks for BMC catalyst at 27.0°, 30.9°, 27.5°, 41.5°, 52.9°, 54.8°, 64.3, and 71.0° 2θ values perfectly match with the $BaMnO_3$ polytype perovskite structure [32,36], which is formed because copper partially replace the manganese in the perovskite lattice and causes a different order of the $MO_6$ units [51]. However, in BMC-A samples, the polytype structure partially changed back into the hexagonal 2H-$BaMnO_3$ structure, thus confirming that A metal has been inserted into the lattice of perovskite. For BMC-La, a peak corresponding to $BaMn_2O_3$ is also detected as a minority phase. The presence of the hexagonal perovskite structure, in addition to the polytype one, could be related to the presence of A metal, which hinders the introduction of copper into the perovskite network. However, as copper species (as CuO) are not clearly detected by XRD, it seems that copper should be inserted into the BM-A perovskites without causing a significant distortion of the hexagonal structure of $BaMnO_3$, so this structure seems to be partially preserved for BMC-A. In fact, the coexistence of both hexagonal and polytype structures was previously observed by the authors for under stoichiometric Ba-Cu-Mn perovskites ($Ba_{0.9}Cu_{0.3}Mn_{0.7}O_3$ and $Ba_{0.8}Cu_{0.3}Mn_{0.7}O_3$) [52]. The transition of the polytype structure to the hexagonal one in BMC-A samples is also evidenced by the decrease in the intensities of the main XRD peak of the former crystalline phase (see values included in Table 3). So, if A metal is present in the catalytic formulation, the crystal growth of the polytype phase seems to be hindered in favour of the formation of the hexagonal structure.

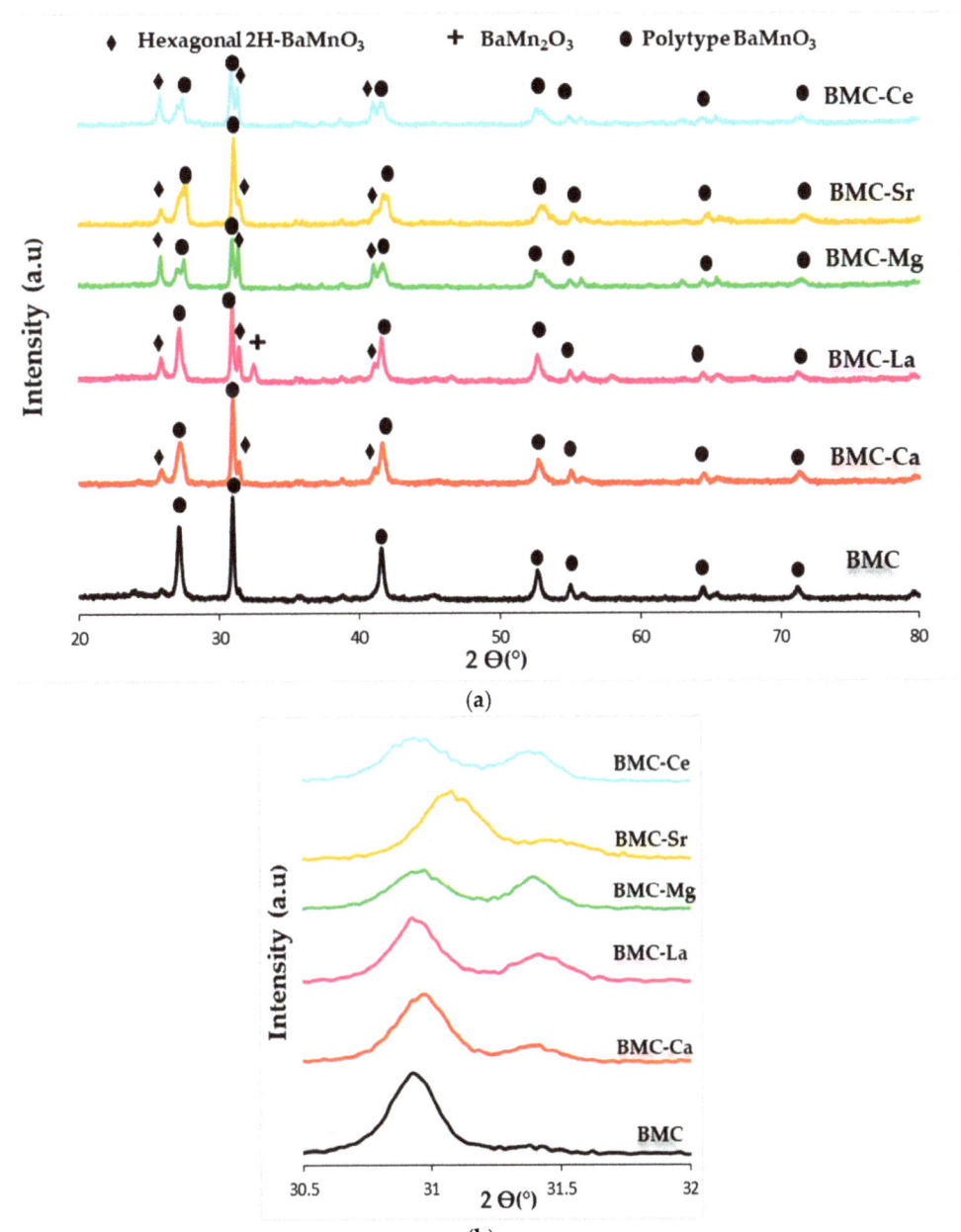

Figure 5. (a) XRD profiles of BMC-A and BMC catalysts and (b) magnification of the 2Θ diffraction angle region corresponding to main diffraction peak of hexagonal and polytype structure of $BaMnO_3$.

To deeply analyze the partial replacement of Ba by A metal, in Figure 5b, the 2θ diffraction angle between 30.5° and 32.5° (where the main peak of hexagonal and polytype perovskite structures appear, respectively) has been magnified. A slight displacement of the polytype structure main peak to higher diffraction angles with respect to BMC is observed for BMC-Sr catalyst, being less pronounced for BMC-Ca. In this line, Albaladejo-Fuentes and co-workers [53] reported a structural distortion for the $Ba_{0.9}Sr_{0.1}Ti_{0.8}Cu_{0.2}O_3$ (BTCu-Sr) perovskite due to the partial replacement of Ba(II) by Sr(II). Also, a shift of the main XRD peak was detected by Fu and coworkers [54] due to the presence of the smaller Ca(II) ions in the Ba-site in the $Ba_{1-x}Ca_xTiO_3$ catalyst series. Concerning BMC-La, BMC-Ce, and BMC-Mg catalysts, as a change in the position of the polytype $BaMnO_3$ structure peak is not clearly detected, it looks as if the A position of the perovskite structure does not include cerium, lanthanum, or magnesium. According to the effect of magnesium on the structure of $BaTiO_3$ [53,55], in which Mg(II) was located in the Ti(IV) position, in BMC-Mg, Mg(II) could be replacing Cu and/or Mn instead of Ba. Thus, based on the ionic radius values shown in Table 4, it is suggested that Mg could be incorporated in the B-site (partially replacing Cu and/or Mn) since the ionic radius of Mg(II) is closer to Mn(III) and Cu(II) ionic radii than to Ba(II) radius. However, Ce and La cations should be placed in the A site, since their ionic radii are closer to those of Ba(II), but it seems that this fact happens without a significant distortion of the hexagonal structure. As for the BM-A series, the Williamson-Hall method [39] was applied to determine the average crystallite size of the polytype perovskite phase included in Table 3. The average crystallite size of perovskite decreases for BMC-A samples, achieving the lowest value for BMC-La. On the contrary, the $a$ and $c$ parameters of BMC-A do not change with respect to the BMC, and only the BMC-Ce catalyst displays a small distortion of the polytype structure.

Table 4. Ionic radii of cation metals using the Goldshmidt correction [16,31,35].

| Metals | Ba(II) | Ca(II) | Mg(II) | La(III) | Ce(IV) | Ce(III) | Sr(II) | Cu(II) | Mn(IV) | Mn(III) |
|---|---|---|---|---|---|---|---|---|---|---|
| Ionic radii (pm) | 146.4 | 115.5 | 65.0 | 107.3 | 90.6 | 105.2 | 129.9 | 73.0 | 53.0 | 65.0 |

The XPS spectra for the O 1s, Mn $2p_{3/2}$, and Cu $2p_{3/2}$ transitions are shown in Figure 6.

The O 1s spectra of BMC and BMC-A (Figure 6a) feature the three contributions previously described for BM and BM-A, and Table 5 displays some relevant XPS data. Focusing attention on the binding energies corresponding to the maximum of the $O_L$ and $O_{ads}$ deconvoluted peaks, BMC-Mg shows the highest chemical shift towards higher binding energies for the $O_L$ peak, even though it is very low (0.2 eV). It is important to remember that a shift of the deconvoluted band towards lower binding energies indicates the existence of a richer electronic environment, whereas a displacement towards higher binding energies indicates the opposite. So, the displacement of the $O_L$ peak to higher binding energies reveals a poor electronic environment created because of the loss of oxygen from the $MnO_6$ octahedra, which, in order to achieve electroneutrality, takes place when Mg(II) occupies the Mn sites. As the $O_L$/(Ba + Mn + Cu + A) nominal ratio (1.5) is higher than the XPS one for all perovskites, oxygen surface vacancies exist on all samples. As discussed above for the BM-A series, this fact is caused by the coexistence of surface Mn(III) and Cu(II) [32]. Note that all BMC-A catalysts (except for BMC-Ca) present a slightly higher value of $O_L$/(Ba + Mn + Cu + A) ratio than BMC, so it seems that the partial substitution of Ba by A metal in BMC slightly decreases the presence of surface oxygen vacancies.

Figure 6b presents the Mn $2p_{3/2}$ XPS spectra of BMC and BMC-A samples, where Mn(III) and Mn(IV) signals and Mn(III) satellite contributions were indexed, with Mn(IV) located at 642.4 eV, Mn(III) at 641.2 eV, and the Mn(III) satellite at 644.0 eV [56]. Table 5 provides the binding energy corresponding to the maximum of these decovoluted peaks and, also, the Mn(IV)/Mn(III) ratios. Note that the presence of A metal does not significantly modify the binding energy for Mn(III) and Mn(IV) peaks, being the highest change observed for Mn(III) in BMC-Mg, which, in turn, is low (0.2 eV) and seems to be related to the location of Mg(II) in the Mn site. After doping with A metals, a significant change in the Mn(IV)/Mn(III) ratio is not observed, as the Mn(IV)/Mn(III) ratio decreases from 1.3 to

1.2 for most BMC-A samples, from 1.3 to 1.1 for BMC-La, and is not modified for BMC-Sr. However, as the ratios are higher than 1, Mn(IV) is the main oxidation state on the surface.

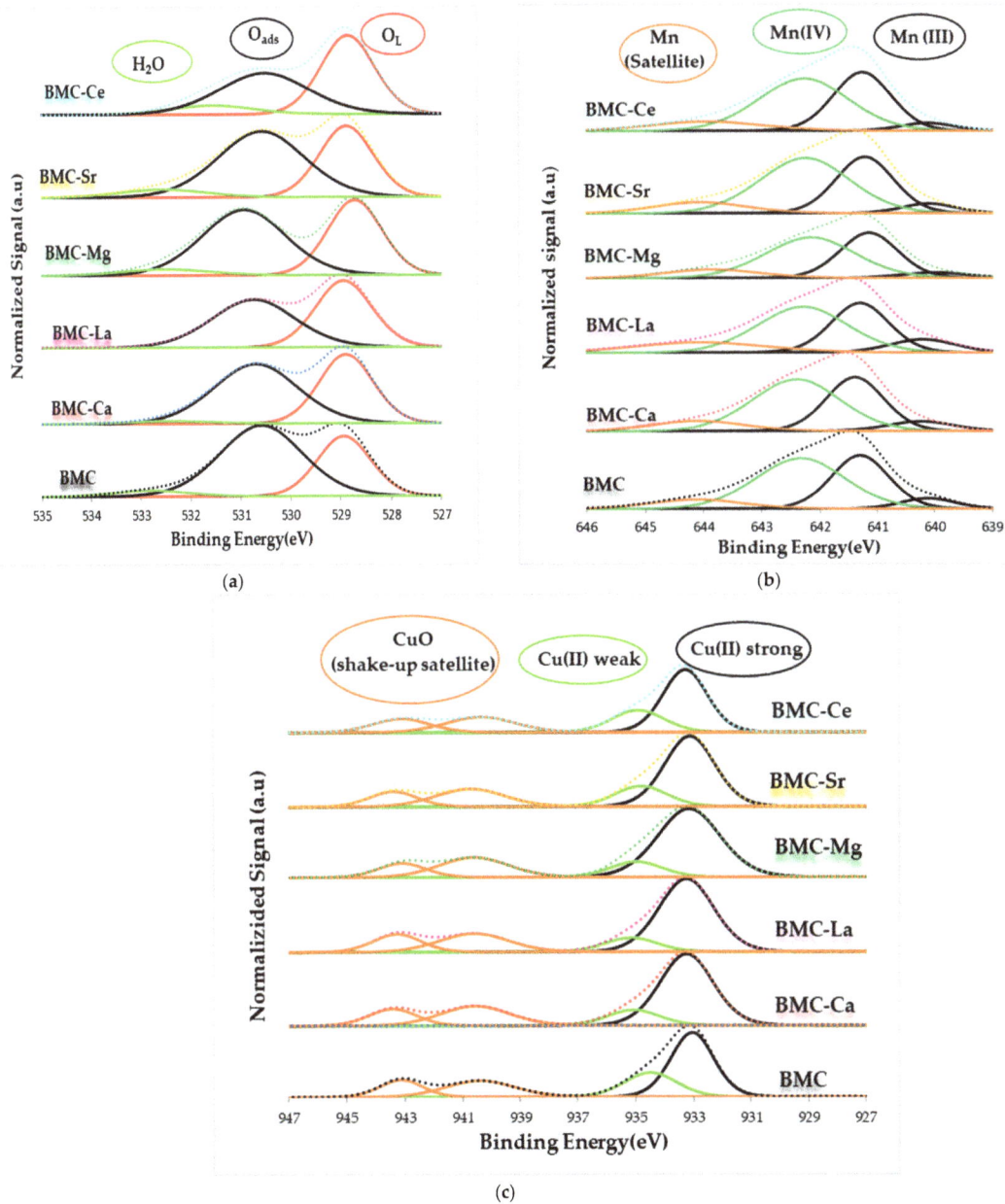

**Figure 6.** XPS spectra of BMC-A and BMC catalysts in the (a) O 1s and (b) Mn $2p_{3/2}$ and (c) Cu $2p_{3/2}$ core level regions.

Table 5. XPS data of BMC-A and BMC catalysts.

| Catalyst | B.Emax Cu(II)$_s$ [a] (eV) | B.Emax Cu(II)$_w$ [b] (eV) | B.Emax Mn(III) (eV) | B.Emax Mn(IV) (eV) | B.Emax O$_L$ (eV) | B.Emax O$_{ads}$ (eV) | $\frac{Mn(IV)}{Mn(III)}$ | $\frac{Cu}{M}$ [c] (Nominal = 0.15) | $\frac{O_L}{M}$ [c] (Nominal = 1.5) |
|---|---|---|---|---|---|---|---|---|---|
| BMC | 933.1 | 934.5 | 641.3 | 642.3 | 528.9 | 530.6 | 1.3 | 0.09 | 0.8 |
| BMC-Ca | 933.3 | 934.9 | 641.4 | 642.4 | 528.9 | 530.7 | 1.2 | 0.07 | 0.8 |
| BMC-La | 933.2 | 935.1 | 641.3 | 642.3 | 528.9 | 530.7 | 1.1 | 0.10 | 0.9 |
| BMC-Mg | 933.1 | 935.0 | 641.1 | 642.2 | 528.7 | 530.9 | 1.2 | 0.09 | 0.9 |
| BMC-Sr | 933.2 | 935.2 | 641.2 | 642.3 | 528.9 | 530.6 | 1.3 | 0.09 | 0.9 |
| BMC-Ce | 933.1 | 934.8 | 641.3 | 641.3 | 528.8 | 530.5 | 1.2 | 0.07 | 0.9 |

[a] s = strong, [b] w = weak, [c] M = Ba + Mn + Cu.

The Cu 2p$_{3/2}$ XPS profiles featured in Figure 6c reveal that Cu(II) exists on the surface, as the XPS peaks close to 933.0 eV appear and because the satellite peaks expected for Cu(II) (at 940.0 eV and 943.0 eV) are also present [32]. Additionally, after deconvolution, two contributions are detected at ca. 933.0 eV and 934.5 eV (see Table 5), which can be attributed to Cu(II) with strong (Cu(II)$_s$) and weak (Cu(II)$_w$) interactions with the perovskites, respectively [23,32,38]. After A doping, an increase in the binding energy corresponding to the maximum of Cu(II)$_w$ is detected, which indicates the presence of a poorer electronic environment than in raw BMC caused by the insertion of A metal. In Table 5, the Cu/(Ba + Mn + Cu + A) ratios are lower than the nominal ones (0.15), so it seems that Cu(II) has been inserted into the perovskite structure for all samples. Note that, after the addition of Ce or Ca, the ratio decreases to 0.07, and, for La, it increases to 0.1. Thus, the distribution of copper is only modified with respect to BMC for Ce, Ca, and La metals: Ce and Ca seem to promote the introduction of copper into the structure, and La seems to favour a slightly higher proportion of surface copper [23].

H$_2$-TPR profiles are shown in Figure 7a, where the corresponding MnO$_2$ and CuO profiles (divided by 4 to be comparable with the catalyst profiles) have also been included as references. The H$_2$-TPR profile for MnO$_2$, as above discussed, features two overlapping reduction peaks at around 400 and 500 °C, corresponding to the MnO$_2$/Mn$_2$O$_3$ reduction of Mn$_3$O$_4$ and to Mn$_3$O$_4$ to MnO reduction [45]. CuO displayed at ca. 320 °C a single broad reduction peak due to the reduction Cu(II) to Cu(0). The H$_2$-TPR profiles of BMC-A samples suggest that the metal reduction occurs in multiple steps:

- Between approximately 200 and 400 °C, the reduction of Mn(IV) and Mn(III) to Mn(II), and also of Cu(II) to Cu(0), takes place.
- Between 700 °C and 800 °C, a small peak attributed to the desorption/reduction of oxygen species is featured.
- Between 900 °C and 1000 °C, a third peak with very low intensity, corresponding to the bulk Mn(III) reduction, seems to be present.

A decrease in the temperature for the reduction of Mn(IV)/Mn(III) to Mn(II) and of Cu(II) to Cu(0) in the H$_2$-TPR profiles of BMC-A (Figure 7a) is exclusively detected after Ce doping. In fact, after Mg doping, an increase in temperature is detected, which could be related to the different location of Mg in this sample. The experimental hydrogen consumption per gram of catalyst, determined between 200 °C and 450 °C using the hydrogen consumption profiles shown in Figure 7a, has been compared with the theoretical hydrogen consumption determined considering the total reduction of manganese and copper (as Mn(III) + Cu(II) in the blue line or Mn(IV) + Cu(II) in the red line) in Figure 7b. For BMC-A samples, the experimental H$_2$ consumptions are between the nominal ones for Mn(IV) + Cu(II) and Mn(III) + Cu(II), corresponding to BMC-Mg and BMC-Ce being closer to Mn(IV) + Cu(II), while for BMC, BMC-Ca, BMC-La, and BMC-Sr are closer to Mn(III) + Cu(II). Thus, it appears that both Mn(III) and Mn(IV) exist in the bulk of perovskites, as detected by XPS on the surface.

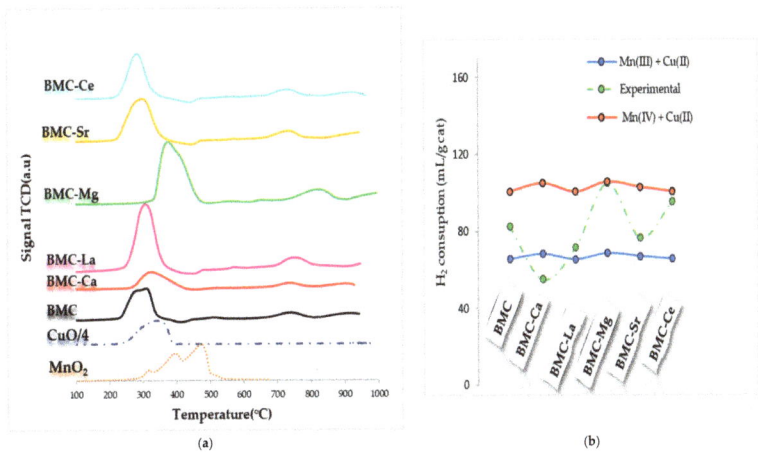

**Figure 7.** $H_2$-TPR profiles of BMC-A and BMC catalysts (**a**) and $H_2$ consumption (mL/g of catalyst) (**b**).

$O_2$-TPD tests were also performed, and the obtained profiles are illustrated in Figure 8. It is observed that, as previously described for the BM-A series, the catalysts exclusively evolve β-$O_2$, which is linked to the reduction of Mn(IV) to Mn(III) and of Cu(II) to Cu(I) [57–61] and, for BMC-Ce, also to the reduction of Ce(IV) to Ce(III) [50]. After A metal is included in the formulation of samples, a shift towards higher temperatures is detected for the temperature of the maximum oxygen emission. This fact is directly related to the A-O bond energy, which is expected to be higher than the Ba-O bond energy, as Ba(II) presents a larger ionic radius than the A metal (see Table 4). As shown in Figure 8b, the partial substitution of Ba causes an increase in the amount of β-$O_2$ released, which is evidence of the improved mobility of the bulk oxygen, according to the following trend: BMC-Ce > BMC-Ca > BMC-Mg > BMC-Sr > BMC-La > BMC. This finding seems to be related, in general terms, to the reducibility of the samples, as BMC-Ce is the most reducible catalyst and has also evolved the highest amount of oxygen.

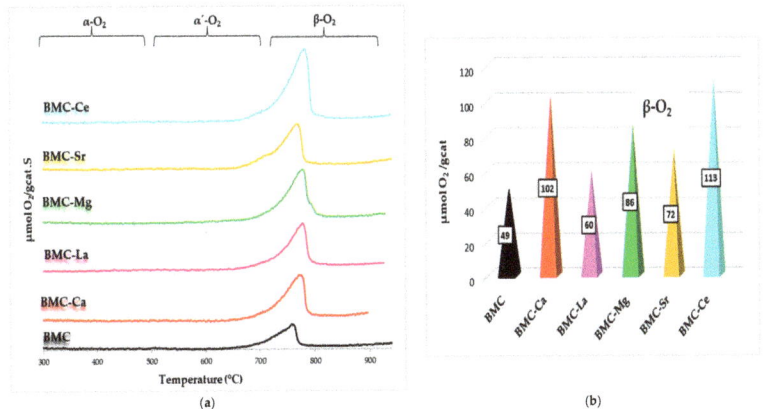

**Figure 8.** $O_2$-TPD profiles of BMC and BMC-A catalysts (**a**) and β-$O_2$ amount emitted during $O_2$-TPD experiments (**b**).

In summary, by comparing the characterization results for the two series of A-doped catalysts (BM-A and BMC-A), it seems that:

(i) The hexagonal structure is favoured in the presence of A dopant, as it is the main phase detected for BM-A, and the polytype structure detected in the BMC sample (formed by distortion of the hexagonal perovskite due to copper insertion into the lattice) is not favoured in BMC-A perovskites, presenting a mixture of the two structures. Also, Mn(IV) and Mn(III) coexist on the surfaces of all samples.
(ii) Oxygen vacancies are present on the surfaces of all perovskites.
(iii) The partial substitution of Ba in BM and BMC enhances the reducibility and the lattice oxygen mobility, making Ce the most efficient A metal due to the contribution of the Ce(IV)/Ce(III) redox pair.

## 3.2. Catalytic Activity

### 3.2.1. BM-A Series

To assess the role of perovskites as GPF catalysts to be used for soot removal in simulated GDI engine exhaust conditions, soot-TPR tests in the two gaseous mixtures previously described (0% and 1% $O_2$ in He) were carried out [23,24,31], being the soot conversion profiles featured in Figure 9. Table 6 contains the $T_{10\%}$ and $T_{50\%}$ values, that are the temperatures for the 10% and 50% of soot conversion, respectively, and the selectivity to $CO_2$ during the reaction.

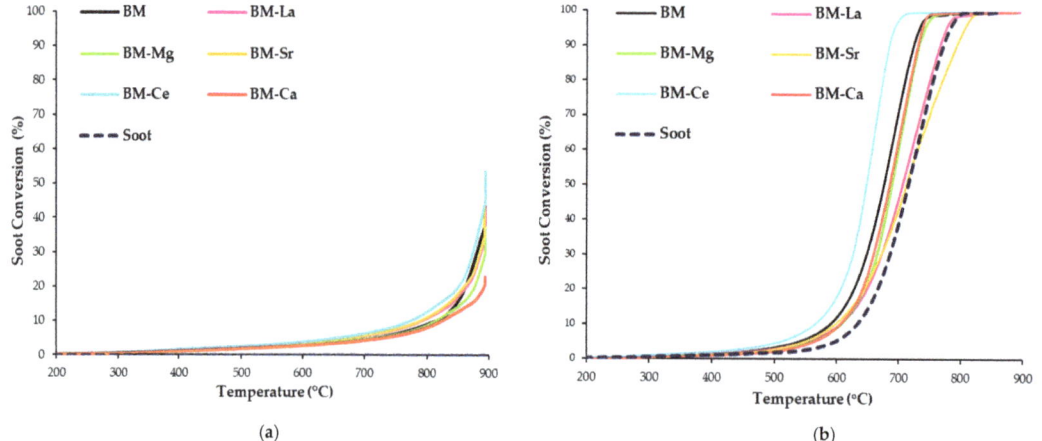

**Figure 9.** Soot-TPR conversion profiles as a function of temperature of BM and BM-A catalysts in 100% He (**a**) and in 1% $O_2$/He (**b**).

**Table 6.** $T_{10\%}$, $T_{50\%}$, and selectivity to $CO_2$ ($S_{CO2}$) for soot oxidation of BM and BM-A catalysts.

| Catalysts | | 1% $O_2$/He | | 100% He |
|---|---|---|---|---|
| | $S_{CO2}$ (%) | $T_{50\%}$ (°C) | $T_{10\%}$ (°C) | $T_{10\%}$ (°C) |
| Soot | 44 | 714 | 631 | - |
| BM | 73 | 710 | 610 | 813 |
| BM-La | 93 | 708 | 606 | 791 |
| BM-Mg | 92 | 684 | 589 | 813 |
| BM-Sr | 93 | 711 | 591 | 789 |
| BM-Ce | 90 | 641 | 548 | 772 |
| BM-Ca | 91 | 680 | 584 | 823 |

Note that almost all samples catalyze the soot oxidation reaction, as most of the profiles are shifted to lower temperatures with respect to the uncatalyzed reaction (soot in Figure 9) in the two atmospheres tested. The soot conversion in the 100% He begins to be significant at above 700 °C (Figure 9a), while in the 1% $O_2$, the soot conversion starts being relevant at around 450 °C (Figure 9b). So, as previously concluded [23,31], soot oxidation is improved in the presence of oxygen in the gaseous mixture used for the reaction. Thus, in the He atmosphere, soot oxidation does not occur in the absence of a catalyst, as there is no oxygen available. However, in the presence of perovskites, the reaction takes place with the oxygen coming from the bulk of mixed oxides (β-$O_2$), whose emission was promoted by the Mn(IV) to Mn(III) reduction [31], as well as by the Ce(IV) to Ce(III) reduction in the case of the BM-Ce catalyst [50]. In the 1% $O_2$ atmosphere, BM-Ca, BM-Ce, and BM-Mg show a higher soot conversion than BM, giving BM-Ce the best catalytic performance.

Thus, the data in Table 6 reveal that the partial replacement of Ba by A metal in BM perovskite improves the catalytic performance for soot removal in simulated GDI engine conditions, as $T_{10\%}$ and $T_{50\%}$ values decrease with respect to BM for most A metals. BM-Ce is the best catalyst in the two conditions tested, as it presents the lowest $T_{10\%}$ and $T_{50\%}$ values. This sample presents the highest oxygen mobility due to the contribution of Ce(IV)/Ce(III) along with the Mn(IV)/Mn(III), which enhances the redox reaction and the oxygen emission, which is directly involved in the soot oxidation. Additionally, oxygen vacancies exist on the surface (as $O_L$/(Ba + Mn + Cu) XPS ratios are lower than 1.5) that work as active sites where the oxygen from the gas phase is activated and participates in the soot oxidation reaction. So, the more efficient activation of oxygen on the active sites and the higher release of oxygen from the catalyst make BM-Ce the most active sample of the BM-A series.

Finally, regarding $CO_2$ selectivity, all BM-A catalysts improve this parameter with respect to BM, so these catalysts seem to be also promising for CO oxidation reactions.

### 3.2.2. BMC-A Series

In order to determine the impact of A metal doping (A = La, Mg, Sr, Ce, Ca) on the catalytic performance of BMC samples for soot removal, as for the BM-A series, soot-TPR tests have been developed using conditions similar to those of GDI engine exhaust. The soot conversion profiles are displayed in Figure 10a,b, and Table 7 summarizes the temperatures needed to achieve 10% and 50% soot conversion, respectively, as well as the selectivity to $CO_2$.

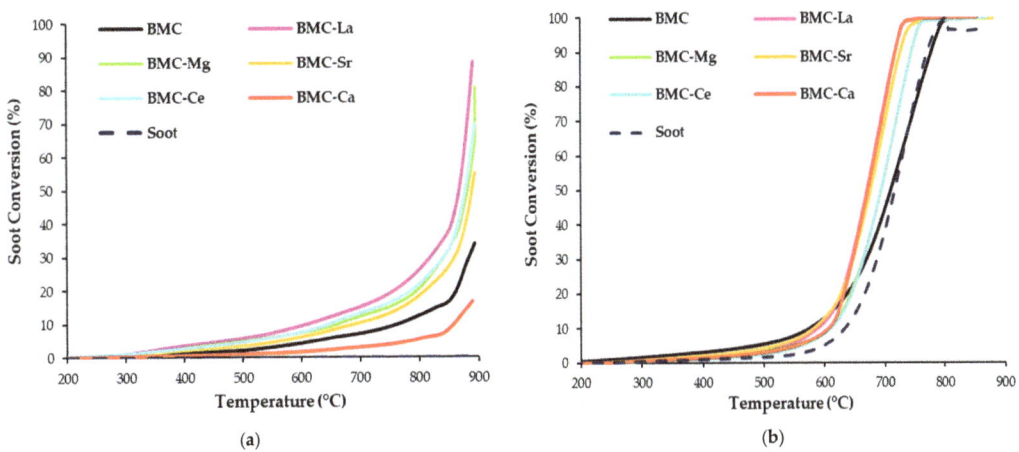

**Figure 10.** Soot conversion profiles as a function of temperature of BMC and BMC-A catalysts in: (a) 100% He and (b) 1% $O_2$/He.

**Table 7.** $T_{10\%}$, $T_{50\%}$ and selectivity to $CO_2$ ($S_{CO2}$) for soot oxidation in the two tested atmospheres of BMC and BMC-A catalysts.

| Catalysts | 1% $O_2$/He | | | 100% He | |
|---|---|---|---|---|---|
| | $S_{CO2}$ (%) | $T_{50\%}$ (°C) | $T_{10\%}$ (°C) | $T_{10\%}$ (°C) | $S_{CO2}$ (%) |
| Soot | 44 | 756 | 631 | - | - |
| BMC | 70 | 732 | 599 | 879 | 33 |
| BMC-Ca | 93 | 671 | 606 | 859 | 37 |
| BMC-La | 94 | 671 | 588 | 611 | 40 |
| BMC-Mg | 92 | 695 | 605 | 660 | 25 |
| BMC-Sr | 88 | 712 | 582 | 689 | 27 |
| BMC-Ce | 94 | 693 | 610 | 646 | 32 |

First, focusing on the results obtained using 100% He (Figure 10a), all perovskites are active for soot oxidation since the uncatalyzed reaction (denoted as soot in Figure 10) does not take place as there is no oxygen available for the reaction. Note that, after doping BMC with A metal, all the soot conversion profiles are shifted to lower temperatures with respect to BMC. However, the catalytic activity is notably lower than in the 1% $O_2$ atmosphere, as the oxygen involved in soot oxidation only comes from the bulk of samples. The BMC-La sample exhibits the lowest $T_{10\%}$, probably due to the highest amount of copper on the surface (see Table 5), which is active for soot oxidation [33]. According to the literature, this surface copper species are present as $BaO_x$–$CuO_x$ phases, forming Cu–O–Ba units in the interface between CuO and the perovskite [24]. As the Cu–O bond energy is lower than that of the Mn–O bond, the release of oxygen from Cu–O–Ba units is easier, thus allowing a high oxygen release rate that improves the catalytic performance for soot oxidation [62]. Aneggi et al. [5], using ceria-zirconia Cu-based catalysts, also detected an improvement in the catalytic activity for soot removal in the absence of oxygen due to the well-known high Oxygen Storage Capacity (OSC) of these mixed oxides. In the 1% $O_2$/He reactant mixture, the addition of A metal also enhances the catalytic activity for soot oxidation, and $T_{50\%}$ values for BMC-A samples are lower than those corresponding to the raw BMC. In these conditions, BMC-Ca and BMC-La are the most active formulations, as these catalysts increase the amount of oxygen evolved (according to $O_2$-TPD results) and present more oxygen vacancies (that allow the activation of oxygen from the gas phase) and surface copper species with a poorer electronic environment than raw BMC (see XPS results); this gives BMC-La the highest proportion of surface copper. In this sense, W.Y. Hernández et al. [63] also found that the use of $La_{0.6}Sr_{0.4}FeO_3$ and $La_{0.6}Sr_{0.4}MnO_3$ allows a notable decrease in the light-off temperatures in the 1% $O_2$/He atmosphere. On the other hand, according to other papers [24,46], the $CO_2$ selectivity values are higher when oxygen is present in the reaction atmosphere, and the generation of $CO_2$ is highly favoured by A doping, being around 90% for all samples. Thus, as well as the BM-A series, these samples are promising catalysts for the CO oxidation reaction.

Finally, by comparing the $T_{10\%}$ and $T_{50\%}$ values of the best catalysts of the two series analyzed (BM-Ce and BMC-La, selected because they featured the highest selectivity to $CO_2$), it could be concluded that:

(i) In 100% He, BMC-La is the most active catalyst, as $T_{10\%}$ is lower than the observed for BMC-Ce (611 °C and 646 °C, respectively), mainly because more copper (as BaO-CuO species) is present on the surface of BMC-La than on BMC-Ce (Cu/(Ba + Mn + Cu + A) ratios are 0.10 and 0.07, respectively).

(ii) In 1% $O_2$ in He, the best catalyst is BM-Ce, as it presents a lower $T_{50\%}$ value than BMC-La (641 °C and 671 °C, respectively) and a similar $CO_2$ selectivity (around 90% in both catalysts). BM-Ce is the sample with the highest oxygen mobility and the best redox properties due to the participation of the Ce(IV)/Ce(III) redox pair as well as a high amount of oxygen vacancies on the surface.

So, according to the previous discussion, the role of copper seems to be relevant if the oxygen involved in the soot oxidation comes from the perovskite (i.e., in 100% He), and BMC-La is the most active catalyst as it presents the highest fraction of copper on the surface. However, if soot is oxidized using the oxygen present in the reaction atmosphere (i.e., in 1% $O_2$ in He), the presence of copper in the perovskite composition seems not to be significant, as the most active catalyst is BM-Ce. Note that the BM-Ce sample presents a higher fraction of surface Ce(IV) than BMC-Ce (see data in Appendix A), so it has better redox performance.

## 4. Conclusions

In this paper, $Ba_{0.9}A_{0.1}MnO_3$ (BM-A, A = Mg, Ca, Sr, Ce, La) and $Ba_{0.9}A_{0.1}Mn_{0.7}Cu_{0.3}O_3$ (BMC-A, A = Mg, Ca, Sr, Ce, La) perovskite-type mixed oxides were prepared, characterized, and used for soot removal by oxidation in simulated GDI engine exhaust conditions. Considering the results discussed above, the following conclusions can be drawn:

- The hexagonal structure is favoured in the presence of A metal, as it is the main phase detected for BM-A, and the polytype structure found in the BMC sample (formed by distortion of the hexagonal perovskite due to copper insertion into the lattice) is not favoured in BMC-A perovskites that present a mixture of the two structures.
- On the surface of all perovskites, coexisting Mn(IV) and Mn(III) and oxygen vacancies are present.
- The partial substitution of Ba in BM and BMC enhances the reducibility and the lattice oxygen mobility, and Ce is the most efficient due to the contribution of the Ce(IV)/Ce(III) redox pair.
- Almost all samples are active as catalysts for soot removal by oxidation, as most of the conversion profiles are shifted to lower temperatures in the presence of perovskites in the two atmospheres tested (0% and 1% $O_2$ in He).
- The soot conversion is notably lower in the absence of $O_2$ than in the 1% $O_2$ atmosphere, as the oxygen available for soot oxidation exclusively comes from the bulk of samples. In these conditions, BMC-La is the most active catalyst due to its highest proportion of copper on the surface (as Ba-O-Cu species).
- In 1% $O_2$ in He, BM-Ce is the best catalyst as it presents a high amount of oxygen surface vacancies, the highest oxygen mobility, and the best redox properties due to the participation of the Ce(IV)/Ce(III) pair along with the Mn(IV)/Mn(III) pair that promote the $O_2$ emission from perovskite, which is directly involved in the soot oxidation.
- The role of copper seems to be relevant only if the oxygen used for the soot oxidation exclusively comes from the perovskite (i.e., in 100% He), as BMC-La, which presents the highest fraction of surface copper, is the most active catalyst. On the contrary, if soot is oxidized using the oxygen present in the reaction atmosphere (i.e., in 1% $O_2$ in He), the presence of copper in the perovskite composition is not significant, as the most active catalyst is BM-Ce because it presents a higher fraction of surface Ce(IV) than BMC-Ce and, consequently, a better redox performance.

**Author Contributions:** The individual contributions of each author are indicated as follows: N.G.: investigation, resources, data curation, and writing—original draft preparation; Á.D.V.: investigation and resources; M.J.I.G.: conceptualization, methodology, writing—review and editing, visualization, supervision, project administration, and funding acquisition. All authors have read and agreed to the published version of the manuscript.

**Funding:** This research was funded by the Spanish Government (MINCINN: PID2019-105542RB-I00/AEI/10.13039/501100011033 Project), the European Union (FEDER Funds), and Generalitat Valenciana (CIPROM/2021-070 Project). N. Ghezali thanks Argelian Government for her thesis grant and Á. Díaz-Verde of the University of Alicante for his predoctoral contract.

**Conflicts of Interest:** The authors declare no conflict of interest.

## Abbreviations

| | |
|---|---|
| GDI | Gasoline direct injection |
| PM | Particulate matter |
| PM2.5 | Particulate matter with a diameter of 2.5 µm |
| GPF | Gasoline particulate filter |
| TWC | Three-way catalyst |
| $CO_2RR$ | $CO_2$ reduction reaction |
| ORR | Oxygen reduction reaction |
| OER | Oxygen evolution reaction |
| ICP-OES | Inductively coupled plasma optical emission spectroscopy |
| XRD | X-ray diffraction |
| XPS | X-ray photoelectron spectroscopy |
| $H_2$-TPR | Temperature-programmed reduction with $H_2$ |
| $O_2$-TPD | Temperature-programmed desorption of oxygen |
| TCD | Thermal conductivity detector |
| BE | Binding energy |
| KE | Kinetic energy |
| TG-MS | Thermal gravimetric mass spectrometry |
| Soot-TPR | Soot oxidation in temperature-programmed reaction conditions |
| BET | Brunauer-emmett-teller |
| ICDD | International centre of diffraction data |
| $O_L$ | Lattice oxygen |
| $O_{ads}$ | Adsorbed oxygen species |
| OSC | Oxygen storage capacity |
| $SCO_2$ | Selectivity to $CO_2$ |

## Appendix A

The deconvoluted Ce 3d spectra and the corresponding XPS data for BM-Ce and BMC-Ce catalysts are shown in Figure A1 and Table A1, respectively. The deconvoluted spectra presents eight sub-peaks: the v quadruplets in the Ce $3d_{5/2}$ area and the u quadruplets in the Ce $3d_{3/2}$ region. The distinctive identifiers for the Ce(III) oxidation state are two subpeaks, $v_1$ and $u_1$ (indicated by regions with deeper colours in Figure A1), whereas the other subpeaks are attributed to the Ce(IV) oxidation state [64,65]. The surface Ce(IV) amount in the BM-Ce sample is comparable to that of the $CeO_2$ sample used as a reference, while the BMC-Ce sample exhibits a larger abundance of Ce(III). This implies that the presence of Ce(III) species seems to be favoured for the copper-containing sample.

$$Ce(III) = u_1 + v_1 \tag{A1}$$

$$Ce(IV) = u_0 + u_2 + u_3 + v_0 + v_2 + v_3 \tag{A2}$$

**Table A1.** XPS data of $CeO_2$, BM-Ce, and BMC-Ce catalysts.

| Catalyst | $CeO_2$ | BM-Ce | BMC-Ce |
|---|---|---|---|
| $\dfrac{Ce(III)}{Ce(IV)}$ | 0.54 | 0.93 | 1.33 |

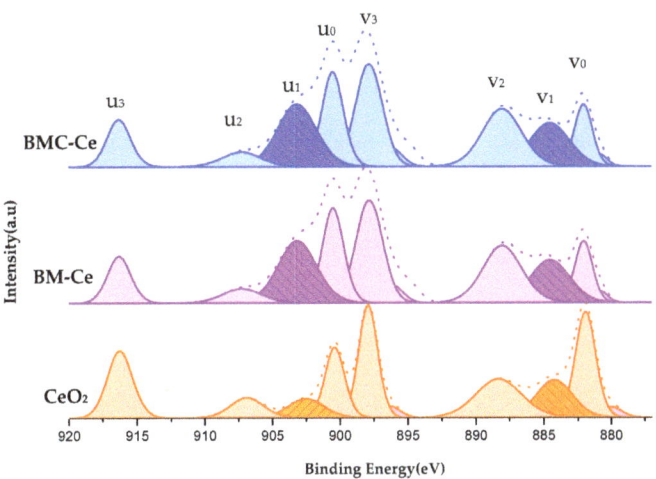

**Figure A1.** Deconvoluted X-ray photoelectron spectra of Ce 3d core levels for BMC-Ce (blue colour), BM-Ce (purple colour) and CeO$_2$ (orange colour).

## References

1. Hamanaka, R.B.; Mutlu, G.M. Particulate Matter Air Pollution: Effects on the Cardiovascular System. *Front. Endocrinol.* **2018**, *9*, 680. [CrossRef] [PubMed]
2. Xing, J.; Shao, L.; Zhang, W.; Peng, J.; Wang, W.; Hou, C.; Shuai, S.; Hu, M.; Zhang, D. Morphology and Composition of Particles Emitted from a Port Fuel Injection Gasoline Vehicle under Real-World Driving Test Cycles. *J. Environ. Sci.* **2019**, *76*, 339–348. [CrossRef]
3. Wang, J.; Zhao, B.; Wang, S.; Yang, F.; Xing, J.; Morawska, L.; Ding, A.; Kulmala, M.; Kerminen, V.-M.; Kujansuu, J.; et al. Particulate Matter Pollution over China and the Effects of Control Policies. *Sci. Total Environ.* **2017**, *584–585*, 426–447. [CrossRef]
4. Liu, H.; Li, Z.; Zhang, M.; Xu, H.; Ma, X.; Shuai, S. Exhaust Non-Volatile Particle Filtration Characteristics of Three-Way Catalyst and Influencing Factors in a Gasoline Direct Injection Engine Compared to Gasoline Particulate Filter. *Fuel* **2021**, *290*, 120065. [CrossRef]
5. Aneggi, E.; Trovarelli, A. Potential of Ceria-Zirconia-Based Materials in Carbon Soot Oxidation for Gasoline Particulate Filters. *Catalysts* **2020**, *10*, 768. [CrossRef]
6. Varga, T.; Ballai, G.; Vásárhelyi, L.; Haspel, H.; Kukovecz, Á.; Kónya, Z. Co$_4$N/Nitrogen-Doped Graphene: A Non-Noble Metal Oxygen Reduction Electrocatalyst for Alkaline Fuel Cells. *Appl. Catal. B Environ.* **2018**, *237*, 826–834. [CrossRef]
7. Asset, T.; Atanassov, P. Iron-Nitrogen-Carbon Catalysts for Proton Exchange Membrane Fuel Cells. *Joule* **2020**, *4*, 33–44. [CrossRef]
8. Srinivasu, K.; Ghosh, S.K. Transition Metal Decorated Graphyne: An Efficient Catalyst for Oxygen Reduction Reaction. *J. Phys. Chem. C* **2013**, *117*, 26021–26028. [CrossRef]
9. Matarrese, R. Catalytic Materials for Gasoline Particulate Filters Soot Oxidation. *Catalysts* **2021**, *11*, 890. [CrossRef]
10. Peña, M.A.; Fierro, J.L.G. Chemical Structures and Performance of Perovskite Oxides. *Chem. Rev.* **2001**, *101*, 1981–2018. [CrossRef]
11. Yamazoe, N.; Teraoka, Y. Oxidation Catalysis of Perovskites—Relationships to Bulk Structure and Composition (Valency, Defect, Etc.). *Catal. Today* **1990**, *8*, 175–199. [CrossRef]
12. Qi, G.; Li, W. Pt-Free, LaMnO$_3$ Based Lean NO$_x$ Trap Catalysts. *Catal. Today* **2012**, *184*, 72–77. [CrossRef]
13. Royer, S.; Duprez, D.; Can, F.; Courtois, X.; Batiot-Dupeyrat, C.; Laassiri, S.; Alamdari, H. Perovskites as Substitutes of Noble Metals for Heterogeneous Catalysis: Dream or Reality. *Chem. Rev.* **2014**, *114*, 10292–10368. [CrossRef] [PubMed]
14. Xu, J.; Liu, J.; Zhao, Z.; Zheng, J.; Zhang, G.; Duan, A.; Jiang, G. Three-Dimensionally Ordered Macroporous LaCo$_x$Fe$_{1-x}$O$_3$ Perovskite-Type Complex Oxide Catalysts for Diesel Soot Combustion. *Catal. Today* **2010**, *153*, 136–142. [CrossRef]
15. Uppara, H.P.; Dasari, H.; Singh, S.K.; Labhsetwar, N.K.; Murari, M.S. Effect of Copper Doping Over GdFeO$_3$ Perovskite on Soot Oxidation Activity. *Catal. Lett.* **2019**, *149*, 3097–3110. [CrossRef]
16. Koolen, C.D.; Luo, W.; Züttel, A. From Single Crystal to Single Atom Catalysts: Structural Factors Influencing the Performance of Metal Catalysts for CO$_2$ Electroreduction. *ACS Catal.* **2023**, *13*, 948–973. [CrossRef]
17. Jiang, J.; Huang, B.; Daiyan, R.; Subhash, B.; Tsounis, C.; Ma, Z.; Han, C.; Zhao, Y.; Effendi, L.H.; Gallington, L.C.; et al. Defective Sn-Zn Perovskites through Bio-Directed Routes for Modulating CO$_2$RR. *Nano Energy* **2022**, *101*, 107593. [CrossRef]
18. Zhang, J.; Pham, T.H.M.; Ko, Y.; Li, M.; Yang, S.; Koolen, C.D.; Zhong, L.; Luo, W.; Züttel, A. Tandem Effect of Ag@C@Cu Catalysts Enhances Ethanol Selectivity for Electrochemical CO$_2$ Reduction in Flow Reactors. *Cell Rep. Phys. Sci.* **2022**, *3*, 100949. [CrossRef]
19. Yadav, P.; Yadav, S.; Atri, S.; Tomar, R. A Brief Review on Key Role of Perovskite Oxides as Catalyst. *ChemistrySelect* **2021**, *6*, 12947–12959. [CrossRef]

20. Pan, Y.; Xu, X.; Zhong, Y.; Ge, L.; Chen, Y.; Veder, J.-P.M.; Guan, D.; O'Hayre, R.; Li, M.; Wang, G.; et al. Direct Evidence of Boosted Oxygen Evolution over Perovskite by Enhanced Lattice Oxygen Participation. *Nat. Commun.* **2020**, *11*, 2002. [CrossRef]
21. Zhang, Y.; Chang, T.-R.; Zhou, B.; Cui, Y.-T.; Yan, H.; Liu, Z.; Schmitt, F.; Lee, J.; Moore, R.; Chen, Y.; et al. Direct Observation of the Transition from Indirect to Direct Bandgap in Atomically Thin Epitaxial $MoSe_2$. *Nat. Nanotechnol.* **2014**, *9*, 111–115. [CrossRef]
22. Laursen, A.B.; Kegnæs, S.; Dahl, S.; Chorkendorff, I. Molybdenum Sulfides—Efficient and Viable Materials for Electro—And Photoelectrocatalytic Hydrogen Evolution. *Energy Environ. Sci.* **2012**, *5*, 5577–5591. [CrossRef]
23. Torregrosa-Rivero, V.; Moreno-Marcos, C.; Albaladejo-Fuentes, V.; Sánchez-Adsuar, M.-S.; Illán-Gómez, M.-J. $BaFe_{1-x}Cu_xO_3$ Perovskites as Active Phase for Diesel (DPF) and Gasoline Particle Filters (GPF). *Nanomaterials* **2019**, *9*, 1551. [CrossRef] [PubMed]
24. Moreno-Marcos, C.; Torregrosa-Rivero, V.; Albaladejo-Fuentes, V.; Sánchez-Adsuar, M.S.; Illán-Gómez, M.J. $BaFe_{1-x}Cu_xO_3$ Perovskites as Soot Oxidation Catalysts for Gasoline Particulate Filters (GPF): A Preliminary Study. *Top. Catal.* **2019**, *62*, 413–418. [CrossRef]
25. Cimino, S.; Lisi, L.; De Rossi, S.; Faticanti, M.; Porta, P. Methane combustion and CO oxidation on $LaAl_{1-x}Mn_xO_3$ perovskite-type oxide solid solutions. *Appl. Catal. B Environ.* **2003**, *43*, 397–406. [CrossRef]
26. Teraoka, Y.; Nii, H.; Kagawa, S.; Jansson, K.; Nygren, M. Influence of the simultaneous substitution of Cu and Ru in the perovskite-type (La,Sr)$MO_3$ (M = Al, Mn, Fe, Co) on the catalytic activity for CO oxidation and CO-NO reactions. *Appl. Catal. A Gen.* **2000**, *194–195*, 35–41. [CrossRef]
27. Petrolekas, P.D.; Metcalfe, I.S. Solid Electrolyte Potentiometric Study of La(Sr)$MnO_3$ Catalyst During Carbon-Monoxide Oxidation. *J. Catal.* **1995**, *152*, 147–163. [CrossRef]
28. Roy, C.; Budhani, R.C. Raman, Infrared and x-Ray Diffraction Study of Phase Stability in $La_{1-x}Ba_xMnO_3$ Doped Manganites. *J. Appl. Phys.* **1999**, *85*, 3124–3131. [CrossRef]
29. Caignaert, V.; Hervieu, M.; Domengès, B.; Nguyen, N.; Pannetier, J.; Raveau, B. $BaMn_{1-x}Fe_xO_{3-\delta}$, an Oxygen-Deficient 6H' Oxide: Electron Microscopy, Powder Neutron Diffraction, and Mössbauer Study. *J. Solid State Chem.* **1988**, *73*, 107–117. [CrossRef]
30. Patcas, F.; Buciuman, F.C.; Zsako, J. Oxygen Non-Stoichiometry and Reducibility of B-Site Substituted Lanthanum Manganites. *Thermochim. Acta* **2000**, *360*, 71–76. [CrossRef]
31. Torregrosa-Rivero, V.; Sánchez-Adsuar, M.-S.; Illán-Gómez, M.-J. Modified $BaMnO_3$-Based Catalysts for Gasoline Particle Filters (GPF): A Preliminary Study. *Catalysts* **2022**, *12*, 1325. [CrossRef]
32. Torregrosa-Rivero, V.; Albaladejo-Fuentes, V.; Sánchez-Adsuar, M.-S.; Illán-Gómez, M.-J. Copper Doped $BaMnO_3$ Perovskite Catalysts for NO Oxidation and $NO_2$-Assisted Diesel Soot Removal. *RSC Adv.* **2017**, *7*, 35228–35238. [CrossRef]
33. Albaladejo-Fuentes, V.; López-Suárez, F.E.; Sánchez-Adsuar, M.S.; Illán-Gómez, M.J. Tailoring the Properties of $BaTi_{0.8}Cu_{0.2}O_3$ Catalyst Selecting the Synthesis Method. *Appl. Catal. A Gen.* **2016**, *519*, 7–15. [CrossRef]
34. Özkan, D.Ç.; Türk, A.; Celik, E. Synthesis and characterizations of $LaMnO_3$ perovskite powders using sol-gel method. *J. Mater. Sci.* **2021**, *32*, 1554–15562. [CrossRef]
35. Akinlolu, K.; Omolara, B.; Shailendra, T.; Abimbola, A.; Kehinde, O. Synthesis, Characterization and Catalytic Activity of Partially Substituted $La_{1-x}Ba_xCoO_3$ ($x \geq 0.1 \leq 0.4$) Nano Catalysts for Potential Soot Oxidation in Diesel Particulate Filters in Diesel Engines. *Int. Rev. Appl. Sci. Eng.* **2020**, *11*, 52–57. [CrossRef]
36. Torregrosa-Rivero, V.; Sánchez-Adsuar, M.-S.; Illán-Gómez, M.-J. Improving the Performance of $BaMnO_3$ Perovskite as Soot Oxidation Catalyst Using Carbon Black during Sol-Gel Synthesis. *Nanomaterials* **2022**, *12*, 219. [CrossRef]
37. Jacobsen, C.J.H.; Dahl, S.; Hansen, P.L.; Törnqvist, E.; Jensen, L.; Topsoe, H.; Prip, D.V.; Moenshaug, P.B.; Chorkendorff, I. Structure sensitivity of supported ruthenium catalysts for ammonia synthesis. *J. Mol. Catal. A Chem.* **2000**, *163*, 19–26. [CrossRef]
38. Díaz-Verde, A.; Torregrosa-Rivero, V.; Illán-Gómez, M.J. Copper Catalysts Supported on Barium Deficient Perovskites for CO Oxidation Reaction. *Top. Catal.* **2023**, *66*, 895–907. [CrossRef]
39. Aarif Ul Islam, S.; Ikram, M. Structural Stability Improvement, Williamson Hall Analysis and Band-Gap Tailoring through A-Site Sr Doping in Rare Earth Based Double Perovskite $La_2NiMnO_6$. *Rare Met.* **2019**, *38*, 805–813. [CrossRef]
40. Merino, N.A.; Barbero, B.P.; Eloy, P.; Cadús, L.E. $La_{1-x}Ca_xCoO_3$ Perovskite-Type Oxides: Identification of the Surface Oxygen Species by XPS. *Appl. Surf. Sci.* **2006**, *253*, 1489–1493. [CrossRef]
41. Tejuca, L.G.; Fierro, J.L.G. XPS and TPD Probe Techniques for the Study of $LaNiO_3$ Perovskite Oxide. *Thermochim. Acta* **1989**, *147*, 361–375. [CrossRef]
42. Khaskheli, A.A.; Xu, L.; Liu, D. Manganese Oxide-Based Catalysts for Soot Oxidation: A Review on the Recent Advances and Future Directions. *Energy Fuel.* **2022**, *36*, 7362–7381. [CrossRef]
43. Yoon, J.S.; Lim, Y.-S.; Choi, B.H.; Hwang, H.J. Catalytic Activity of Perovskite-Type Doped $La_{0.08}Sr_{0.92}Ti_{1-x}M_xO_{3-\delta}$ (M = Mn, Fe, and Co) Oxides for Methane Oxidation. *Int. J. Hydrogen Energy* **2014**, *39*, 7955–7962. [CrossRef]
44. Shen, M.; Zhao, Z.; Chen, J.; Su, Y.; Wang, J.; Wang, X. Effects of Calcium Substitute in $LaMnO_3$ Perovskites for NO Catalytic Oxidation. *J. Rare Earths* **2013**, *31*, 119–123. [CrossRef]
45. Zhang, C.; Wang, C.; Hua, W.; Guo, Y.; Lu, G.; Gil, S.; Giroir-Fendler, A. Relationship between Catalytic Deactivation and Physicochemical Properties of $LaMnO_3$ Perovskite Catalyst during Catalytic Oxidation of Vinyl Chloride. *Appl. Catal. B Environ.* **2016**, *186*, 173–183. [CrossRef]
46. Díaz-Verde, Á.; Montilla-Verdú, S.; Torregrosa-Rivero, V.; Illán-Gómez, M.-J. Tailoring the Composition of $Ba_xBO_3$ (B = Fe, Mn) Mixed Oxides as CO or Soot Oxidation Catalysts in Simulated GDI Engine Exhaust Conditions. *Molecules* **2023**, *28*, 3327. [CrossRef] [PubMed]

47. Najjar, H.; Lamonier, J.-F.; Mentré, O.; Giraudon, J.-M.; Batis, H. Optimization of the Combustion Synthesis towards Efficient LaMnO$_{3+y}$ Catalysts in Methane Oxidation. *Appl. Catal. B Environ.* **2011**, *106*, 149–159. [CrossRef]
48. Lin, H.; Chen, D.; Liu, H.; Zou, X.; Chen, T. Effect of MnO$_2$ Crystalline Structure on the Catalytic Oxidation of Formaldehyde. *Aerosol Air Qual. Res.* **2017**, *17*, 1011–1020. [CrossRef]
49. Buciuman, F.C.; Patcas, F.; Zsakó, J. TPR-Study of Substitution Effects on Reducibility and Oxidative Non-Stoichiometry of La$_{0.8}$A'$_{0.2}$MnO$_{3+\delta}$ Perovskites. *J. Therm. Anal. Calorim.* **2000**, *61*, 819–825. [CrossRef]
50. Levasseur, B.; Kaliaguine, S. Effects of Iron and Cerium in La$_{1-y}$Ce$_y$Co$_{1-x}$Fe$_x$O$_3$ Perovskites as Catalysts for VOC Oxidation. *Appl. Catal. B Environ.* **2009**, *88*, 305–314. [CrossRef]
51. Ulyanov, A.N.; Yu, S.-C.; Yang, D.-S. Mn-Site-Substituted Lanthanum Manganites: Destruction of Electron Pathway and Local Structure Effects on Curie Temperature. *J. Magn. Magn. Mater.* **2004**, *282*, 303–306. [CrossRef]
52. Díaz-Verde, A.; dos Santos-Veiga, E.L.; Beltrán-Mir, H.; Torregrosa-Rivero, V.; Illán-Gómez, M.J.; Cordoncillo-Cordoncillo, E. CO Oxidation Performance in GDI Engine Exhaust Conditions of Ba$_x$Mn$_{0.7}$Cu$_{0.3}$O$_3$ (x = 1, 0.9, 0.8 and 0.7) Perovskite Catalysts. In Proceedings of the 15th European Congress on Catalysis, Prague, Czech Republic, 27 August–1 September 2023. Available online: https://www.europacat2023.cz/Amca-Europacat2021/media/content/Docs/Book_of_abstracts-EuropaCat2023.pdf (accessed on 28 September 2023).
53. Albaladejo-Fuentes, V.; Sánchez-Adsuar, M.-S.; Illán-Gómez, M.-J. Tolerance and Regeneration versus SO$_2$ of Ba$_{0.9}$A$_{0.1}$Ti$_{0.8}$Cu$_{0.2}$O$_3$ (A = Sr, Ca, Mg) LNT Catalysts. *Appl. Catal. A Gen.* **2019**, *577*, 113–123. [CrossRef]
54. Fu, D.; Itoh, M.; Koshihara, S. Invariant Lattice Strain and Polarization in BaTiO$_3$-CaTiO$_3$ Ferroelectric Alloys. *J. Phys. Condens. Matter* **2010**, *22*, 052204. [CrossRef] [PubMed]
55. Su, B.; Button, T.W. Microstructure and Dielectric Properties of Mg-Doped Barium Strontium Titanate Ceramics. *J. Appl. Phys.* **2004**, *95*, 1382–1385. [CrossRef]
56. Chen, J.; Shen, M.; Wang, X.; Qi, G.; Wang, J.; Li, W. The Influence of Nonstoichiometry on LaMnO$_3$ Perovskite for Catalytic NO Oxidation. *Appl. Catal. B Environ.* **2013**, *134–135*, 251–257. [CrossRef]
57. Levasseur, B.; Kaliaguine, S. Effect of the Rare Earth in the Perovskite-Type Mixed Oxides AMnO$_3$ (A=Y, La, Pr, Sm, Dy) as Catalysts in Methanol Oxidation. *J. Solid State Chem.* **2008**, *181*, 2953–2963. [CrossRef]
58. Peron, G.; Glisenti, A. Perovskites as Alternatives to Noble Metals in Automotive Exhaust Abatement: Activation of Oxygen on LaCrO$_3$ and LaMnO$_3$. *Top. Catal.* **2019**, *62*, 244–251. [CrossRef]
59. Tien-Thao, N.; Alamdari, H.; Zahedi-Niaki, M.H.; Kaliaguine, S. LaCo$_{1-x}$Cu$_x$O$_{3-\delta}$ Perovskite Catalysts for Higher Alcohol Synthesis. *Appl. Catal. A Gen.* **2006**, *311*, 204–212. [CrossRef]
60. Zhang, R.; Villanueva, A.; Alamdari, H.; Kaliaguine, S. Catalytic Reduction of NO by Propene over LaCo$_{1-x}$Cu$_x$O$_3$ Perovskites Synthesized by Reactive Grinding. *Appl. Catal. B Environ.* **2006**, *64*, 220–233. [CrossRef]
61. Zhang, R.; Villanueva, A.; Alamdari, H.; Kaliaguine, S. SCR of NO by Propene over Nanoscale LaMn$_{1-x}$Cu$_x$O$_3$ Perovskites. *Appl. Catal. A Gen.* **2006**, *307*, 85–97. [CrossRef]
62. Ashikaga, K.; Murata, K.; Ito, T.; Yamamoto, Y.; Arai, S.; Satsuma, A. Tuning the Oxygen Release Properties of CeO$_2$-Based Catalysts by Metal–Support Interactions for Improved Gasoline Soot Combustion. *Catal. Sci. Technol.* **2020**, *10*, 7177–7185. [CrossRef]
63. Hernández, W.Y.; Tsampas, M.N.; Zhao, C.; Boreave, A.; Bosselet, F.; Vernoux, P. La/Sr-Based Perovskites as Soot Oxidation Catalysts for Gasoline Particulate Filters. *Catal. Today* **2015**, *258*, 525–534. [CrossRef]
64. Konsolakis, M.; Carabineiro, S.A.C.; Marnellos, G.E.; Asad, M.F.; Soares, O.S.G.P.; Pereira, M.F.R.; Órfão, J.J.M.; Figueiredo, J.L. Volatile Organic Compounds Abatement over Copper-Based Catalysts: Effect of Support. *Inorg. Chim. Acta* **2017**, *455*, 473–482. [CrossRef]
65. Amadine, O.; Essamlali, Y.; Fihri, A.; Larzek, M.; Zahouily, M. Effect of Calcination Temperature on the Structure and Catalytic Performance of Copper–Ceria Mixed Oxide Catalysts in Phenol Hydroxylation. *RSC Adv.* **2017**, *7*, 12586–12597. [CrossRef]

**Disclaimer/Publisher's Note:** The statements, opinions and data contained in all publications are solely those of the individual author(s) and contributor(s) and not of MDPI and/or the editor(s). MDPI and/or the editor(s) disclaim responsibility for any injury to people or property resulting from any ideas, methods, instructions or products referred to in the content.

Article

# Influence of the Deposition Parameters on the Properties of TiO₂ Thin Films on Spherical Substrates

Maria Covei [1], Cristina Bogatu [1], Silvioara Gheorghita [1], Anca Duta [1], Hermine Stroescu [2,*], Madalina Nicolescu [2], Jose Maria Calderon-Moreno [2], Irina Atkinson [2], Veronica Bratan [2] and Mariuca Gartner [2,*]

[1] Department of Product Design, Mechatronics and Environment, Transilvania University of Brasov, 29 Eroilor Bd., 500036 Brasov, Romania; maria.covei@unitbv.ro (M.C.)
[2] "Ilie Murgulescu" Institute of Physical Chemistry, Romanian Academy, 202 Splaiul Independentei St., 060021 Bucharest, Romania
* Correspondence: hstroescu@icf.ro (H.S.); mgartner@icf.ro (M.G.)

**Abstract:** Wastewater treatment targeting reuse may limit water scarcity. Photocatalysis is an advanced oxidation process that may be employed in the removal of traces of organic pollutants, where the material choice is important. Titanium dioxide (TiO₂) is a highly efficient photocatalyst with good aqueous stability. TiO₂ powder has a high surface area, thus allowing good pollutant adsorption, but it is difficult to filter for reuse. Thin films have a significantly lower surface area but are easier to regenerate and reuse. In this paper, we report on obtaining sol-gel TiO₂ thin films on spherical beads (2 mm diameter) with high surface area and easy recovery from wastewater. The complex influence of the substrate morphology (etched up to 48 h in concentrated $H_2SO_4$), of the sol dilution with ethanol (1:0 or 1:1), and the number of layers (1 or 2) on the structure, morphology, chemical composition, and photocatalytic performance of the TiO₂ thin films is investigated. Etching the substrate for 2 h in $H_2SO_4$ leads to uniform, smooth surfaces on which crystalline, homogeneous TiO₂ thin films are grown. Films deposited using an undiluted sol are stable in water, with some surface reorganization of the TiO₂ aggregates occurring, while the films obtained using diluted sol are partially washed out. By increasing the film thickness through the deposition of a second layer, the roughness increases (from ~50 nm to ~100 nm), but this increase is not high enough to promote higher adsorption or overall photocatalytic efficiency in methylene blue photodegradation (both about 40% after 8 h of UV-Vis irradiation at 55 W/m²). The most promising thin film, deposited on spherical bead substrates (etched for 2 h in $H_2SO_4$) using the undiluted sol, with one layer, is highly crystalline, uniform, water-stable, and proves to have good photocatalytic activity.

**Keywords:** TiO₂ thin film; photocatalysis; glass bead substrate; sol-gel; dilution ratio; number of layers; morphology; structure; stability

**Citation:** Covei, M.; Bogatu, C.; Gheorghita, S.; Duta, A.; Stroescu, H.; Nicolescu, M.; Calderon-Moreno, J.M.; Atkinson, I.; Bratan, V.; Gartner, M. Influence of the Deposition Parameters on the Properties of TiO₂ Thin Films on Spherical Substrates. *Materials* **2023**, *16*, 4899. https://doi.org/10.3390/ma16144899

Academic Editor: Aleksej Zarkov

Received: 31 May 2023
Revised: 30 June 2023
Accepted: 5 July 2023
Published: 8 July 2023

**Copyright:** © 2023 by the authors. Licensee MDPI, Basel, Switzerland. This article is an open access article distributed under the terms and conditions of the Creative Commons Attribution (CC BY) license (https://creativecommons.org/licenses/by/4.0/).

## 1. Introduction

Due to water scarcity in many parts of the world, wastewater treatment (WWT) targeting its reuse for different applications (e.g., in agriculture, industry, domestic activities, etc.) represents an important aspect of sustainable communities [1,2].

However, there are many organic pollutants (dyes, pesticides, phenol, antibiotics) that cannot be (fully) removed using traditional methods. For the treatment of wastewater loaded with low concentrations of these pollutants (in the ppm range), the use of advanced oxidation processes, such as photocatalysis, may be a viable option [3,4]. However, its applications are mostly reported at the laboratory or pilot plant scale [4].

For up-scaled operation, the associated process cost needs to be reduced by using (a) non-toxic, low-cost, easily recoverable (through filtration) photocatalysts, (b) natural solar-/Vis-radiation, and (c) up-scalable technologies from laboratory to demonstrator level and beyond, when integrated into WWT plants [4–6].

One frequent discussion on the selection of the appropriate photocatalysts is the choice between powder and thin film. Powders have a higher specific surface area and more adsorption sites available for the pollutant molecules, thus being more efficient than thin films; however, their use in suspension reactors is limited by the particles' agglomeration, by the radiation losses (due to scattering) and the limited powder recovery. They can also cause secondary pollution as there is a risk of nanoparticle release in the environment [7]. The use of thin film-photocatalysts can solve these problems, but these types of materials are generally less efficient in pollutant removal as they are characterized by lower specific surface area compared to powders [7–9]. Loading thin photocatalytic films on spherical bead substrates made of glass, silica, aluminum, ceramics, chitosan, alginate, etc., was recently reported as a viable path to prepare efficient and stable photocatalysts with large specific surface areas that are easy to filter, treat and reuse [7,9–19]. Thus, the use of photocatalytic beads in the reactor is much more attractive compared to powders.

In the photocatalytic bead development, titanium dioxide ($TiO_2$) is intensively investigated [7,9–15,17–20] as it has been the main photocatalyst choice for several decades due to its high photocatalytic efficiency and its good aqueous stability over a broad pH range [21,22]. Photocatalytic beads obtained by coating different spherical substrates (glass, silica, alumina, etc.) with a thin film based on commercial titanium dioxide (Degussa P25, Aeroxide 25) or obtained by various techniques such as the sol-gel method, fluidized bed chemical vapor deposition, etc., were successfully employed in the degradation of different organic pollutants (dyes, pesticides, phenols) under UV irradiation [10–19]. In these studies, $TiO_2$ synthesis mainly followed the sol-gel route, and the immersion/dip coating method was further involved in coating the beads. These are versatile, cost-effective techniques (low processing temperature, low-cost equipment) that allow deposition on substrates with different geometry and shapes and allow good control of the photocatalyst properties (particles with good homogeneity, controllable morphology, and crystallinity when an appropriate annealing treatment is applied) [5,11].

The degradation of different dyes such as methylene blue [12], methyl orange [11,14], or reactive red, acid brown [15,17] is reported on $TiO_2$-coated beads. Thus, $TiO_2$ coatings were deposited on various bead substrates (glass, alumina, silica) with a thickness of ~35 nm; the results showed that the morphology and the efficiency in the standard methylene blue removal are influenced by the substrate type and the deposition conditions, and the best results correspond to the $TiO_2$ coated on silica beads [12].

Porous glass beads coated with a thin film of $TiO_2$ showed good efficiency in the methyl orange degradation under UV irradiation: 65% up to entire discoloration of the methyl orange after 30 min of irradiation depending on the Ti (at. %) content in the coating and the catalyst load [11]. Sol-gel S, N co-doped $TiO_2$ nanoparticles immobilized on glass beads using the dip coating technique, tested in the methyl orange photodegradation in both laboratory- and in large-scale photoreactors exhibited up to 95% removal efficiency under solar irradiation for 2 h [14].

The photocatalytic efficiency of $TiO_2$ coated beads using different substrates (glass, silica, clay) in the degradation of phenol, different types of pesticides, or emerging contaminates are reported both at laboratory and pilot plat scale [10,13,18–20]; the removal efficiencies reach values of 70–92% for the investigated pesticides under UV irradiation [10,13,18], while 15 emerging contaminants were successfully degraded to a few µg/L (removal efficiency 70–100% after 60 min of irradiation) by solar photocatalysis in compound parabolic collector (CPC) photoreactors. Moreover, the photoactivity of the investigated $TiO_2$-coated glass beads is preserved even after five testing cycles [20].

Besides the good photocatalytic efficiency in pollutant(s) removal, the photocatalytic beads' stability in the aqueous environment under irradiation is essential. This is correlated with a good adherence of the coating to the spherical substrate, which can be improved by etching the spherical beads (glass, silica) in concentrated acid solutions (nitric acid, sulphuric acid, hydrochloric acid or hydrofluoric acid) [14,17,19] when the bead porosity and specific surface area are expected to increase.

In this study, we report on photocatalytic beads obtained by coating glass spherical substrates with sol-gel TiO$_2$ aimed at the degradation of the standard methylene blue (MB) pollutant (ISO 10678:2010 [23]) from diluted solution (10 ppm) under UV-Vis irradiation at low irradiance value. Methylene blue is frequently used as a target molecule for degradation in the photocatalytic process, mainly due to the simplicity in the assessment of its degradation efficiency that is based on UV-Vis spectrometry. Methylene blue is a dye with a high molar absorbance that allows the UV-Vis detection of the color change from blue up to colorless during photobleaching supported by the photocatalyst in the aqueous medium under irradiation [23,24].

The influence of the substrate morphology (correlated with the duration of the etching treatment) of the sol dilution with ethanol (1:0 or 1:1) and the number of deposited layers (1 or 2) on the structure, morphology, chemical composition, photocatalytic performance and stability of the TiO$_2$ coating was investigated; based on these results, the best TiO$_2$ photocatalytic beads in the experimental conditions were outlined.

## 2. Materials and Methods

Thin TiO$_2$ films obtained following the sol-gel method were deposited on glass beads with a diameter of 2 mm. Before deposition, the substrates were etched in sulfuric acid (Scharlau, 96%) for 0.5 h up to 48 h to remove possible impurities as well as to increase the surface area and activate it. This treatment was followed by washing with water, detergent, and, finally, ethanol (15 min ultra-sonication for each). The impact of the etching duration on the substrate morphology and chemical composition was investigated to select the optimum etching duration.

The sol for the TiO$_2$ layer was obtained by mixing titanium isopropoxide (Aldrich, >97%), ethanol (Chimreactiv, 99.5%), acetylacetone (Scharlau, 99%), acetic acid (Scharlau, 99.8%), and water in a volumetric ratio of 20:16:0.89:0.18:2.4, for 30 min under magnetic stirring, followed by 90 min ultrasonication. The previously etched substrate (1 g of beads) was immersed in the sol (5 mL) and stirred for 30 min, using an orbital stirrer, followed by drying (110 °C—1 h) and annealing (450 °C—3 h) in a furnace Nabertherm B150. The methodology of the thin film deposition is schematically represented in Figure 1.

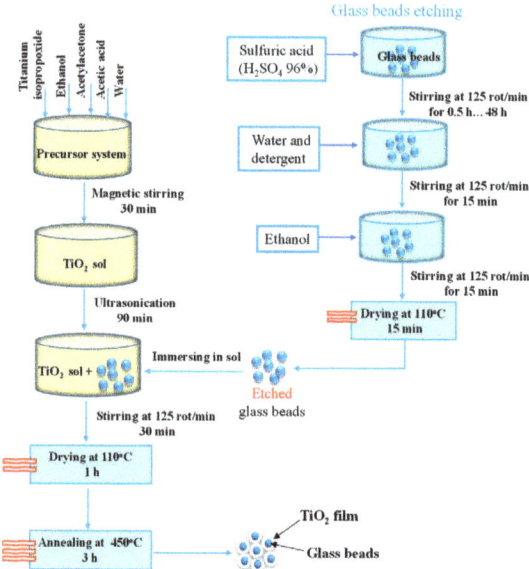

**Figure 1.** Diagram of the thin film deposition methodology.

The crystallinity of the thin films was investigated using a Bruker D8 Discover X-ray Diffractometer (CuKα1 = 1.5406Å, step size 0.025, scan speed 1.5 s/step, 2θ from 5 to 80°). Scanning electron microscopy (SEM) was done on a Hitachi SEM S-3400 N type 121 II apparatus coupled with a Thermo Scientific (Waltham, MA, USA) UltraDry energy dispersive X-ray spectrometer (EDX). Surface topology was analyzed through atomic force microscopy (AFM) using an NT-MDT microscope, model NTGRA PRIMA EC, working in semicontact mode with Si-tips, NSG10, force constant 0.15 N/m, tip radius 10 nm. Roughness was estimated using the AFM software (Nova1138) on a 20 μm × 20 μm surface.

The DR UV-Vis spectra were recorded using a Perkin Elmer Lambda (Shelton, CT, USA) 35 spectrophotometer equipped with an integrating sphere. The measurements were carried out in the 900–200 nm wavelength range using spectralon as a reference and a special holder. The reflectance measurements were converted into absorbance spectra using the Kubelka–Munk function. The optical band gap was calculated using the Tauc plot, which assumes that the absorption coefficient, α, is related to the band gap energy of the semiconductor by applying Equation (1):

$$(\alpha * h\nu)^{1/n} = B * (h\nu - E_g) \quad (1)$$

where h is the Planck constant, ν is the photon frequency, and B is a constant. The factor n has different values, depending on the nature of the electronic transitions, and can be ½ or 2 for the direct or indirect transition, respectively. The band gap is usually measured from diffuse reflectance spectra, by replacing α with F(R). $E_g$ is determined from the x axis intercept of the graphical representation of $(F(R) * h\nu)^{1/n}$ vs. hν, for $n = 2$ [25].

Souri et al. [26] showed that the η value can be determined using the absorption spectra from the slope of the linear part of the graphs of $\ln[A(\lambda)/\lambda]$ versus $1/\lambda$ to determine the optical bandgap without any presumption of the nature of the transition by the absorption spectra [26]. Elemental analysis (XRF) of the films was performed in a vacuum using a Rigaku ZSX Primus II spectrometer (Tokyo, Japan). The results were analyzed using EZ-scan combined with Rigaku SQX (version 5.18) fundamental parameters software capable of automatically correcting all matrix effects, including line overlaps.

Contact angle analysis was performed using OCA20 equipment, with water as the testing liquid (5 μL droplet). Due to the small area of the beads, the contact angle measurements could not be performed directly on the beads and were done on glass slides (2 × 2 cm$^2$) purchased from Citoglas and etched in 98% sulfuric acid, similar to the glass beads for 0.5 up to 48 h.

A circular photo-reactor with two UV light sources (UVA, 340–400 nm, $\lambda_{UV,max}$ = 365 nm, Philips, Amsterdam, The Netherlands) and five VIS light sources (TL-D Super 80 18W/865, 400–700 nm, $\lambda_{VIS,max}$ = 565 nm, Philips), with a total average irradiance of G = 55 W/m$^2$ was used. The UV share in the radiation is 10%, to resemble solar radiation conditions at a much lower irradiance value. The irradiance value was recorded using a Kipp and Zonnen pyranometer.

The photocatalytic experiments used the standard methylene blue (MB, 99.8%, Merck) pollutant at the initial concentration of 10 ppm (ISO 10678:2010 standard determination of photocatalytic activity of surfaces in an aqueous medium by degradation of methylene blue) [16].

The photocatalytic beads (1 g) were immersed in a 20 mL aqueous pollutant solution. Before irradiation, 1 h contact between these was allowed in the dark to reach the adsorption/desorption equilibrium. The photodegradation efficiency, η, was calculated based on the initial absorbance of the pollutant solution ($A_0$) and the absorbance recorded hourly up to 8 h of irradiation (A) at the maximum absorbance wavelength of the pollutant ($\lambda_{MB}$ = 664 nm) using a UV-Vis-NIR spectrophotometer (Perkin Elmer Lambda 950) and applying Equation (2):

$$\eta = \frac{A_0 - A}{A_0} \times 100 \quad (2)$$

To assess the contribution of the adsorption to the entire removal efficiency, experiments were also conducted in the dark using the same set-up and conditions as in the photocatalytic experiments but in the dark, measuring the solution absorbance hourly during 1–9 h, corresponding to contact times equal to those used in the photocatalytic experiments.

## 3. Results and Discussion

The influence of the parameters of interest on the photocatalyst $TiO_2$ thin films deposited on glass beads was investigated, considering: (a) the substrate etching duration, (b) the sol dilution with ethanol, and (c) the number of deposited layers.

### 3.1. Influence of the Etching Duration on the Substrate Properties

Surface structure and morphology can significantly influence thin film growth. In order to increase the specific surface area and create more nucleation sites on the substrate, etching in $H_2SO_4$ was proposed. The etching duration was varied to create a large specific surface without damage from concentrated $H_2SO_4$. It is also expected that a longer etching duration will lead to the formation of micro- or nano-pores and capillaries from which $H_2SO_4$ may be difficult to extract, even after multiple washing cycles.

The surface of the glass bead substrates before and after etching is presented in Figure 2. It can be noticed that the beads come with certain imperfections at the surface but are generally smooth (Figure 2a). After etching, the surface area increases (up to 2 h, Figure 2b–d). After 24 h (Figure 2e) and 48 h (Figure 2f), some craters start to form, and the surface becomes slightly damaged.

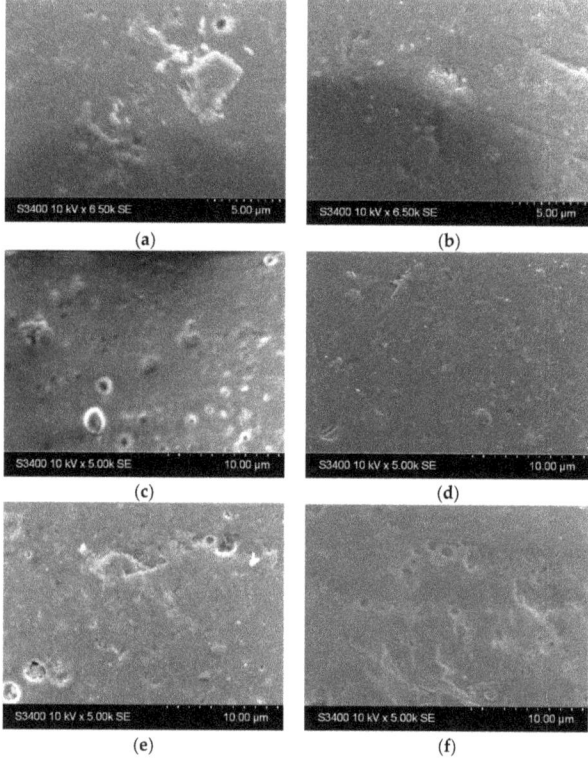

**Figure 2.** SEM images of the bead substrate (a) without etching and after etching with $H_2SO_4$ for (b) 0.5 h, (c) 1 h, (d) 2 h, (e) 24 h, and (f) 48 h.

Etching in sulfuric acid could also lead to increased hydrophilicity of the substrate by the hydrolysis of Si-O bonds to Si-OH. This process could potentially lead to an improved quality of $TiO_2$ thin films deposited on top through the formation of Si-O-Ti bonds. For this reason, we have performed contact angle measurements on glass slides to check the impact of $H_2SO_4$ etching on the substrate hydrophilicity. These results are included in the Supplementary Material. While the water contact angle decreases at a longer etching duration (24 and 48 h), it is not significant enough to promote better quality films deposited on top (Table S1).

To check the impact of chemical etching on the substrate composition, EDX was employed, and the results are inserted in Table 1. These confirm that longer etching durations can lead to the degradation of the surface, and traces of sulfur can be noticed. This may be due either to the infiltration of $H_2SO_4$ into the (nano-)pores or, alternatively, to the formation of metal sulfates from the reaction of sulfuric acid and metal ion impurities from the glass substrate (Na, Mg, Ca, Al).

**Table 1.** Chemical composition of the bead substrates according to the etching duration in $H_2SO_4$.

| Etching Time | Si | O | S | Other Elements (Na, Mg, Ca, Al) |
|---|---|---|---|---|
| 0 h | 18.57 | 69.00 | - | 12.43 |
| 0.5 h | 18.09 | 70.94 | - | 10.97 |
| 1 h | 17.70 | 71.60 | - | 10.70 |
| 2 h | 17.36 | 71.22 | - | 11.42 |
| 24 h | 16.95 | 71.15 | 0.01 | 11.89 |
| 48 h | 17.38 | 66.83 | 0.02 | 15.77 |

The impact of the etching process on the surface roughness was investigated following AFM measurements on the bead substrates. The surface appears smooth, with a root mean square roughness (RMS) of 19 nm before etching (Figure 3a). This roughness maintains almost constant after 0.5 h (Figure 3b) and starts slowly increasing after 1 h (Figure 3c) and 2 h (Figure 3d) of etching. Although the RMS continues to increase to 44 nm or even 58 nm after 24 h or 48 h, respectively, the formation of small crevices/cracks on the substrate that may occur prevents these substrates from being ideal for thin film deposition. It may also be expected that sulfuric acid traces will remain on the substrate, infiltrated in the imperfections created after 24 and 48 h of etching, which would be detrimental to the thin film properties.

Based on these results, the optimum substrate is considered the one that was etched in sulfuric acid for 2 h. This was further used for the deposition of the $TiO_2$ thin films using diluted or undiluted sols with one or two deposition layers. The film thickness was thus varied in order to identify the most efficient and stable photocatalyst for methylene blue degradation.

*3.2. Influence of the Dilution of the Sol on the $TiO_2$ Thin Film Properties*

Two types of samples were investigated: from diluted and from undiluted sol. Their structural and morphological properties and their chemical composition were compared and correlated with their photocatalytic efficiency and stability.

The crystallinity of both types of samples was calculated based on the X-ray diffractograms (Figure 4). The value obtained from the undiluted sol was slightly higher (57%) compared to 54% for those obtained from the diluted sol. Both samples exhibit similar structural properties, with the only observable peak attributed to anatase (101).

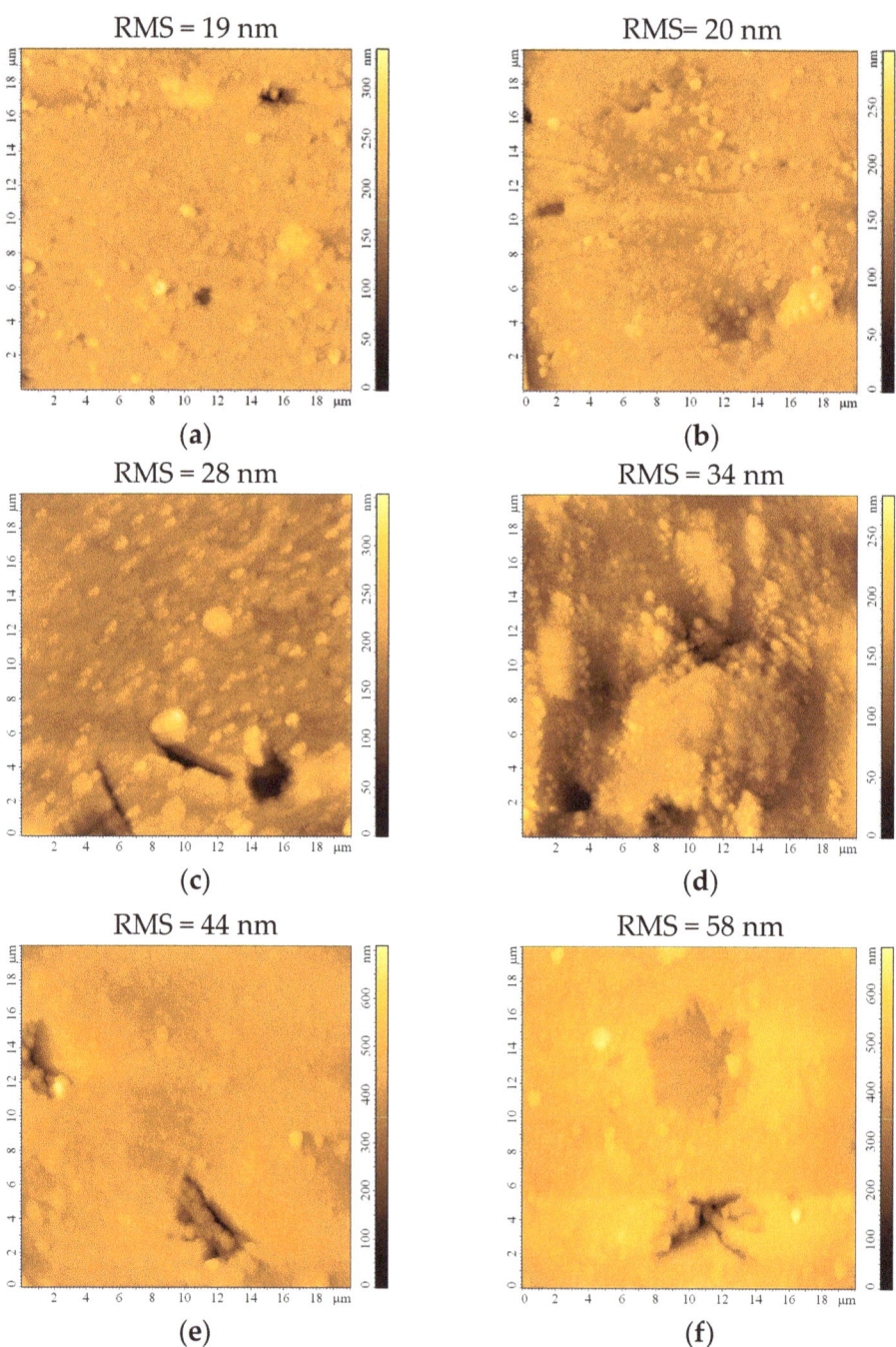

**Figure 3.** AFM images of the bead substrate (**a**) without etching or after etching with $H_2SO_4$ for (**b**) 0.5 h, (**c**) 1 h, (**d**) 2 h, (**e**) 24 h, and (**f**) 48 h.

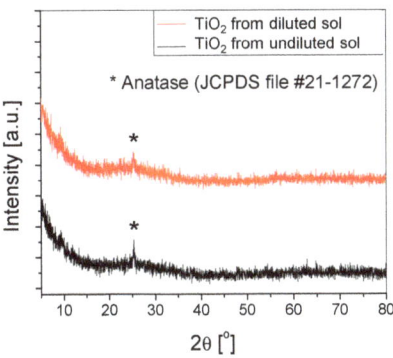

**Figure 4.** XRD of the TiO$_2$ thin films deposited using (un)diluted sol on substrates etched with H$_2$SO$_4$ for 2 h.

Figure 5 shows the morphology of the TiO$_2$ film covering the substrate. There are aggregates of various sizes and shapes that increase the surface area of the films. The average roughness of the films is quite large, about 100 to 150 nm for slightly larger investigated areas (20 × 20 μm$^2$), as seen in Figure 5b,d. The higher roughness may be an advantage in the photocatalytic process, as it promotes increased pollutant adsorption on the photocatalytic surface. However, the stability of these aggregates in water must be confirmed to ensure they do not detach from the film surface. From the EDX analysis, it can be established that the film thickness is not uniform, and some areas of the substrate are covered with a very thin layer of TiO$_2$. The amount of Ti in these areas is very low (or even undetectable), so the amount of Si and other elements (from the substrate) increases, values that can most likely be attributed to the small thickness of the film at the analyzed point.

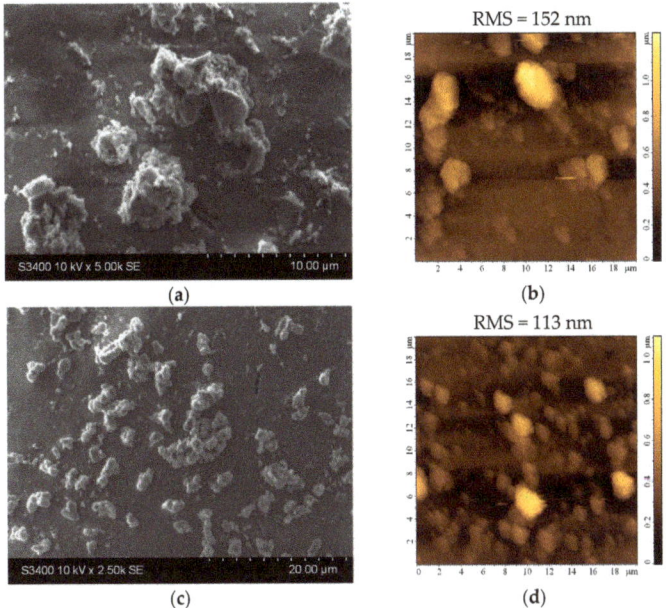

**Figure 5.** (**a,c**) SEM and (**b,d**) AFM images of the TiO$_2$ thin film deposited on the bead substrate etched for 2 h in H$_2$SO$_4$ from (**a,b**) undiluted and (**c,d**) diluted sol.

Stability tests were performed by immersing the samples in distilled water (1 g photocatalytic beads in 20 mL water, under orbital stirring) for 9 h. Surface morphological changes were investigated using SEM analysis (Figure 6a,c), and then the samples were immersed for another 9 h under identical test conditions. The surface changes were investigated once more using SEM (Figure 6b,d).

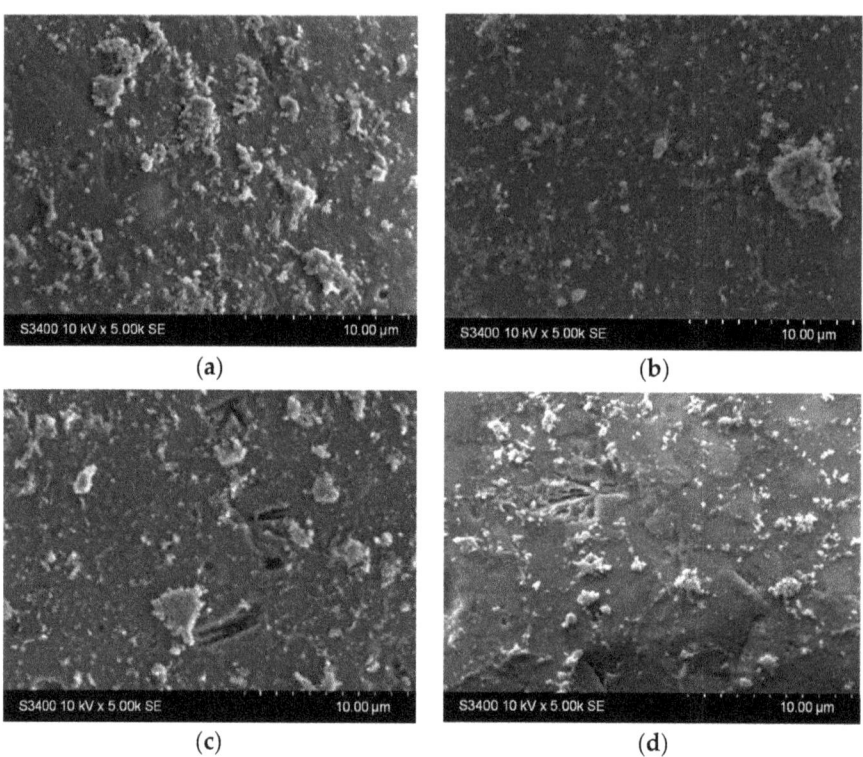

**Figure 6.** SEM images of the $TiO_2$ thin film (**a,c**) after 9 h and (**b,d**) 18 h immersion in water for the samples deposited from undiluted (**a,b**) and diluted (**c,d**) sols.

It can be noticed that there are fewer larger aggregates on the surface before the stability experiments (Figure 5, Table 2). In contrast, the surface of the thin films appears to be covered with smaller, better-dispersed particles that are quite adherent, as they can be observed even after 18 h of testing (Figure 6b,d). Larger aggregates may be removed or even broken down into smaller aggregates during the first stages of immersion (during the first 9 h of testing). The slight decrease in the Ti at% concentration in the films can also be the result of the detachment of some of the larger $TiO_2$ aggregates from the surface of the films, which is more pronounced in the first 9 h of testing (as also shown in Figure S1). The increase in the concentration of substrate elements (Si, Na, Mg, etc.), which can be noticed in Table 3 (correlated to the images shown in Figure S1), can explain the slight decrease in the thickness of the $TiO_2$ film due to these reorganizations in the aggregates at the surface. The decrease of the Ti content after 9 and 18 h is more significant for the sample obtained using the diluted sol. This suggests that aggregate migration is more significant in this case, and therefore, the thin film is less stable in an aqueous medium. The presence of sulfur in the samples before photocatalysis is probably due to the etching treatment with $H_2SO_4$. The amount of this element increases at the surface of the samples after photocatalysis due to the methylene blue or its oxidation products (containing S) that remain adsorbed on the film surface at the end of the photocatalytic test.

Table 2. Chemical composition of the TiO₂ thin film obtained from undiluted and diluted sols.

| | Sample | Si | O | Ti | S | Other Elements (Ca, Mg, Na) |
|---|---|---|---|---|---|---|
| Undiluted sol | Pt1 | - | 84.36 | 15.64 | - | - |
| | Pt2 | 1.36 | 72.11 | 26.53 | 0.00 | - |
| | Pt3 | 22.44 | 68.76 | - | 0.00 | 8.81 |
| | Pt4 | 19.40 | 70.25 | - | 0.15 | 10.21 |
| | Pt5 | 4.56 | 75.16 | 6.48 | 0.59 | 12.85 |
| | Pt6 | 11.49 | 75.56 | - | 0.25 | 12.70 |
| | Pt7 | 12.98 | 78.15 | 0.33 | 0.10 | 8.50 |
| | Pt8 | 20.32 | 70.07 | 2.15 | - | 7.47 |
| Diluted sol | Pt1 | 17.85 | 66.72 | 15.36 | 0.08 | - |
| | Pt2 | 5.82 | 70.07 | 13.37 | - | 10.74 |
| | Pt3 | 20.50 | 67.90 | 3.35 | - | 8.25 |
| | Pt4 | 16.45 | 71.59 | 0.41 | - | 11.55 |
| | Pt5 | 14.28 | 77.27 | - | - | 8.45 |

Table 3. Chemical composition (at. %) of the surface of the TiO₂ layers after 9 h and 18 h of stability testing for the samples obtained using undiluted and diluted sols.

| Sol Type | Stability Testing Time | Si | O | Ti | S | Other Elements (Na, Mg, Ca, Al) |
|---|---|---|---|---|---|---|
| Undiluted | initial | 13.99 | 74.93 | 1.72 | 0.06 | 9.30 |
| | 9 h | 17.02 | 71.48 | 1.43 | 0.06 | 9.99 |
| | 18 h | 16.73 | 70.37 | 1.55 | 0.11 | 11.22 |
| Diluted | initial | 16.00 | 72.01 | 1.46 | 0.02 | 10.51 |
| | 9 h | 16.57 | 71.03 | 1.03 | 0.00 | 11.36 |
| | 18 h | 15.96 | 75.32 | 0.67 | 0.00 | 11.05 |

Photocatalytic experiments were run using the 10 ppm methylene blue aqueous solution under UV-Vis, UV, and no irradiation (in the dark) to check the thin films' photodegradation and adsorption efficiency. The results are presented in Figure 7.

As can be noticed in Figure 7, the samples obtained using undiluted sol had improved photocatalytic efficiency over the testing period compared to the samples using the diluted sol. This could be due to the higher amount of TiO₂ in the film, but also following the improved stability of these films, as previously discussed.

A thicker film is preferable for enhanced photocatalytic activity since it contains more photocatalytic mass that may absorb more radiant energy. However, in the photocatalytic film, the transport of reagents relies on diffusion, which may become a limiting process if the film is relatively thick and the diffusivity is low. Moreover, the thickness of the film plays a role in determining the distribution of radiant energy intensity within it. The availability of photons at a local level, necessary for activating the reaction, depends on the depth they need to penetrate. As a result, when the optical path is relatively long, the intensity of photons can be reduced, and the reaction does not take place inside the whole film efficiently [27]. However, photocatalysis is fundamentally a surface process, and the surface morphology of the two samples plays an important role. Higher roughness means higher surface area, leading to improved photocatalytic efficiency.

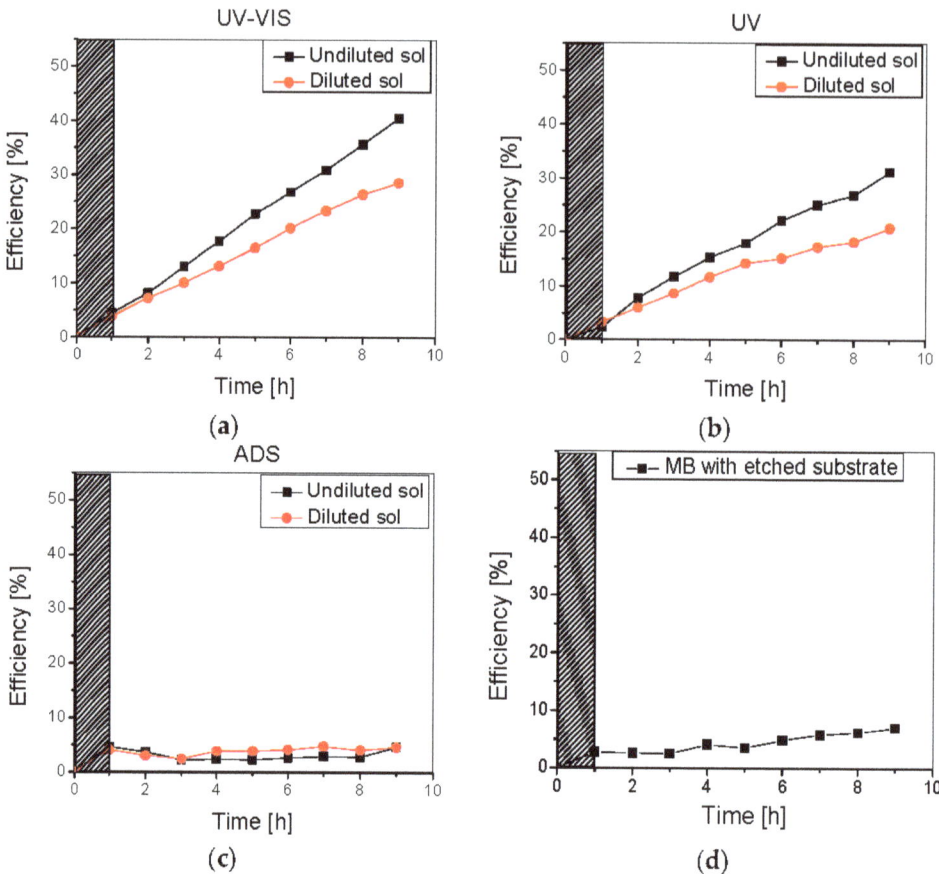

**Figure 7.** Photocatalytic efficiency of the $TiO_2$ thin film obtained using undiluted and diluted sol under (**a**) UV-Vis, (**b**) UV, and (**c**) no irradiation, as well as (**d**) Methylene blue photolysis with an uncoated substrate.

Moreover, the slight differences between the photocatalytic efficiency recorded on the samples under UV-Vis (Figure 7a) and UV (Figure 7b) can be correlated to the dye-sensitization of the metal oxide thin film, as $TiO_2$ is only active under UV irradiation. Finally, it is worth mentioning that while the adsorption of methylene blue on the $TiO_2$ surface represents the first step in photocatalysis, it has a less significant contribution to pollutant removal as an individual process. The adsorption on both samples was almost constant over the testing period (9 h) at 3–5% (Figure 6c).

Photodegradation experiments using the methylene blue solution, with no addition of (photocatalytic) beads, were also performed in order to establish the amount of MB which is degraded through irradiation. After 1 h in the dark and 8 h of UV-Vis irradiation, following the same experimental conditions as those previously mentioned, an efficiency of 5.5% was recorded. Therefore, the difference up to 30% (for the sample obtained using the diluted sol) and up to 40% (for the sample obtained using the undiluted sol) can be mainly attributed to photodegradation and only partly to the adsorption of MB by the $TiO_2$-covered beads. Moreover, during photolysis tests performed with the etched substrate (1 g) but uncoated with $TiO_2$ thin film (using the same conditions as described above), a degradation efficiency of 7% was reached (Figure 7d). This increase from 5.5% to 7% was most likely due to additional MB adsorption on the glass beads substrate.

This confirms that both samples are promising photocatalysts that could potentially lead to methylene blue mineralization under solar radiation (with higher irradiance of 1000 W/m$^2$). Thus, it is recommended that no dilution of the sol should be applied when aiming at depositing uniform, photoactive, and stable TiO$_2$ photocatalyst on bead substrates.

*3.3. Influence of the Number of Deposition Layers on the TiO$_2$ Thin Film Properties*

Another path to modify the film thickness is to sequentially deposit multiple layers. Therefore, a second TiO$_2$ layer was further deposited on the glass beads previously coated with TiO$_2$ using undiluted sol (previously determined as optimal).

The photocatalytic performance on methylene blue removal under UV-Vis, UV, and no irradiation was investigated (Figure 8) and compared to the TiO$_2$ sample with one layer (Figure 7).

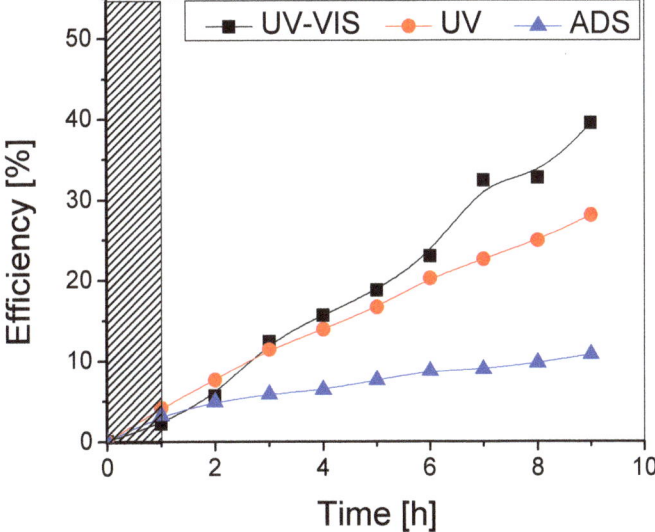

**Figure 8.** Photocatalytic (UV-VIS and UV) and adsorption (ADS) efficiency of the TiO$_2$ thin film with two layers obtained from undiluted sol.

It can be noticed that the addition of a second TiO$_2$ layer does not significantly improve the photocatalytic performance of the samples, as the overall maximum efficiencies after the testing period (9 h) are very similar (~40% for both samples under UV-Vis and ~28% under UV irradiation). Although the double-layered sample has a higher adsorption efficiency (6% compared to 3% for the single-layered one), this did not improve the photocatalytic efficiency. Also, it is likely that some aggregates containing the adsorbed pollutant are removed from the photocatalytic surface during the testing period. In both cases, there was no flattening of the curve even after higher testing durations, which indicates that the photocatalyst surface is not affected by clogging with the pollutant or degradation by-products and may potentially lead to mineralization if (a) the process duration is extended, while maintaining the same irradiance or (b) the irradiance is increased (natural solar radiation), thus decreasing the required process duration.

To test the stability of the samples with one or two layers, their morphology and chemical composition were evaluated before and after photocatalysis. The highest impact on the samples was considered under maximum irradiance; therefore, the samples were investigated before and after testing under UV-Vis irradiation.

The results presented in Figure 9 show that both samples host surface reorganization processes as the larger aggregates stacked before photocatalysis (Figure 9a,c) detach from the surface and/or are broken down into smaller aggregates after photocatalysis (Figure 9b,d). This behavior is similar to the one noticed during the stability experiments on the previously discussed films in water. Moreover, there were not any areas where film washout could be detected, either for the single-layer or for the two-layered sample.

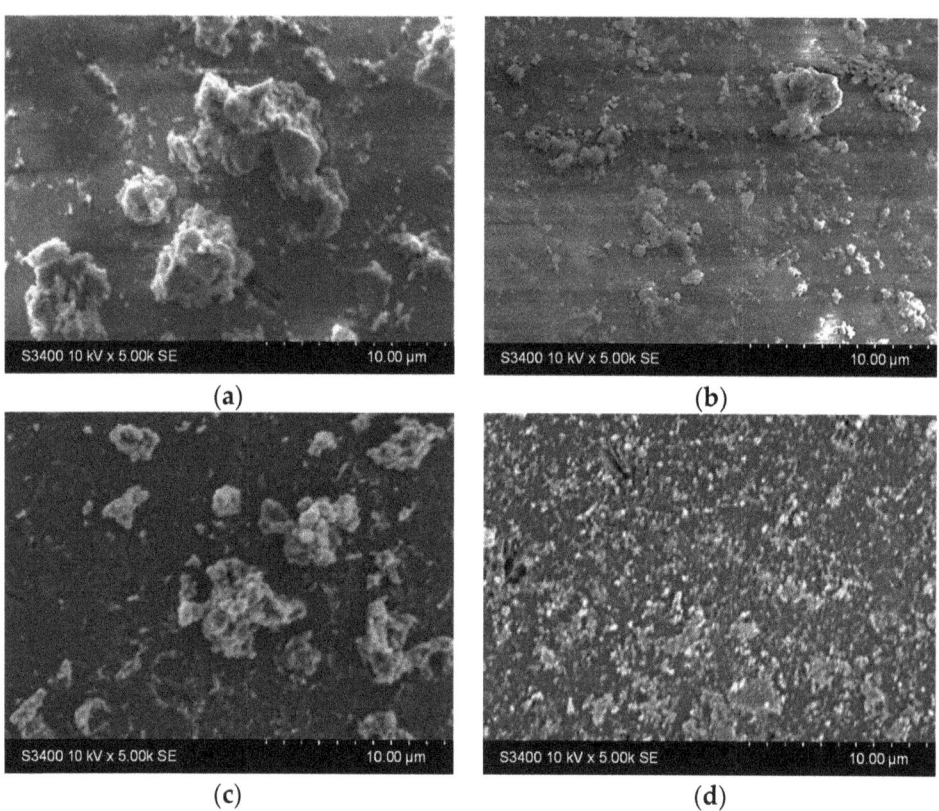

**Figure 9.** SEM images of the TiO$_2$ thin films with (**a,b**) one and (**c,d**) two layers (**a,c**) before and (**b,d**) after photocatalysis.

The chemical composition at the sample surface was evaluated through EDX mapping, and the results are inserted in Table 4. The higher amount of Ti in the double-layered sample compared to the single-layered one can be due to the increased film thickness. This matches the lower substrate (Si and Na, Mg, Ca, and Al) elements contained in the double-layered film. Before photocatalysis, sulfur presence in the single-layered sample could be attributed to traces left over from the substrate etching step. After photocatalysis, sulfur content increases which could be attributed to the presence of some photodegradation by-products that were not desorbed from the photocatalyst surface. Also, after photocatalysis, the decrease in the Ti amount may be correlated with the decrease in the film thickness as a result of thin film reorganization at the surface and the partial transfer of the TiO$_2$ aggregates into the solution (which is also supported by the images in Figure S2). The SEM images of the films presented before and after the photocatalytic tests (Figure 9 and Figure S2) give an indication of the entire film surface on the beaded substrate, as the process of aggregates transfer could be observed all over the surface of these samples.

Table 4. Chemical composition (at. %) of the surface of the TiO$_2$ thin films with one or two layers before and after UV-Vis photocatalysis (PC) of methylene blue.

| No. of Layers | Stability Testing Time | Si | O | Ti | S | Other Elements (Na, Mg, Ca, Al) |
|---|---|---|---|---|---|---|
| 1 | Before PC | 16.84 | 71.69 | 1.20 | 0.02 | 10.25 |
| 1 | After PC | 14.16 | 74.68 | 0.73 | 0.12 | 10.31 |
| 2 | Before PC | 13.12 | 73.72 | 3.90 | - | 9.26 |
| 2 | After PC | 17.38 | 70.85 | 1.71 | 0.12 | 9.95 |

The differences between the single- and double-layered TiO$_2$ samples from a morphological, chemical, and photocatalytic point of view were not significant enough to recommend a multilayered structure. The increase in cost that the double layer involves offsets any slight improvement that the second layer has on the overall film properties that were investigated.

XRF measurements were performed to verify the chemical composition of the most promising sample obtained on a substrate etched for 2 h in sulfuric acid from an undiluted sol with a single-layer deposition (Table 5). The XRF spectra of titanium and oxygen elements of the optimized TiO$_2$ thin film on the bead surface illustrated in Figure 10 reveal the spectral lines of Ti$_{K\alpha1}$ (4.50 eV), Ti$_{K\beta1}$ (4.92 eV) (Figure 9a), and O$_{K\alpha}$ (0.52 eV) (Figure 10b). As shown in Table 5, the experimental results are in good agreement with the theoretical ones for the film, confirming that a stoichiometric TiO$_2$ was deposited.

Table 5. Chemical composition determined by XRF for the optimized TiO$_2$ thin film on bead surface.

| Chemical Element | Theoretical Chemical Composition (wt.%) | Experimental Chemical Composition (wt.%) | Absolute Error (wt.%) |
|---|---|---|---|
| Ti | 59.93 | 60.32 | +0.39 |
| O | 40.07 | 39.67 | +0.39 |

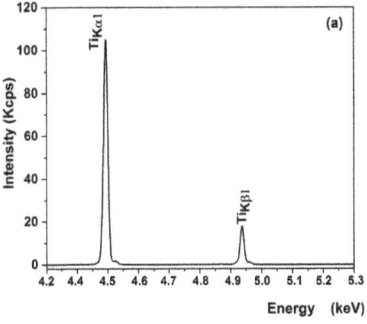
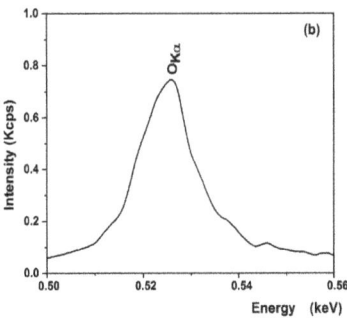

Figure 10. XRF spectra of (a) titanium and (b) oxygen elements of the optimized TiO$_2$ thin film on bead surface.

The indirect bandgap energy, characteristic of anatase [28], was evaluated starting from the UV-Vis absorption spectrum (Figure 11a) and applying the Tauc method (Figure 11b). The obtained value was 3.15 eV and matches well with the reference literature values [29,30]; the slight differences may be the result of the surface agglomerations.

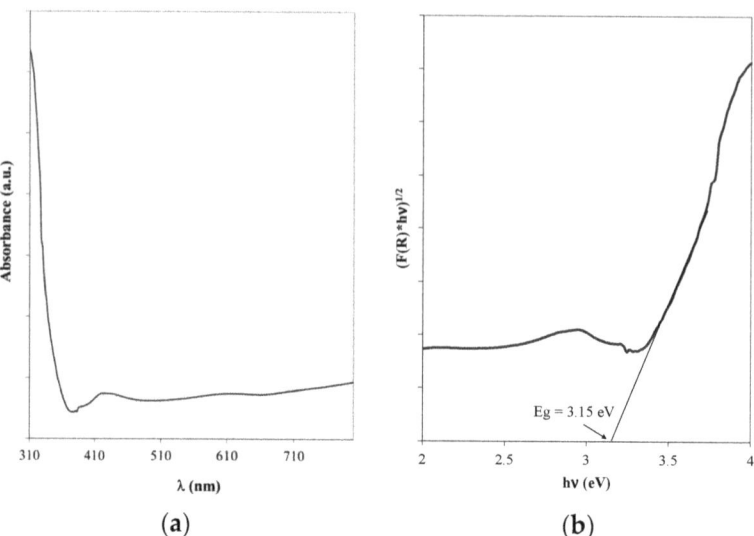

**Figure 11.** (a) UV-Vis spectrum and (b) Tauc representation for the optimized TiO$_2$ thin film on bead surface.

Contact angle measurements on the thin films deposited on glass beads could not be made due to the small diameter of the substrate, which offers only a curved surface for baseline determination. The wetting properties were determined for a corresponding TiO$_2$ thin film obtained from undiluted sol, with one layer on 2 h etched planar glass substrate. The contact angle was ~15°, decreasing to 9.5° after UV conditioning for 9 h, as previously reported in the literature [31]. However, the differences between the chemical composition and morphology of the two substrate types do not guarantee that the values obtained on the films deposited on the planar surface would match those of the films deposited on the bead surface. The morphology and the chemical composition (or even crystallinity) of the two substrates can lead to different film growth mechanisms, thus changing their morphology and their chemical composition and, therefore, their wetting properties.

## 4. Conclusions

The study investigated the influence of various parameters on photocatalytic TiO$_2$ thin films deposited on glass beads applied in the degradation of methylene blue from aqueous solutions, focusing on substrate etching duration, sol dilution, and the number of deposited layers.

The substrate etching duration in H$_2$SO$_4$ was adjusted to enhance surface area and nucleation sites. Longer etching time led to crater formation after 24 and 48 h, as observed in the SEM images. Chemical composition analysis using EDX confirmed sulfur traces and other elements, likely caused by H$_2$SO$_4$ infiltration into the pores or reactions with metal ion impurities from the glass substrate. Longer etching durations increased surface roughness (58 nm after 48 h) but hindered thin film ideal deposition due to crevices and cracks on the substrate. Residual sulfuric acid traces after etching could negatively impact thin film properties. Based on these findings, the optimum etching duration was determined to be 2 h.

This substrate was used for the deposition of TiO$_2$ thin films using diluted or undiluted sols, with variations in the number of deposition layers. XRD analysis indicated similar structural properties for the thin films from undiluted and diluted sols. SEM and AFM images showed varied aggregates on the film surface, potentially enhancing photocatalytic efficiency through increased pollutant adsorption. However, the stability of these aggregates in water needed to be confirmed to ensure they did not detach from the film surface. Chemical composition analysis indicated a non-uniform film thickness, with some areas having a very thin layer of $TiO_2$. Stability tests in distilled water outlined the removal or breakdown of larger aggregates into smaller ones, as shown by SEM images, particularly in the diluted sol samples, indicating reduced stability in an aqueous environment.

Photocatalysis is an advanced oxidation process that can remove trace concentration (ppm or ppb) of organic pollutants and (ideally) using solar or UV-Vis irradiation. This makes it an excellent option for advanced wastewater treatment. Photocatalytic experiments with methylene blue solution revealed that undiluted sol samples exhibited superior photocatalytic efficiency compared to diluted sol samples under various irradiation conditions. Thicker films generally exhibited enhanced photocatalytic activity due to increased photocatalytic mass, but diffusion limitations could occur in relatively thick films with low diffusivity. Higher roughness and higher surface area also supported an improved photocatalytic efficiency. On the other hand, titanium dioxide powder has a high specific surface area, thus allowing good pollutant adsorption. However, using powders as photocatalyst raises challenges related to their full recovery for reuse and limiting their dispersion in the environment. Although thin films have a smaller specific surface area than powders, they are easier to regenerate and reuse.

In conclusion, the study provides valuable insights into the impact of etching duration and sol dilution on the properties and photocatalytic performance of $TiO_2$ thin films on glass beads. The findings can guide the optimization of these parameters for more efficient photocatalysts in applications such as pollutant degradation.

**Supplementary Materials:** The following supporting information can be downloaded at: https://www.mdpi.com/article/10.3390/ma16144899/s1, Figure S1: SEM images of the $TiO_2$ layers before (a, b) and after 9 h (c, d) and 18 h (e, f) of stability testing for the samples obtained using undiluted (a, c, e) and diluted sols (b, d, f), matching the points in Table 3; Figure S2: SEM images of the $TiO_2$ thin films with one or two layers, before and after UV-Vis photocatalysis (PC) of methylene blue, matching the points in Table 4; Table S1: Water contact angle on flat microscopic glass slides.

**Author Contributions:** Conceptualization: M.C., A.D. and M.G.; methodology: M.C. and A.D.; validation: J.M.C.-M., I.A. and V.B.; formal analysis: M.C., C.B., S.G., I.A. and V.B.; investigation: M.C., C.B. and S.G.; data curation: S.G., M.C., I.A., V.B. and J.M.C.-M.; writing—original draft preparation: M.C., H.S. and M.N.; writing—review and editing: M.G. and A.D.; supervision: A.D. and M.G.; project administration: M.C. and M.G.; funding acquisition: M.C. and M.G. All authors have read and agreed to the published version of the manuscript.

**Funding:** This research was funded by a grant from the Ministry of Research and Innovation, CNCS-UEFISCDI, project number PN-III-P2-2.1-PED-2021-2928, Contract no. 598PED/2022.

**Institutional Review Board Statement:** Not applicable.

**Informed Consent Statement:** Not applicable.

**Data Availability Statement:** The data presented in this study are available on request from the corresponding author.

**Conflicts of Interest:** The authors declare no conflict of interest.

# References

1. Mannina, G.; Gulhan, H.; Ni, B.-J. Water reuse from wastewater treatment: The transition towards circular economy in the water sector. *Bioresour. Technol.* **2022**, *363*, 127951. [CrossRef]
2. Cagno, E.; Garrone, P.; Negri, M.; Rizzuni, A. Adoption of water reuse technologies: An assessment under different regulatory and operational scenarios. *J. Environ. Manag.* **2022**, *317*, 115389. [CrossRef]
3. Wang, H.; Li, X.; Zhao, X.; Li, C.; Song, X.; Zhang, P.; Huo, P.; Li, X. A review on heterogeneous photocatalysis for environmental remediation: From semiconductors to modification strategies. *Chin. J. Catal.* **2022**, *43*, 178–214. [CrossRef]

4. Malato, S.; Maldonado, M.I.; Fernández-Ibáñez, P.; Oller, I.; Polo, I.; Sánchez-Moreno, R. Decontamination and disinfection of water by solar photocatalysis: The pilot plants of the Plataforma Solar de Almeria. *Mater. Sci. Semicond. Process.* **2016**, *42*, 15–23. [CrossRef]
5. Wetchakun, K.; Wetchakun, N.; Sakulsermsuk, S. An overview of solar/visible light-driven heterogeneous photocatalysis for water purification: $TiO_2$- and ZnO-based photocatalysts used in suspension photoreactors. *J. Ind. Eng. Chem.* **2019**, *71*, 19–49. [CrossRef]
6. Zhang, S.; Zhang, J.; Sun, J.; Tang, Z. Capillary microphotoreactor packed with $TiO_2$-coated glass beads: An efficient tool for photocatalytic reaction. *Chem. Eng. Process. Process Intensif.* **2020**, *147*, 107746. [CrossRef]
7. Wang, L.; Fei, X.; Zhang, L.; Yu, J.; Cheng, B.; Ma, Y. Solar fuel generation over nature-inspired recyclable $TiO_2/g$-$C_3N_4$ S-scheme hierarchical thin-film photocatalyst. *J. Mater. Sci. Technol.* **2022**, *112*, 1–10. [CrossRef]
8. Dell'Edera, M.; Lo Porto, C.; De Pasquale, I.; Petronella, F.; Curri, M.L.; Agostiano, A.; Comparelli, R. Photocatalytic $TiO_2$-based coatings for environmental applications. *Catal. Today* **2021**, *380*, 62–83. [CrossRef]
9. Xing, Z.; Zhang, J.; Cui, J.; Yin, J.; Zhao, T.; Kuang, J.; Xiu, Z.; Wan, N.; Zhou, W. Recent advances in floating $TiO_2$-based photocatalysts for environmental application. *Appl. Catal. B Environ.* **2018**, *225*, 452–467. [CrossRef]
10. Shifu, C.; Gengyu, C. Photocatalytic degradation of organophosphorus pesticides using floating photocatalyst $TiO_2 \cdot SiO_2$/beads by sunlight. *Sol. Energy* **2005**, *79*, 1–9. [CrossRef]
11. Shen, C.; Wang, Y.J.; Xu, J.H.; Luo, G.S. Facile synthesis and photocatalytic properties of $TiO_2$ nanoparticles supported on porous glass beads. *Chem. Eng. J.* **2012**, *209*, 478–485. [CrossRef]
12. Ha, J.-W.; Do, Y.-W.; Park, J.-H.; Han, C.-H. Preparation and photocatalytic performance of nano-$TiO_2$-coated beads for methylene blue decomposition. *J. Ind. Eng. Chem.* **2009**, *15*, 670–673. [CrossRef]
13. Balakrishnan, A.; Appunni, S.; Gopalram, K. Immobilized $TiO_2$/chitosan beads for photocatalytic degradation of 2,4-dichlorophenoxyacetic acid. *Int. J. Biol. Macromol.* **2020**, *161*, 282–291. [CrossRef] [PubMed]
14. Khalilian, H.; Behpour, M.; Atouf, V.; Hosseini, S.N. Immobilization of S, N-codoped $TiO_2$ nanoparticles on glass beads for photocatalytic degradation of methyl orange by fixed bed photoreactor under visible and sunlight irradiation. *Sol. Energy* **2015**, *112*, 239–245. [CrossRef]
15. Isik, Z.; Bilici, Z.; Adiguzel, S.K.; Yatmaz, H.C.; Dizge, N. Entrapment of $TiO_2$ and ZnO powders in alginate beads: Photocatalytic and reuse efficiencies for dye solutions and toxicity effect for DNA damage. *Environ. Technol. Innov.* **2019**, *14*, 100358. [CrossRef]
16. Hui, J.; Pestana, C.J.; Caux, M.; Gunaratne, H.Q.N.; Edwards, C.; Robertson, P.K.J.; Lawton, L.A.; Irvine, J.T.S. Graphitic-$C_3N_4$ coated floating glass beads for photocatalytic destruction of synthetic and natural organic compounds in water under UV light. *J. Photochem. Photobiol. A Chem.* **2021**, *405*, 112935. [CrossRef]
17. Sakthivel, S.; Shankar, M.V.; Palanichamy, M.; Arabindoo, B.; Murugesan, V. Photocatalytic decomposition of leather dye: Comparative study of $TiO_2$ supported on alumina and glass beads. *J. Photochem. Photobiol. A Chem.* **2002**, *148*, 153–159. [CrossRef]
18. Sraw, A.; Kaur, T.; Pandey, Y.; Sobti, A.; Wanchoo, R.K.; Toor, A.P. Fixed bed recirculation type photocatalytic reactor with $TiO_2$ immobilized clay beads for the degradation of pesticide polluted water. *J. Environ. Chem. Eng.* **2018**, *6*, 7035–7043. [CrossRef]
19. Chand, R.; Obuchi, E.; Katoh, K.; Luitel, H.N.; Nakano, K. Effect of transition metal doping under reducing calcination atmosphere on photocatalytic property of $TiO_2$ immobilized on $SiO_2$ beads. *J. Environ. Sci.* **2013**, *25*, 1419–1423. [CrossRef]
20. Miranda-García, N.; Maldonado, M.I.; Coronado, J.M.; Malato, S. Degradation study of 15 emerging contaminants at low concentration by immobilized $TiO_2$ in a pilot plant. *Catal. Today* **2010**, *151*, 107–113. [CrossRef]
21. Ohtani, B. Photocatalysis A to Z—What we know and what we do not know in a scientific sense. *J. Photochem. Photobiol. C Photochem. Rev.* **2010**, *11*, 157–178. [CrossRef]
22. Lee, S.-Y.; Park, S.-J. $TiO_2$ photocatalyst for water treatment applications. *J. Ind. Eng. Chem.* **2013**, *19*, 1761–1769. [CrossRef]
23. ISO 10678:2010; Fine Ceramics (Advanced Ceramics, Advanced Technical Ceramics): Determination of Photocatalytic Activity of Surfaces in an Aqueous Medium by Degradation of Methylene Blue. International Organization for Standardization: Geneva, Switzerland, 2010.
24. Mills, A.; Hill, C.; Robertson, P.K.J. Overview of the current ISO tests for photocatalytic materials. *J. Photochem. Photobiol. A Chem.* **2012**, *237*, 7–23. [CrossRef]
25. Makuła, P.; Pacia, M.; Macyk, W. How to correctly determine the band gap energy of modified semiconductor photocatalysts based on UV–Vis spectra. *J. Phys. Chem. Lett.* **2018**, *9*, 6814–6817. [CrossRef] [PubMed]
26. Souri, D.; Tahan, Z.E. A new method for the determination of optical band gap and the nature of optical transitions in semiconductors. *Appl. Phys. B* **2015**, *119*, 273–279. [CrossRef]
27. Camera-Roda, G.; Santarelli, F. Optimization of the thickness of a photocatalytic film on the basis of the effectiveness factor. *Catal. Today* **2007**, *129*, 161–168. [CrossRef]
28. Zhang, J.; Zhou, P.; Liu, J.; Yu, J. New understanding of the difference of photocatalytic activity among anatase, rutile and brookite $TiO_2$. *Phys. Chem. Chem. Phys.* **2014**, *16*, 20382–20386. [CrossRef]
29. Essalhi, Z.; Hartiti, B.; Lfakir, A.; Siadat, M.; Thevenin, P. Optical properties of $TiO_2$ thin films prepared by sol gel method. *J. Mater. Environ. Sci.* **2016**, *7*, 1328–1333.

90. Möls, K.; Aarik, L.; Mändar, H.; Kasikov, A.; Niilisk, A.; Rammula, R.; Aarik, J. Influence of phase composition on optical properties of $TiO_2$: Dependence of refractive index and band gap on formation of $TiO_2$-II phase in thin films. *Opt. Mater.* **2019**, *96*, 109335. [CrossRef]
91. Hashimoto, K.; Irie, H.; Fujishima, A. $TiO_2$ photocatalysis: A historical overview and future prospects. *Jpn. J. Appl. Phys.* **2005**, *44*, 8269. [CrossRef]

**Disclaimer/Publisher's Note:** The statements, opinions and data contained in all publications are solely those of the individual author(s) and contributor(s) and not of MDPI and/or the editor(s). MDPI and/or the editor(s) disclaim responsibility for any injury to people or property resulting from any ideas, methods, instructions or products referred to in the content.

Article

# High NIR Reflectance and Photocatalytic Ceramic Pigments Based on M-Doped Clinobisvanite BiVO$_4$ (M = Ca, Cr) from Gels

Guillermo Monrós *, Mario Llusar and José A. Badenes

Department of Inorganic and Organic Chemistry, Jaume I University of Castelló, 12006 Castelló de la Plana, Spain; mllusar@uji.es (M.L.); jbadenes@uji.es (J.A.B.)
* Correspondence: monros@uji.es

**Abstract:** Clinobisvanite (monoclinic scheelite BiVO$_4$, S.G.I2/b) has garnered interest as a wide-band semiconductor with photocatalyst activity, as a high NIR reflectance material for camouflage and cool pigments and as a photoanode for PEC application from seawater. BiVO$_4$ exists in four polymorphs: orthorhombic, zircon-tetragonal, monoclinic, and scheelite-tetragonal structures. In these crystal structures, V is coordinated by four O atoms in tetrahedral coordination and each Bi is coordinated to eight O atoms from eight different VO$_4$ tetrahedral units. The synthesis and characterization of doped bismuth vanadate with Ca and Cr are studied using gel methods (coprecipitated and citrate metal–organic gels), which are compared with the ceramic route by means of the UV–vis–NIR spectroscopy of diffuse reflectance studies, band gap measurement, photocatalytic activity on Orange II and its relation with the chemical crystallography analyzed by the XRD, SEM-EDX and TEM-SAD techniques. The preparation of bismuth vanadate-based materials doped with calcium or chromium with various functionalities is addressed (a) as pigments for paints and for glazes in the chrome samples, with a color gradation from turquoise to black, depending on whether the synthesis is by the conventional ceramic route or by means of citrate gels, respectively; (b) with high NIR reflectance values that make them suitable as fresh pigments, to refresh the walls or roofs of buildings colored with them; and (c) with photocatalytic activity.

**Keywords:** clinobisvanite; ceramic pigment; high NIR reflectance; photocatalysis; sol–gel

## 1. Introduction

Monoclinic BiVO$_4$ (I2/b) is a direct band gap (2.4 eV) semiconductor with proper alignment with water oxidation potential that is applied to seawater splitting in the presence of a high ion concentration and in a corrosive environment [1]. BiVO$_4$ exists in four polymorphs: orthorhombic (o-BiVO$_4$) or pucherite, tetragonal-zircon type (t-BiVO$_4$, I41/amd), monoclinic-scheelite or clinobisvanite (m-BiVO$_4$, group I2/b) and scheelite-tetragonal. Although orthorhombic is the most common phase in nature (mineral pucherite), it has not been synthesized in the laboratory. The low-temperature synthesis of BiVO$_4$ produces the zircon-tetragonal phase with a band gap of 2.9 eV, which, at 528 K, transforms into the monoclinic phase, reversible to tetragonal by adjusting the temperature. In these crystal structures, V is coordinated by four O atoms in a tetrahedral coordination and each Bi is coordinated to eight O atoms [2]. The BiO$_8$ polyhedron known as bisdisphenoid or dodecadeltahedron is rather familiar in this case because all polymorphs of described BiVO$_4$ show this typical coordination in zircon and scheelite; this polyhedron has 12 triangular faces, 18 edges and, of course, 8 vertices with symmetry D$_{2d}$ in Schönflies notation [3].

Among non-titania (TiO$_2$)-based visible-light-driven photocatalysts, monoclinic BiVO$_4$ has proved to be an excellent material for photocatalytic water splitting and the photocatalytic degradation of organic compounds [4], as a high NIR reflectance material for camouflage and cool pigments and as a photoanode for PEC application from seawater [5].

BiVO$_4$ is not classified as hazardous and is used to replace the highly toxic yellow cadmium sulfoselenide in ecofriendly paints applications [6]. In general, bismuth compounds do not show classified harmful properties and are used in medical application,, e.g., the so-called milk of bismuth [7] and or the bismuth subsalicylate (or hydrolyzed bismuth salicylate (Bi(C$_6$H$_4$(OH)CO$_2$)$_3$)), an antacid and anti-diarrheal agent [8].

BiVO$_4$ can be stabilized in different crystal structures, yet high PEC efficiency is found only for the n-type-doped monoclinic phase. The theoretical description of the monoclinic polymorph is difficult because of spontaneous transformation to a tetragonal phase. The cause of instability of the monoclinic phase which spontaneously transforms to a tetragonal phase has not solved yet. Laraib et al. [9] suggest a crucial role of doping in the structure and, thus, the photoelectrochemical performance of BiVO$_4$ because m-BiVO$_4$ is related to the higher photoelectrochemical activity.

A simple review of the literature about the studied systems has been carried out analyzing the entries in WOS (Web of Science) and Google Academics. For the topic "calcium-doped BiVO$_4$", there are three entries (two in WOS and another one in Google): one for visible-light-activated (VLA) photocatalysis applied to the degradation of pharmaceutical contaminants of emerging concern and two for photoanodes for water splitting, all based in the semiconductor characteristics of Ca-BiVO$_4$ solid solutions. Fatwa et al. [10] introduces for the first time calcium as an acceptor-type dopant into BiVO$_4$ photoelectrodes, and the resulting Ca-doped BiVO$_4$ photoelectrodes show anodic photocurrents with an enhanced carrier separation efficiency. BiVO$_4$ and Ca-BiVO$_4$ show indirect semiconductor behavior with band gap measured by a Tauc plot of 2.48 and 2.44 eV, respectively. Li et al. [11] prepare micropowders of M$^{II}_x$Bi$_{1-x}$V$_{1-x}$Mo$_x$O$_4$ (M$^{II}$ = Ca, Sr, x = 0.1 to 0.9) by the ceramic method; the samples x = 0.1 in the cases of samples containing calcium and strontium afford the highest water-splitting performance.

In the case of "chromium-doped BiVO$_4$", twelve entries are detected—nine in WOS and two in Google. Seven are related to different photocatalytic systems applied to the photoreduction of chromium (VI) in wastewater and the other four to photoelectrochemical water splitting. The Cr-doped BiVO$_4$ are synthesized using chromates via liquid–solid state reaction obtaining Cr(VI) solid solutions in BiVO$_4$, where Cr(VI) replaces V(V). Krysiak et al. [12] studied photoanodes for water splitting prepared by spray printing of molybdenum-doped BiVO$_4$ using a liquid–solid state reaction in ethanol media by means of multicomposite catalysts containing nickel, iron, and chromium introduced as nitrates. The formed catalyst particles were not uniformly incorporated into the Mo-BiVO$_4$ structure, but rather creating agglomerates. This suggests phase segregation, which may be beneficial for photoelectrochemical activity. Okuno et al. [13] prepared Cr$^{6+}$-BiVO$_4$ solid solutions from Bi(NO$_3$)$_3$·5H$_2$O, V$_2$O$_5$ and CaCrO$_4$ in HNO$_3$. The resulting solution was stirred for 26–168 h in the dark and the resulting precipitate was washed and dried at 80 °C for 5 h. The single phase of monoclinic scheelite BiVO$_4$ was obtained with up to 2% Cr doping while mixtures of monoclinic scheelite and tetragonal-zircon BiVO$_4$ were obtained for samples with larger amounts of doping (3–5%) even with a prolonged reaction time of up to 4 weeks.

Therefore, the literature related to Ca- and Cr-doped BiVO$_4$ applies to the photoelectrochemical splitting of water and uses solid-state liquid reactions in organic media to preserve high valence ions such as Cr(VI). Therefore, it is not applicable for the synthesis of ceramic pigments that need high temperatures to be used in ceramic glazes and earthenware (1000–1300 °C) in construction applications.

This work addresses the synthesis and characterization of bismuth vanadate doped with calcium or chromium with various functionalities (a) as pigments for paints and enamels; (b) with high NIR reflectance values that make them suitable as cool pigments, to refresh the walls or floors of buildings colored with them; and (c) with photocatalytic activity.

## 2. Materials and Methods

### 2.1. Synthesis Methods

The synthesis methodology will be based on the ceramic method CE, the ammonia coprecipitation method CO and the metal–organic decomposition method MOD.

Ceramic samples (CE) were synthesized from $Cr_2O_3$, $\alpha$-$Bi_2O_3$ and $NH_4VO_3$ 99.9 wt% supplied by Sigma-Aldrich. These precursors with a particle size between 0.3 and 5 μm were mechanically homogenized in an electric grinder (20,000 rpm) for 5 min and the mixture fired at the corresponding temperature and soaking time.

CO and MOD samples were synthesized from $Cr(NO_3)_3 \cdot 9H_2O$, $NH_4VO_3$ (previously dissolved in $HNO_3$ 30 wt%) and $Bi(NO_3)_3 \cdot 5H_2O$ (99.9 wt%, Sigma-Aldrich, St. Louis, MO, USA). For 5 g of the product, these precursors were dissolved in 200 mL of water, then citric acid is added with a molar ratio of Bi:Acid = 1:x (x = 0 or CO sample, 0.25, 1, 2). This solution was continuously stirred at 70 °C and ammonia 17.5 wt% was dropped until gelation occurred at approximatively pH 7.5. The gel was dried at 110 °C and fired at the corresponding temperature and soaking time (500 °C/1 h for charring CO and MOD gels, and 600 °C/3 h for CE and the previously charred CO and MOD gels).

## 2.2. Characterization Methods

X-ray diffraction (XRD) was carried out on a Siemens D5000 diffractometer (München, Germany) using Cu $K_\alpha$ radiation (10–70°2θ range, scan step 0.02°2θ, 4 s per step and 40 kV and 20 mA conditions).

L*a*b* and C*h* color parameters of glazed samples were measured following the CIE-L*a*b* (Commission Internationale de l'Éclairage) colorimetric method using a X-Rite SP60 spectrometer, with standard lighting D65 and a 10° observer [14]. L* measures the lightness (100 = white, 0 = black) and a* and b* the chroma (−a* = green, +a* = red, −b* = blue, +b* = yellow). C* (chroma) and h* (hue angle) can be estimated from a* and b* parameters by the Equations (1) and (2), respectively:

$$C^* = (a^2 + b^2)^{1/2} \tag{1}$$

$$h^* = \arctan(b^*/a^*) \tag{2}$$

We can point out that the optimal chroma is obtained when lightness L* (measured by the L*a*b* method) is high for yellow (80–90), cyan (75–85) and green (70–80), middle for magenta (50–60) and red (45–55) and low for blue (25–35) hues.

UV–vis–NIR spectra of fired powder and the applications of the pigments were carried out by a Jasco V670 spectrometer (Madrid, Spain) through the diffuse reflectance technique, measuring absorbance (A in arbitrary units) or reflectance (R(%)) through the Kubelka–Munk transformation model. A band gap energy of semiconductors was calculated by a Tauc plot from the UV-vis–NIR diffuse reflectance spectra [15].

The solar reflectance, R or total solar reflectance, $R_{NIR}$ or solar reflectance in the NIR and $R_{Vis}$ solar reflectance in the visible spectrum are evaluated from UV–vis–NIR spectra, through the diffuse reflectance technique, as the integral of the measured spectral reflectance and the solar irradiance divided by the integral of the solar irradiance in the range of 350–2500 nm for R, 700–2500 nm for $R_{NIR}$ and 350–700 nm for $R_{Vis}$ as in the Equation (3):

$$R = \frac{\int_{350}^{2500} r(\lambda) i(\lambda) d\lambda}{\int_{350}^{2500} i(\lambda) d\lambda} \tag{3}$$

where (a) $r(\lambda)$ is the spectral reflectance ($Wm^{-2}$) measured from UV–vis–NIR spectra and (b) $i(\lambda)$ is the standard solar irradiation ($Wm^{-2} nm^{-1}$) according to the American Society for Testing and Materials (ASTM) Standard G173-03.

Microstructure characterization of powders was carried out by scanning electron microscopy (SEM) using a JEOL 7001F electron microscope (Akishima, Japan) and transmission electron microscopy (TEM) using a HITACHI electron microscope and selected area electron diffraction (SADP) (following conventional preparation and imaging techniques).

The photocatalytic tests were performed using a dispersion of 500 mg/L of powder added to a solution $0.6 \cdot 10^{-5}$ M of Orange II in pH 7.42 phosphate buffer media

($NaH_2PO_4 \cdot H_2O$ 3.31 g and $Na_2HPO_4 \cdot 7H_2O$ 33.77 g solved in 1.000 mL of water). The degradation of Orange II in buffer media was followed by colorimetry at $\lambda$ = 485. The UV irradiation source was a mercury lamp of 125 W emitting in the range 254–365 nm. The suspension was first stirred in the dark for 15 min to reach the equilibrium sorption of the dye. Aliquot samples were taken every 15 min to measure the change in the dye concentration. Control experiments with Orange II solution and without catalyst, were conducted before the photocatalytic experiments (CONTROL). Commercial anatase P25 from Degussa was used as a reference to compare its photocatalytic activity with the studied samples [16].

The photodegradation curves were analyzed following the Langmuir–Hinshelwood model [16] measuring parameters such as the degradation half-time $t_{1/2}$ and the correlation of kinetic data $R^2$.

In order to analyze the pigmenting capacity of powders as ceramic pigments, the powders were 3 wt% mixed with a double-firing frit (composition in oxides (wt%); $SiO_2$(62.3), $B_2O_3$(9.1), $Al_2O_3$(10.3), CaO(5.9), $K_2O$(9.3), ZnO(3.1), $ZrO_2$(0.1)) supplied by Alfarben S.A., with a maturation point at 1050 °C that devitrifies zircon. The mixture of the frit and the pigment were fired applying a standard firing cycle used in the ceramic tile industry, maintaining the maximum of temperature (1050 °C) for 5 min.

## 3. Discussion

A cooperative transition (COT) mechanism is assumed [3]. In this mechanism, the chromophore enters the solid solution, replacing an isovalent cation but with dissimilar size; if the chromophore is bigger than the replaced ion, it becomes "compressed" even in small concentrations and even more if the incorporation of the chromophore ion M occurs, with increasing covalence and polarizability of the M-O bond. The M-O distance decreases and the high tetragonal distortion of d orbitals of the transition ion enhances the crystalline field over de ion. Consequently, absorption bands shift to higher frequencies and the color shifts to blue. If the chromophore is smaller than the replaced ion, it becomes "relaxed" in the lattice site (in this case, Shannon–Prewitt Crystal Radii [17]: $Bi^{3+}$(VIII) = 1.31 Å, $Ca^{2+}$(VIII) = 1.26 Å, $Cr^{3+}$(VIII) = 0.7 Å) and the crystalline field over the d orbitals goes down; therefore, the absorption bands shift to higher wavelengths. In short, in this mechanism, there is a cooperative transition between the chromophore and the replaced ion that modifies the crystalline field over the chromophore ion.

In this case, the replacement of $Bi^{3+}$ by the smaller $Ca^{2+}$ or $Cr^{3+}$ should relax the dopant ions, which shifts its absorption to higher wavelengths, but aliovalent replacement with $Ca^{2+}$ probably increases the concentration of charge-carrying mobile defects of oxygen [13,18].

### 3.1. The Effect of the Dopant Concentration

Figure 1 shows the powder samples $(Ca_xBi_{1-x})VO_4$, x = 0, 0.1, 0.2, 0.4 (600 °C/3 h): (a) aspect by binocular lens ($\times$40), and its L*a*b*, and (b) 3 wt% washed samples glazed in a double-firing frit (1050 °C). The entrance of $Ca^{2+}$ replacing $Bi^{3+}$ is associated with a slight decrease in color intensity (lightness L* and yellow hue b* increase, but red hue a* decreases). The pigments are dissolved, unstabilized in the glaze, and do not produce coloration (Figure 1b).

Figure 2 shows the UV–vis–NIR spectra of calcium-doped samples $(Ca_xBi_{1-x})VO_4$, x = 0.1, 0.2, 0.4 (600 °C/3 h), showing a transference charge band centered at 400 nm with a typical semiconductor performance. Table 1 shows the main characterization of $(Ca_xBi_{1-x})VO_4$ (600 °C/3 h) samples: the band gap measured from UV–vis–NIR spectra by the Tauc procedure (Figure 3) indicates that the semiconductors show a direct performance and the band gap increases slowly from 2.25 eV for pure $BiVO_4$ (x = 0) to 2.31 eV for the x = 0.4 doped sample (10). Thereby, NIR reflectance also increases continuously with x: from 53% for pure $BiVO_4$ to 62% for the x = 0.4 sample, improving the cool performance of $BiVO_4$.

**Figure 1.** Powder samples $(Ca_xBi_{1-x})VO_4$, x = 0, 0.1, 0.2, 0.4 (600 °C/3 h): (**a**) aspect by binocular lens (×40), and L*a*b* of powders; (**b**) 3 wt% washed samples glazed in a double-firing frit (1050 °C).

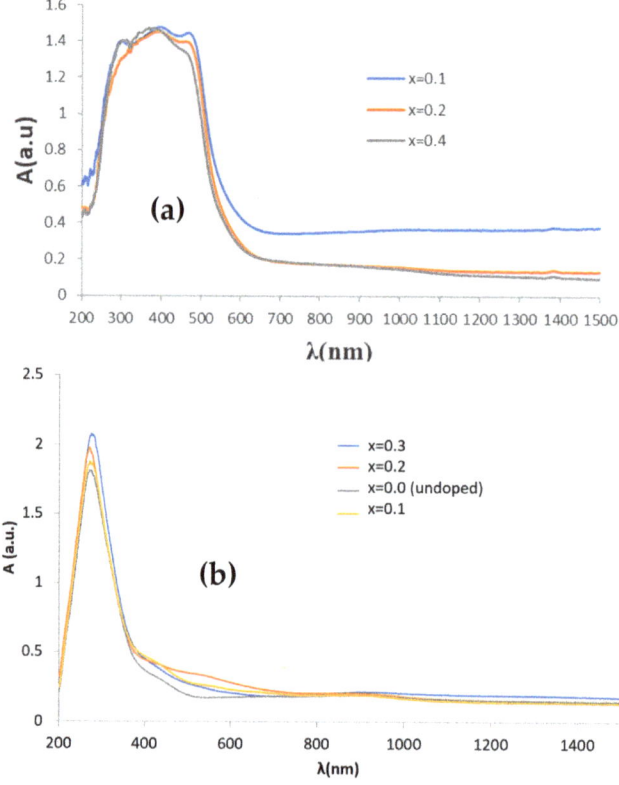

**Figure 2.** UV–vis–NIR spectra of $(Ca_xBi_{1-x})VO_4$, x = 0.1, 0.2, 0.4: (**a**) (600 °C/3 h) powders; (**b**) 3 wt% glazed samples.

**Table 1.** Characterization of $(M_xBi_{1-x})VO_4$, M = Ca, Cr, x = 0, 0.1, 0.2, 0.4 (600 °C/3 h) samples.

| | x = 0 | x = 0.1 | x = 0.2 | x = 0.4 |
|---|---|---|---|---|
| $(Ca_xBi_{1-x})VO_4$ powder | | | | |
| L*a*b* | 74.2/15.0/36.0 | 75.1/11.2/38.2 | 78.6/10.6/38.7 | 80.0/9.5/39.4 |
| Eg (eV) (Tauc) | 2.25 | 2.26 | 2.30 | 2.31 |
| $R_{Vis}/R_{NIR}/R$ | 30/53/41 | 31/55/43 | 32/60/46 | 33/62/47 |
| $(Cr_xBi_{1-x})VO_4$ powder | | | | |
| L*a*b* | 74.2/15/36 | 50.3/12.2/37.2 | 47.6/11.6/35.1 | 46.0/10.5/34.7 |
| Eg (eV) (Tauc) | 2.25 | 2.18 | 2.13 | 2.13 |
| $R_{Vis}/R_{NIR}/R$ | 30/53/41 | 27/47/37 | 23/44/32 | 19/41/30 |
| $(Cr_xBi_{1-x})VO_4$ glazed | | | | |
| L*a*b* | white | 59/−7.9/3.3 | 52.2/−10.1/−2 | 53.8/−5.8/−8.9 |
| Eg (eV) (Tauc) | 3.26 | 1.50 | 1.53 | 1.56 |
| $R_{Vis}/R_{NIR}/R$ | 77/89/82 | 33/76/54 | 26/70/47 | 22/67/44 |

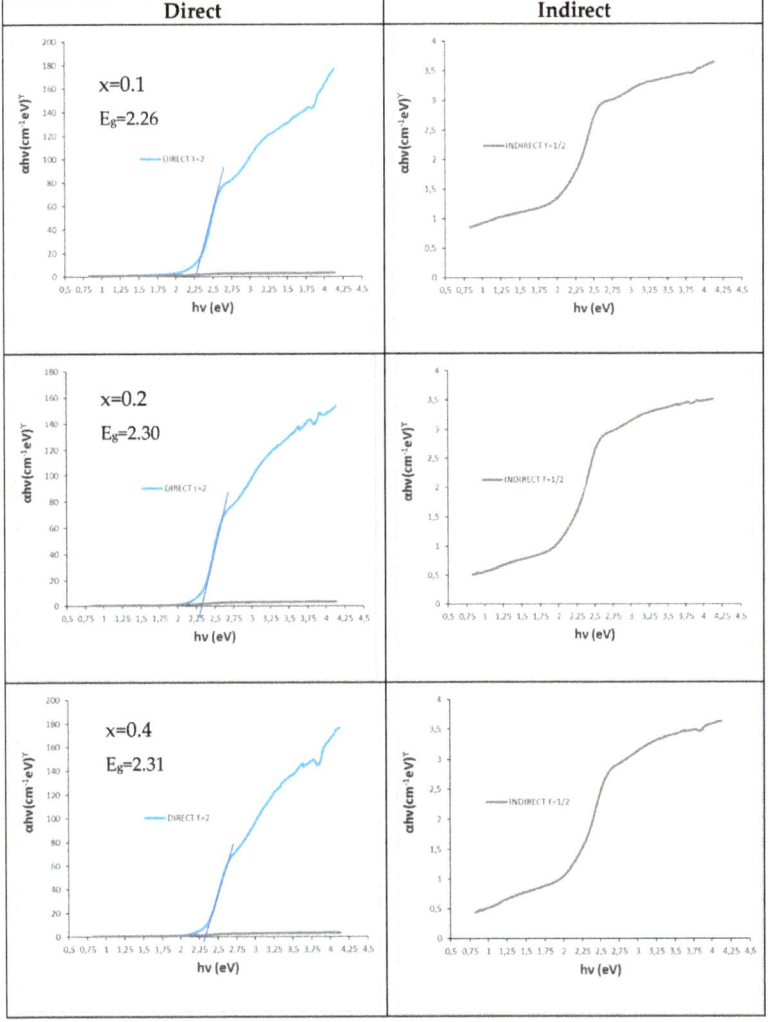

**Figure 3.** Tauc plots of $(Ca_xBi_{1-x})VO_4$, x = 0.1, 0.2, 0.4 (600 °C/3 h) powders.

All XRD of these $Bi_{1-x}Ca_xVO_4$ ceramic samples show clinobisvanite (V) as a single phase, except the x = 0.4 sample, which shows residual peaks of $CaV_2O_6$ (see Figure 4). Therefore, it can be pointed out that the limit of the solid solution of $Ca^{2+}$ in $Bi_{1-x}Ca_xVO_{4-x/2}$ is lower than x = 0.4 and stabilizes the monoclinic polymorph (11).

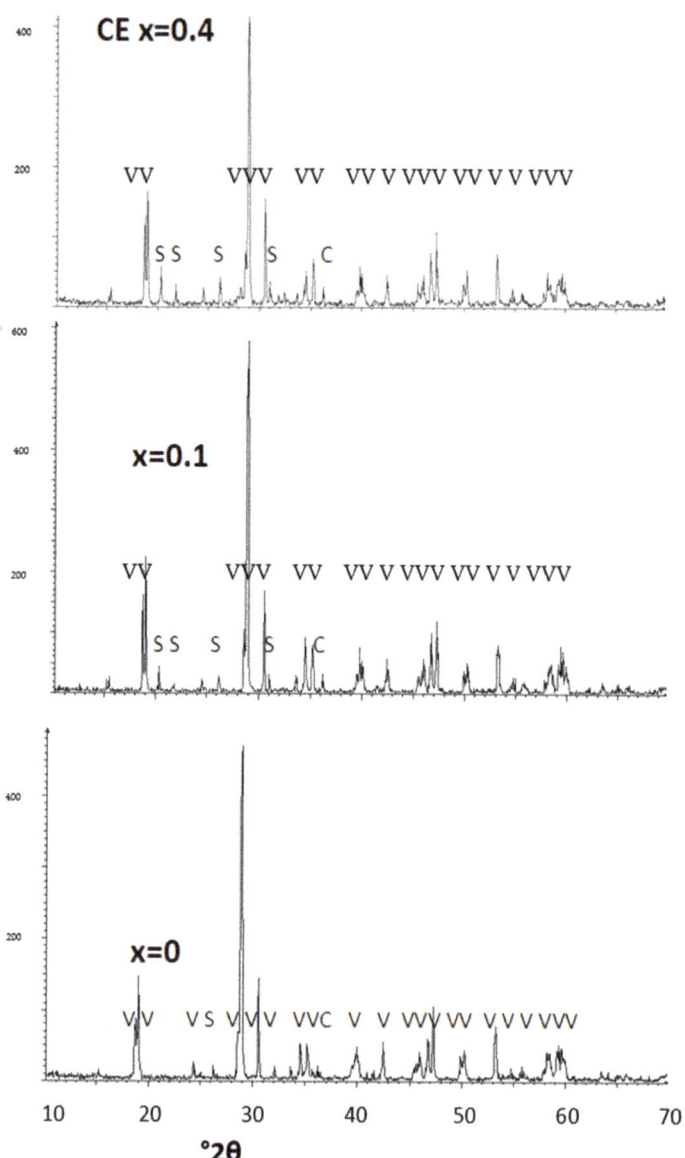

**Figure 4.** XRD of $Bi_{1-x}Cr_xVO_4$ ceramic samples. Crystalline phases: V (m-BiVO$_4$), C (hexagonal Cr$_2$O$_3$), and S (Scherbinite V$_2$O$_5$). The homologous $Bi_{1-x}Ca_xVO_4$ ceramic samples show clinobisvanite (V) as a single phase except the x = 0.4 sample, which shows residual peaks of CaV$_2$O$_6$.

Figure 5 shows the ceramic CE powders with chromium $(Cr_xBi_{1-x})VO_4$, x = 0, 0.1, 0.2, 0.4 fired at 600 °C/3 h). Figure 5a shows the aspect through binocular lens (×40) and

L*a*b* of yellow–green-colored samples; macroaggregates of homogeneous microcrystals can be observed and the entrance of chromium produces a more intense color (L* decreases with x) and less red and yellow hues (both a* and b* parameters decrease with x).

Figure 5. Samples $(Cr_xBi_{1-x})VO_4$, x = 0, 0.1, 0.2, 0.4 (600 °C/3 h): (a) aspect through binocular lens (×40), and L*a*b* of powders; (b) 3 wt% washed samples glazed in a double-firing frit (1050 °C).

When washed powders were 3 wt% glazed in a double-firing frit (1050 °C) (Figure 5b) an interesting turquoise color is developed by chromium-doped samples, more intense with the chromium amount x: L* decreases, green (negative a*) decreases and blue (negative b*) increases, respectively, resulting x = 0.4 (L*a*b* = 53.8/−5.8/−8.9). The optimum turquoise sample. Figure 5 shows the XRD of chromium doped $Bi_{1-x}Cr_xVO_4$ ceramic samples: all samples show clinobisvanite (V on figure) as majoritarian phase with residual peaks associated with hexagonal $Cr_2O_3$ (C) and Scherbinite $V_2O_5$ (S), which slightly increase in intensity with the amount of doping agent x (best observed in the highly doped sample (x = 0.4)).

The UV–vis–NIR spectra of $(Cr_xBi_{1-x})VO_4$, x = 0.1, 0.2, 0.4 (600 °C/3 h) powders (see spectra on Sections 3.2 and 3.3 for sample x = 0.4) show optical absorption bands that can be associated with $Cr^{3+}$ ($3d^3$) in the dodecadeltahedral position ($D_{2d}$ symmetry group), with a multiband at 220–520 nm integrated almost for the overlap of bands centered at 270, 380 and 480 nm, respectively, and a shoulder at 750 nm (that confers a semiconductor appearance to the spectrum) with bands at 680 and a shoulder at 760 nm. The electronic configuration for $Cr^{3+}(d^3)$ may be described as $d_{x^2-y^2}^{21} d_{z^2}^{21} d_{xz,yz}^{1} d_{xz} d_{xy}$ [19]. Table 1 shows the characterization of these chromium-doped $(Cr_xBi_{1-x})VO_4$ (600 °C/3 h) samples: the band gap measured from UV–vis–NIR spectra by the Tauc procedure indicates that the semiconductor shows a direct performance, decreasing its band gap slowly from 2.25 eV for pure $BiVO_4$ (x = 0) to 2.13 eV for the x = 0.4 doped sample. Likewise, NIR reflectance also decreases continuously with x: from 53% for pure $BiVO_4$ to 41% for the x = 0.4 sample, reducing the cool performance of $BiVO_4$.

Figure 6 shows the UV–vis–NIR spectra of chromium-doped $(Cr_xBi_{1-x})VO_4$, x = 0.1, 0.2, 0.4 (600 °C/3 h) glazed samples. It is detected as an intense band centered at 270 nm to the charge transfer band of the frit used in the glaze, and relatively intense bands in the visible range at 380, 500 and 620 with a shoulder at 720 nm; a minimum of absorbance at 520 nm is detected, indeed a maximum of light reflected, that explains the turquoise color of the sample. These bands can be considered the same as that of the powders shifting to lower wavelengths.

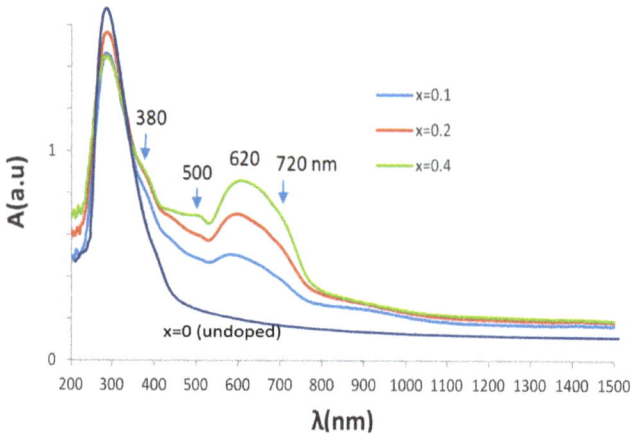

**Figure 6.** UV–vis–NIR spectra of $Cr_xBi_{1-x}VO_4$ glazed samples.

Figure 7 shows the Tauc plots of chromium-doped $(Cr_xBi_{1-x})VO_4$, x = 0, 0.1, 0.2, 0.4 (600 °C/3 h) glazed samples showing a direct type semiconductor behavior. For the pure $BiVO_4$ (x = 0) that unstabilizes in the glaze and produces white color associated with the glaze (see Figures 1b and 2b), its Tauc plot indicates a direct band gap of 3.26 eV in the UV range due to the frit mixed with the pigment in the glaze. The band gap of turquoise glazes with chromium-doped $BiVO_4$ increases slowly from 1.50 eV for x = 0.1 to 1.56 eV for the x = 0.4 sample in the vis–NIR range.

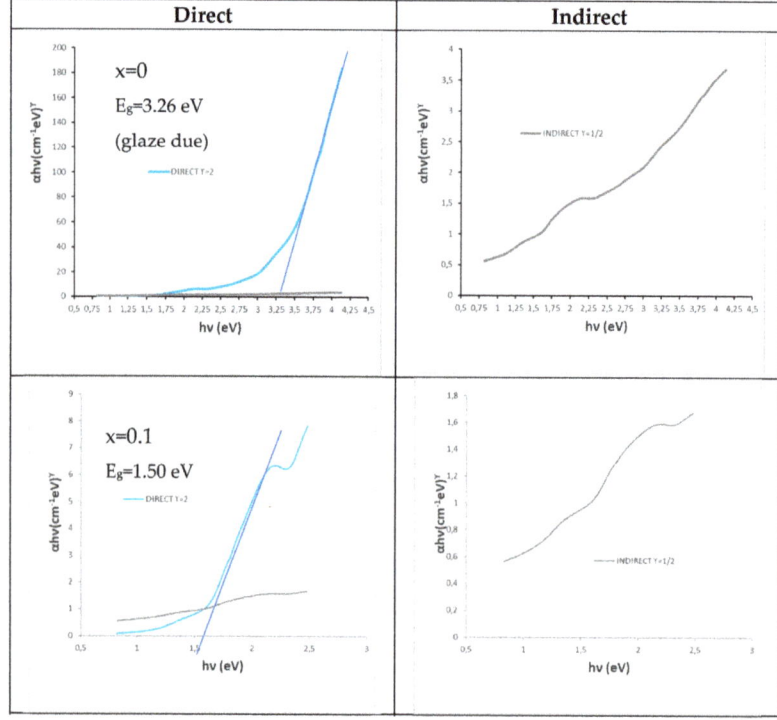

**Figure 7.** Cont.

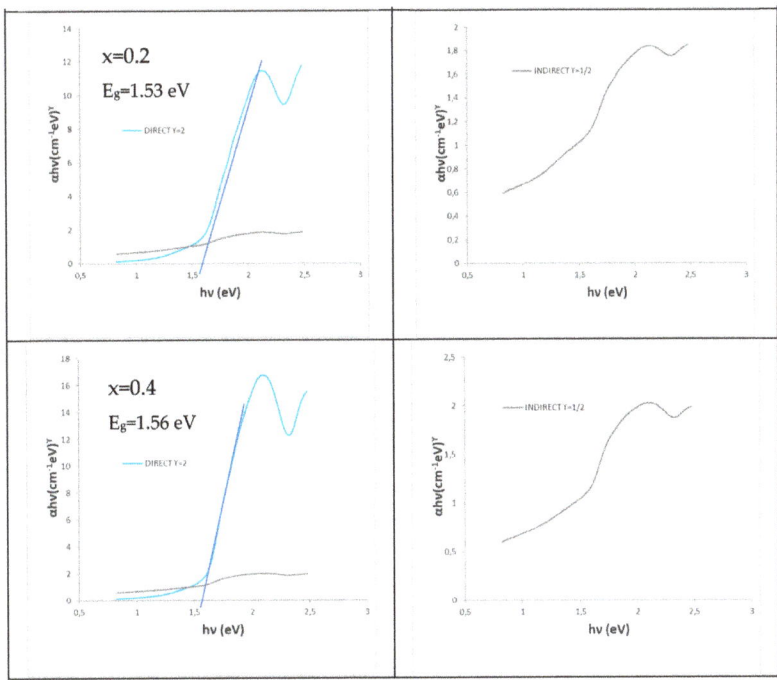

**Figure 7.** Tauc plots of $(Cr_xBi_{1-x})VO_4$, x = 0, 0.1, 0.2, 0.4 (600 °C/3 h) glazed samples.

Since Ca-doped samples do not produce coloration and the solid solution is complete in this case, optimization studies with mineralizers and the use of the coprecipitation and MOD routes to obtain inks for glazes were performed only on chromium-doped samples.

### 3.2. The Effect of the Addition of Mineralizers in Cr-Doped Samples

Figure 8 shows the effect of the addition of three mineralizers in order to activate the synthesis improving the ionic diffusion on the system [20,21] of the optimum composition $Cr_{0.4}Bi_{0.6}VO_4$: 3 wt% $NH_4Cl$ (min 1), 3 wt% $3NaF \cdot 2MgF_2 \cdot Li_2CO_3$ (min 2) and 2 wt% $CaCO_3 \cdot KCl$ +1 wt% $NaF \cdot Na_4B_2O_7$ (min 3). The DTA and TG analyses of the employed mineralizers (Figure 9) chosen with thermal activity at progressive temperatures show: (a) (min 1) melts and decomposes at 290 °C, showing an intense endothermic band (a very week endothermic band at 100 °C is detected also associated with water elimination) with an associated weight loss at TG of 100%; (b) (min 2) endothermic bands at 140, 460 and 625 °C as main DTA bands, and the band at 460 °C is associated with a weight loss of approximately 20% in TG pattern, due to carbonate decomposition, and finally (c) (min 3) shows the main endothermic bands at 140, 580 and 640 °C as main DTA bands with a weight loss in TG approximately 10% at 580 °C associated with carbonate decomposition.

Figure 8b shows the photographs of powders by binocular lens (×40) and the corresponding 3 wt% washed powders glazed in a double-firing frit (1050 °C) (L*a*b*parameters are included); the low-temperature mineralizer $NH_4Cl$ (min 1) is greenish (L* increases, a* decreases and b* increases) but the turquoise shade obtained in glazed samples is similar to the unmineralized sample, the middle mineralizer 2 produces green shades in both powder and glazed samples, and the higher-temperature mineralizer (min 3) is also greenish in powder but shows a high turquoise intensity in glazes (L* decreases and negative b* increases). The addition of mineralizers only shows a significant improvement of the turquoise shade of the glazed sample with mineralizer 3 with a near thermal activity (580 °C) to the firing employed temperature (600 °C) (20).

**Figure 8.** Effect of mineralizers on the $Cr_{0.4}Bi_{0.6}VO_4$ sample: (**a**) photos of powders by binocular lens (×40); (**b**) 3 wt% washed samples glazed in a double-firing frit (1050 °C) (L*a*b* parameters are included).

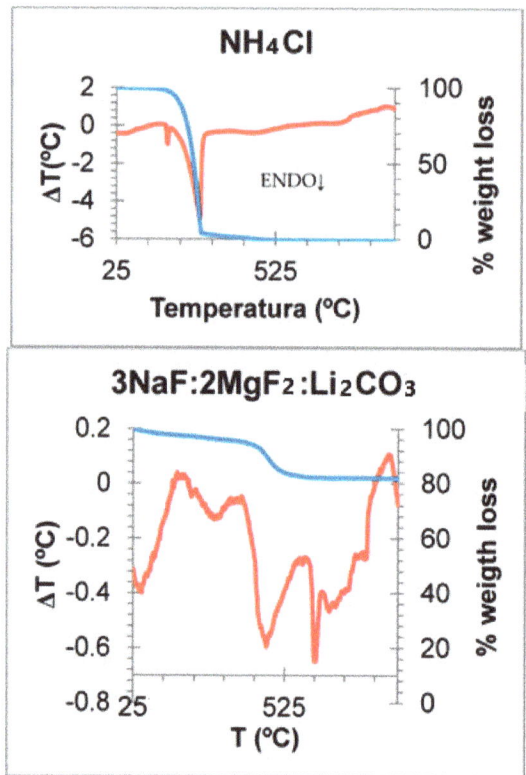

**Figure 9.** *Cont.*

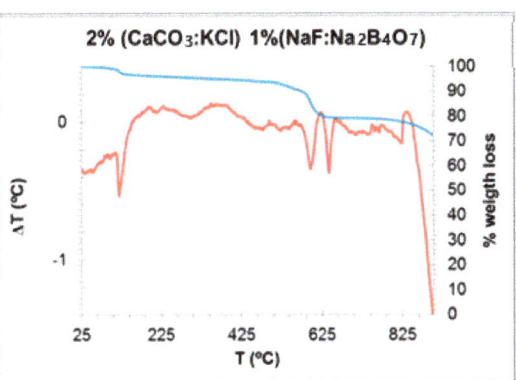

**Figure 9.** DTA (red) and TG (blue) of employed mineralizers.

Figure 10 shows the XRD diffractogram of some mineralized samples: mineralizer 1 shows m-BiVO$_4$ and residual peaks of S (scherbinite V$_2$O$_5$) and C (hexagonal Cr$_2$O$_3$). In the case of mineralizer 2, C (hexagonal Cr$_2$O$_3$) and very weak peaks of W (CrVO$_4$) are detected; finally, for mineralizer 3 only residual peaks of W (CrVO$_4$) are detected. Therefore, the relative high-temperature mineralizer 3 shows the cleaner diffractogram (with minimal residual phases) and the best turquoise result in the glaze.

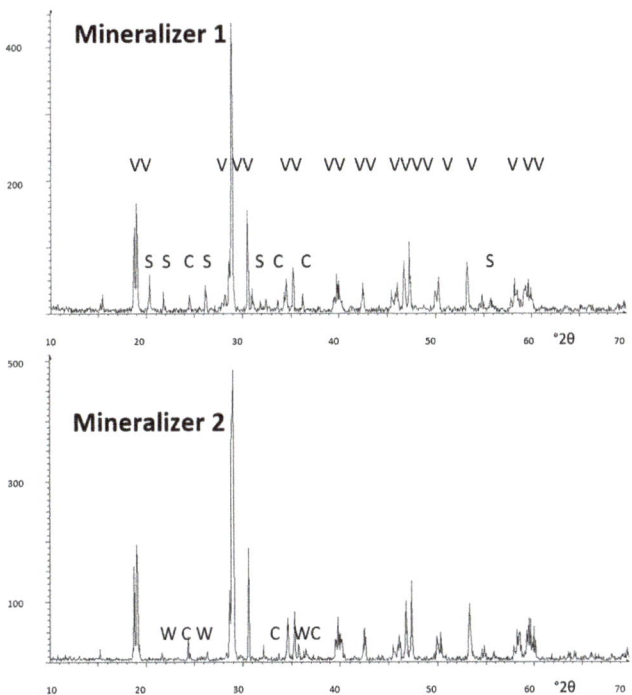

**Figure 10.** XRD of the Cr$_{0.4}$Bi$_{0.6}$VO$_4$ mineralized samples. Crystalline phases: V (m-BiVO$_4$), S (Scherbinite V$_2$O$_5$), C (hexagonal Cr$_2$O$_3$), and W (CrVO$_4$).

Figure 11 shows the UV–vis–NIR of Cr$_{0.4}$Bi$_{0.6}$VO$_4$ powders and glazed mineralized samples, respectively. As described above, powders show the optical absorption bands associated with Cr$^{3+}$ (3d$^3$) in the dodecadeltahedral position but the absorbance is higher in mineralized samples; and in the case of min 3, the shoulder at 680 nm practically disappears. Likewise, glazed samples show an intense band centered at 300 nm due to the charge transfer band of the frit used in the glaze and relatively intense bands in the visible range at 500, 610 and the shoulder at 680 nm; a minimum of absorbance at 500 nm is detected; indeed a maximum of light reflected, which explains the turquoise color of the sample. The band gap of powders (measured by the Tauc method) decreases for mineralized samples; 2.11 eV for min 1 and 2.09 eV for min 2 and min 3 (Table 2). On the other hand, the band gap associated with the shoulder at 680 nm for glazed samples indicates similar values for mineralizer 2 (1.56 eV) whereas those for mineralizers 1 and 3 are slightly lower (1.55 and 1.54 eV, respectively) (Table 2).

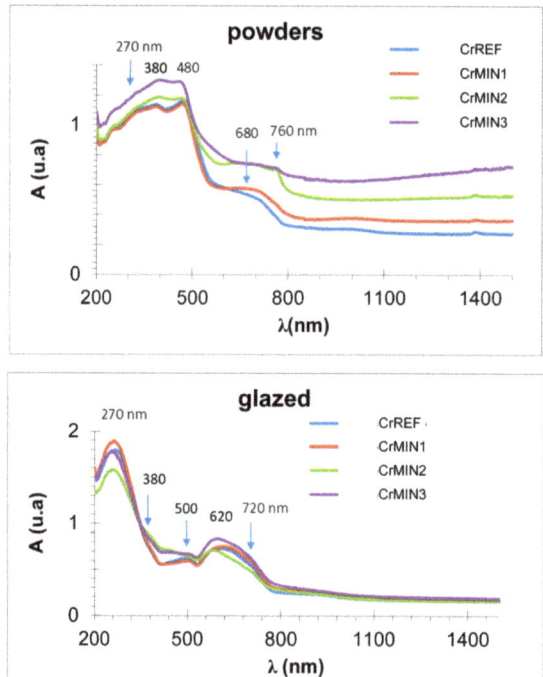

**Figure 11.** UV–vis–NIR spectra of the $Cr_{0.4}Bi_{0.6}VO_4$ mineralized samples.

**Table 2.** Characterization of mineralized samples of $Cr_{0.4}Bi_{0.6}VO_4$ composition (600 °C/3 h).

|  | CE | min 1 | min 2 | min 3 |
|---|---|---|---|---|
| **POWDERS** |  |  |  |  |
| L*a*b* | 46/10.5/34.7 | 49.3/6.6/37.2 | 41.3/4.9/22.1 | 39.8/4.3/24.5 |
| $E_g$ (eV) (Tauc) | 2.13 | 2.11 | 2.11 | 2.09 |
| $R_{Vis}/R_{NIR}/R$ | 19/41/30 | 18/39/28 | 16/37/26 | 13/35/21 |
| **GLAZED** |  |  |  |  |
| L*a*b* | 53.8/−5.8/−8.9 | 54.9/−6.2/−7.3 | 49.3/−8.2/−0.7 | 49.6/−9.4/−9.4 |
| $E_g$ (eV) (Tauc) | 1.56 | 1.55 | 1.56 | 1.56 |
| $R_{Vis}/R_{NIR}/R$ | 22/67/44 | 22/68/44 | 21/67/44 | 23/66/43 |

Finally, the reflectance $R_{Vis}$ and $R_{NIR}$ of powders decrease in all mineralized samples associated with a greenish turn and as the intensity of the color increases (lower L*) (Table 2). On the other hand, for glazed samples, the total reflectance is very similar in all mineralized samples and the CE sample (R = 43–44) (Table 2), in agreement with the very similar band gap and absorbance spectra on Figure 11.

## 3.3. The Effect of Non-Conventional Methods and Ink Application in Cr-Doped Samples

Figure 12 shows the coprecipitated sample (CO) and the effect of citric acid (MOD): (I,II) show the photographs of powders by binocular lens (×40) charred at 500 °C/1 h and successively fired at 600 °C/3 h, respectively, (III,IV) show the screen printing (48 threads/cm) of direct colloidal emulsion and the 500 °C/1 h powder mixed with diethylenglycol (weight ratio 2:5), and (V,VI) show the 3 wt% washed samples glazed in a double-firing frit (1050 °C) of powders fired at 500 °C/1 h and 600 °C/3 h, respectively (L*a*b*parameters are included).

| CO (x=0) | Citric 0.25 | Citric 1 | Citric 1.5 | Citric 2 |
|---|---|---|---|---|
| (I) 500 °C/1h<br>46.5/11.7/35.8 | 46.6/10.7/34.4 | 44.1/11.2/30.2 | 42.3/6.6/22.1 | 40.2/9.7/16.3 |
| (II) 600 °C/3h<br>48.7/12.0/33.7 | 49.7/11.2/36.7 | 46.7/9.6/30.2 | 47.0/10.9/27.4 | 45.8/12.0/25.2 |
| (III) Direct print screen of raw emulsions<br>71.7/-1.7/16.1 | 76.6/-0.7/17.2 | 70.1/-2.8/14.4 | 69.5/-2.2/8.5 | 73.8/-2.9/12 |
| (IV) 500°C/1h Diethylenglycol (2:5) medium print screen<br>68.3/-2/17.2 | 68.3/-2.1/15.1 | 65.9/-2.6/14.2 | 69.8/-2.8/13.7 | 68.6/-2.3/13.1 |
| (V) 500°C/3h single firing 1050°C<br>49.4/0.6/10.7 | 45.9/-0.2/8.7 | 43.3/1.8/9.5 | 41.4/1.0/6.6 | 41.6/-0.4/0.6 |
| (VI) 600°C/3h single firing 1050°C<br>46.6/-1.3/9.1 | 48.9/-0.5/9.8 | 50.4/-0.5/9.5 | 48.4/0.4/8.9 | 44.7/1.1/3.9 |

**Figure 12.** Coprecipitated sample (CO) and effect of citric acid (MOD) of $Cr_{0.4}Bi_{0.6}VO_4$ composition: (I,II) photos of powders by binocular lens (×40) fired at 500 °C/1 h and 600 °C/3 h, respectively, (III,IV) print screen (48 threads/cm) of direct colloidal emulsion and the 500 °C/1 h powder mixed with diethylenglycol (weight ratio 2:5), and (V,VI) 3 wt% washed samples glazed in a double-firing frit (1050 °C) fired at 500 °C/1 h and 600 °C/3 h, respectively (L*a*b*parameters are included).

The charred powders at 500 °C/1 h (Figure 12I) are of dark green shades in MOD samples associated with the remaining graphitic residue in the material. The stabilized samples fired at 600 °C/3 h (Figure 12II), free of carbon, are of yellow–green shades with L*a*b* of approximately 47/10/35, similar in all samples. The screen-printing depositions on ceramic support (Figure 12III,IV) show brown shades for CO and citric 0.25 samples and greenish shades for samples citric 1, 1.5 and 2 (b* approximately −2.5). Finally, glazed samples (Figure 12V,VI) show grey shades which darken with x: the sample with x = 2 shows a dark-grey (black) shade with very low chroma C* (L*a*b* = 41.6/−0.4/0.6 for 500 °C/1 h and 44.7/1.1/3.9 for stabilized sample 600 °C/3 h).

Figure 13 shows the XRD diffractograms of CE, CO (x = 0) and citric MOD samples fired at 500 °C/1 h and 600 °C/3 h, respectively. At 500 °C, only broad peaks of m-BiVO$_4$ are detected. At 600 °C, as in the CE sample, the CO and MOD samples show clinobisvanite m-BiVO$_4$ with residual peaks associated with S (scherbinite V$_2$O$_5$) and C (hexagonal Cr$_2$O$_3$); in the CO and MOD samples, the intensity of residual phases slightly decreases.

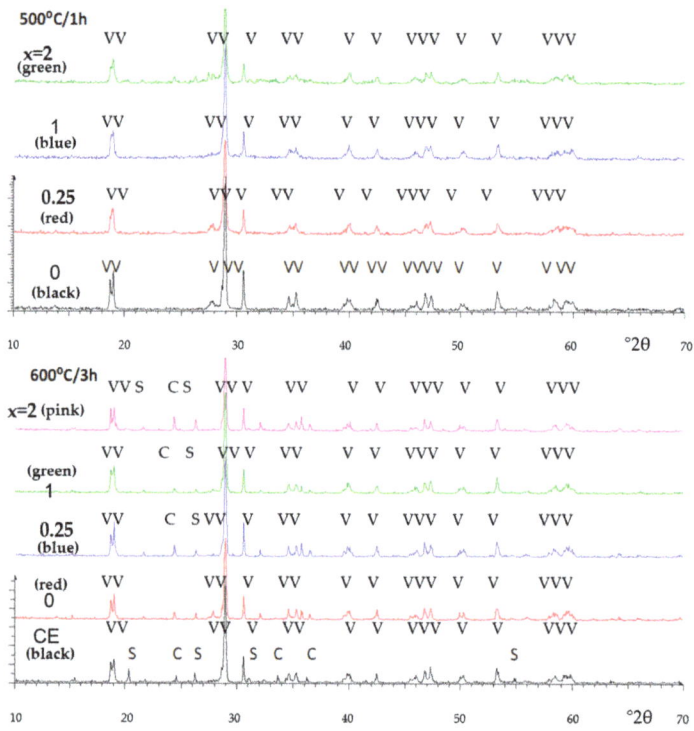

**Figure 13.** XRD spectra of CITRIC MOD samples of Cr$_{0.4}$Bi$_{0.6}$VO$_4$ composition. Crystalline phases: V (m-BiVO$_4$), S (Scherbinite V$_2$O$_5$), C (hexagonal Cr$_2$O$_3$).

CE and mineralized yellow powders produce turquoise colors in the glaze and greenish CO and MOD powders give grey-black colors in the glaze. Figure 14 shows the UV–vis–NIR representative spectra of a yellow powder (CE sample, turquoise in glaze) and a greenish powder (citric x = 2 MOD sample, grey-black in glaze). As described above, the optical absorption bands associated with Cr$^{3+}$ (3d$^3$) in the dodecadeltahedral position (D$_{2d}$ symmetry group) are detected in the case of the yellow powder (CE sample, turquoise in glaze). For the greenish powder (citric x = 2 MOD sample, grey-black in glaze), the spectrum is similar, but the absorbance in the 580–1500 nm range increases. Figure 15 shows the Tauc plots of these representative samples; the band gap increases from 2.13 eV for yellow powder to 2.20 eV for greenish MOD powder (Table 3).

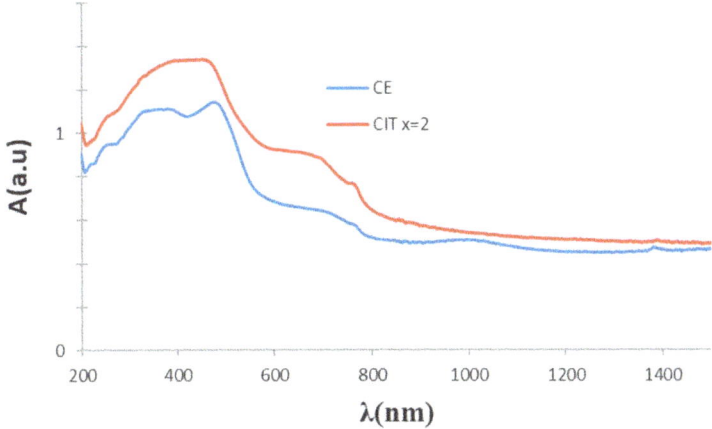

**Figure 14.** UV–vis–NIR spectra of CE and CITRIC x = 2 powder samples of $Cr_{0.4}Bi_{0.6}VO_4$ composition.

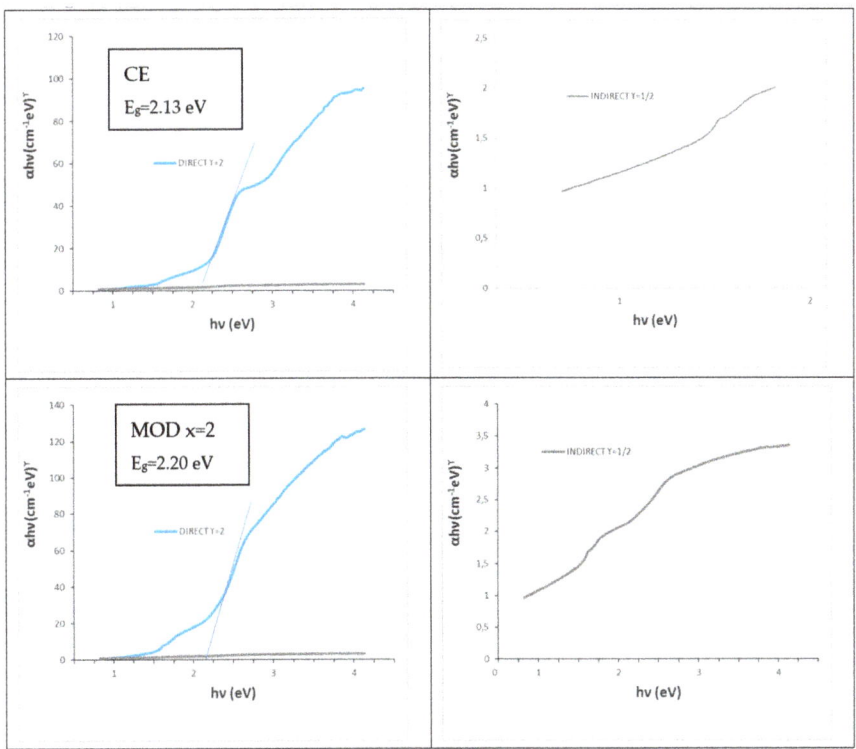

**Figure 15.** Tauc plots of CE and CITRIC x = 2 powders of $Cr_{0.4}Bi_{0.6}VO_4$ composition.

Figure 16 shows UV–vis–NIR spectra of glazed samples of $Cr_{0.4}Bi_{0.6}VO_4$ composition: (a) CITRIC MOD glazed samples 500 °C/1 h; (b) CE and CITRIC x = 2 samples 600 °C/3 h. (including sample CE). Bands at 270, 320, 420, 580 and 670 nm can be detected, showing a shift at lower wavelengths relative to ceramic samples and a significant increase in absorption to approximately 500 nm associated with the darkening of glazes. The direct band gap of the semiconductors measured from Tauc plots, as shown in Figure 17, for the edge of absorbance in the NIR range (3 in Figure 16b) shows that all MOD samples show a band gap of 1.53 eV slightly lower than the homologous CE sample (1.56 eV) (Table 3).

**Table 3.** Characterization of CO and MOD samples of $Cr_{0.4}Bi_{0.6}VO_4$ composition (600 °C/3 h).

|  | CE | CO | MOD 0.25 | MOD 1 | MOD 1.5 | MOD 2 |
|---|---|---|---|---|---|---|
| **POWDERS** | | | | | | |
| L*a*b* | 46/10.5/34.7 | 46.5/11.7/35.8 | 46.6/10.7/34.4 | 44.1/11.2/30.2 | 42.3/6.6/22.1 | 40.2/9.7/16.3 |
| $E_g$ (eV) (Tauc) | 2.13 | 2.14 | 2.14 | 2.16 | 2.18 | 2.20 |
| $R_{Vis}/R_{NIR}/R$ | 19/41/30 | 18/38/28 | 17/39/28 | 16/37/27 | 15/38/27 | 14/38/26 |
| **GLAZED (600 °C)** | | | | | | |
| L*a*b* | 53.8/−5.8/−8.9 | 46.6/−1.3/9.1 | 48.9/−0.5/9.8 | 50.4/−0.5/9.5 | 48.4/0.4/8.9 | 44.7/1.1/3.9 |
| $E_g$ (eV) (Tauc) | 1.56 | 1.53 | 1.53 | 1.53 | 1.53 | 1.53 |
| $R_{Vis}/R_{NIR}/R$ | 22/67/44 | 20/71/46 | 20/71/46 | 19/70/45 | 19/69/44 | 20/69/45 |

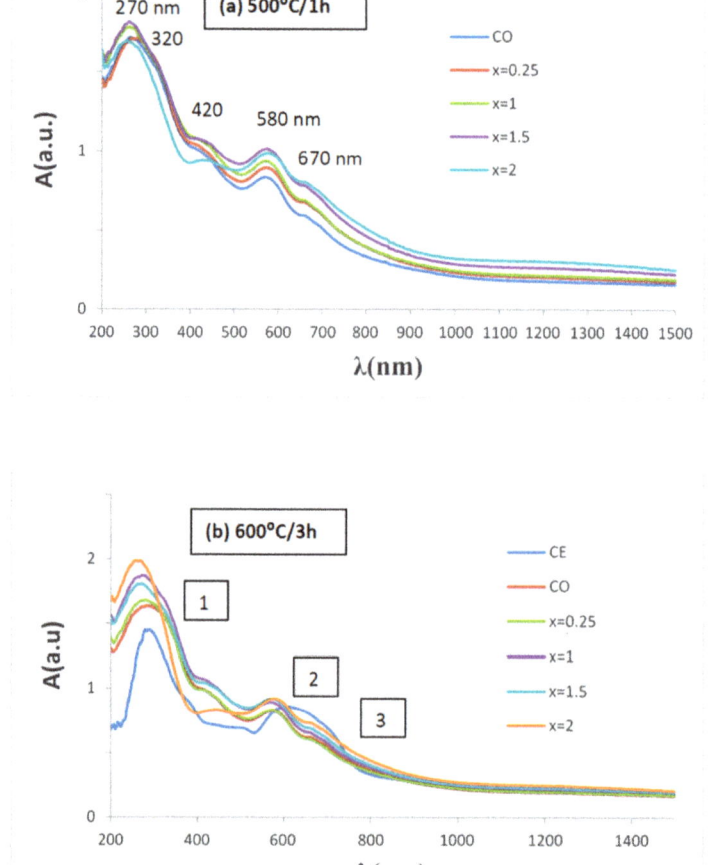

**Figure 16.** UV–vis–NIR spectra of glazed samples of $Cr_{0.4}Bi_{0.6}VO_4$ composition: (**a**) CITRIC MOD glazed samples 500 °C/1 h; (**b**) CE and CITRIC x = 2 samples 600 °C/3 h. (the analyzed absorbance inflection points are shown in the numbered boxes).

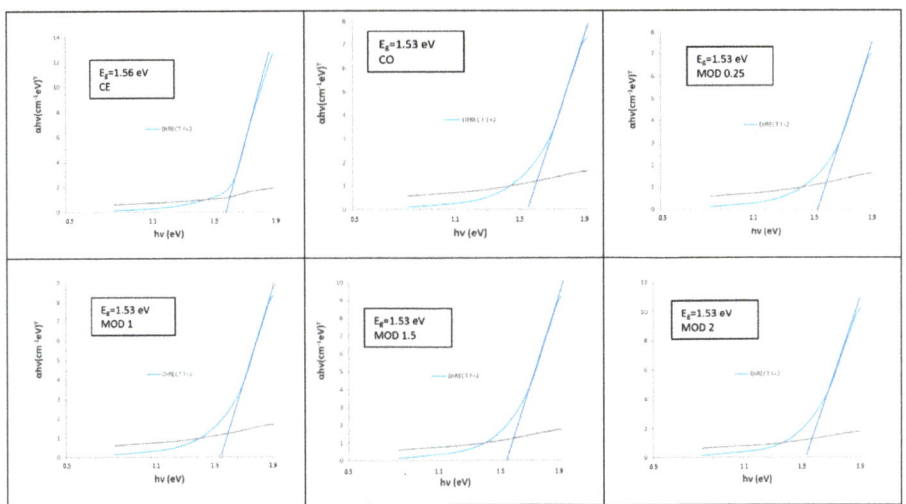

**Figure 17.** Tauc plots of glazed samples of $Cr_{0.4}Bi_{0.6}VO_4$ composition considering the absorption 3 on Figure 16b.

Finally, the $R_{Vis}$ and $R_{NIR}$ reflectance of the powders decrease slightly in CO and MOD samples associated with greenish hue (lower L*) (Table 3). On the other hand, for glazed samples, the reflectance is very similar in all CO and MOD samples, with a lower $R_{Vis}$ than the CE sample associated with the darkening of samples, but higher $R_{NIR}$ (Table 3), in agreement with the absorbance spectra on Figure 16 and despite the darkening of samples.

Figure 18 presents the photodegradation test over Orange II of powders ($Cr_{0.4}Bi_{0.6}VO_4$) fired at 600 °C/3 h (CE and CITRATE (x = 2) samples) with the Langmuir–Hinshelwood kinetics parameters $t_{1/2}$ (degradation half-time and data correlation $R^2$) which for the commercial reference P25 of Degussa and for the control (without powder addition) are $t_{1/2}/R^2$= 25 min/0.898 and 497 min/0.945, respectively. The ceramic samples show an interesting degradation half-time of 85 min that increases to 150 min for the citrate sample (lower than 497 min for the simple photolysis of the CONTROL test).

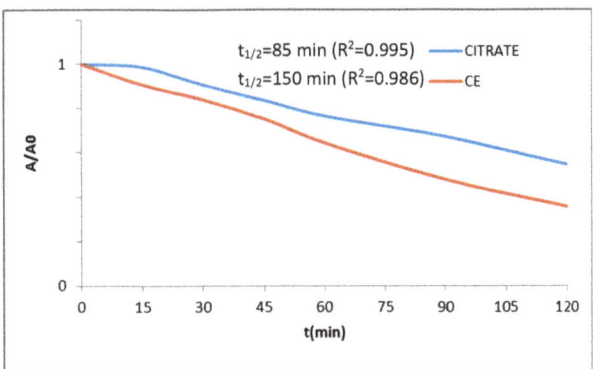

**Figure 18.** Photodegradation test over Orange II of powders ($Cr_{0.4}Bi_{0.6}VO_4$) fired at 600 °C/3 h (CE and CITRATE (x = 2) samples) with the Langmuir–Hinshelwood kinetics parameters (degradation half-time $t_{1/2}$ and data correlation $R^2$).

Figure 19 shows the SEM micrographs of representative powders of $Cr_{0.4}Bi_{0.6}VO_4$ composition: the CE powder, which produces a turquoise color on glazes, shows well-developed prismatic crystals of approximately 3 µm in length together with other irregular particles of 0.5 µm in size. The coprecipitated sample CO, which produces a grey color on glazes, shows very fine nanoparticles, forming agglomerates of 3–15 µm of size, finally MOD x = 0.25, which produces a grey color on glazes, shows microcrystals between 0.5 and 1 µm, forming big agglomerates of 10–20 µm. The decrease in the active surface due to the agglomeration of the fine particles of $BiVO_4$ of the CO and MOD methods, observed by SEM, is probably associated with the decrease in the degradation half-time detected in the photodegradation test of these samples (12).

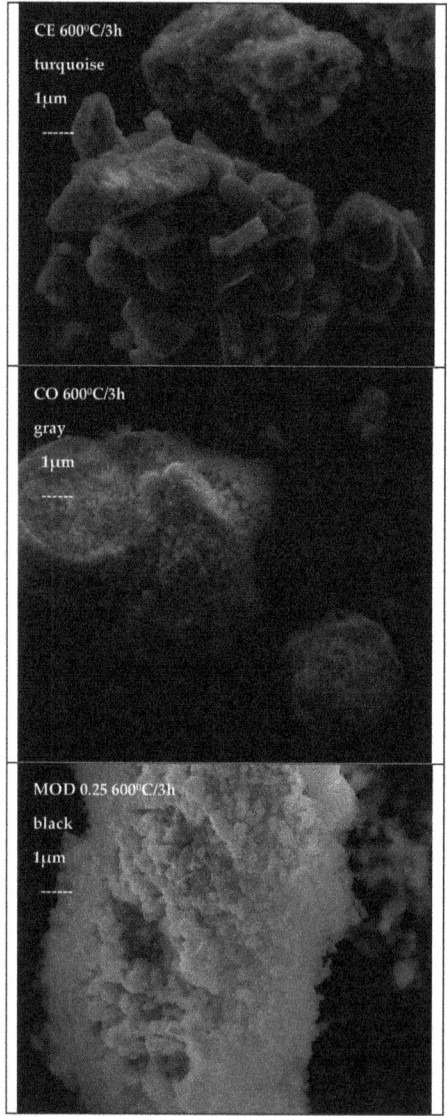

**Figure 19.** SEM micrographs of representative samples of $Cr_{0.4}Bi_{0.6}VO_4$ composition.

Figure 20 shows the TEM micrographs and selected area electron diffraction (SADP) of powders MOD x = 0.25 charred at 500 °C/1 h used in the preparation of inks IV;

nanoparticles of 10–80 nm and a SADP compatible with polycrystals of clinobisvanite m-BiVO$_4$ are detected.

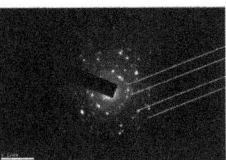

**Figure 20.** TEM micrographs and selected area electron diffraction (SADP) of powders MOD x = 0.25 (Cr$_{0.25}$Bi$_{0.75}$VO$_4$) charred at 500 °C/1 h used in the preparation of inks IV.

## 4. Conclusions

Doped bismuth vanadate was synthesized using gel methods (coprecipitated and citrate metal–organic gels) and compared with the ceramic route.

The limit of the solid solution of Ca$^{2+}$ in Bi$_{1-x}$Ca$_x$VO$_{4-x/2}$ is lower than x = 0.4, in the case of Cr$^{3+}$ fired at 600 °C, and residual Cr$_2$O$_3$ and V$_2$O$_5$ always remained, indicating an incomplete solid solution of chromium in clinobisvanite (m-BiVO$_4$), but both cations stabilize the monoclinic polymorph. NIR reflectance increases continuously with Ca doping from 53% for pure BiVO$_4$ to 62% (Eg = 2.31 eV) for Ca$_{0.4}$Bi$_{0.6}$VO$_4$ samples, improving the cool performance of BiVO$_4$; conversely, the entrance of chromium decreases the NIR reflectance of clinobisvanite. The band gap of clinobisvanite increases for Ca-doped samples and decreases for Cr-doped samples in agreement with NIR reflectance evolution.

The yellow samples doped with Ca do not produce coloration in glazes but the yellow–green Cr-doped samples produce interesting turquoise shades in a double-firing frit (1050 °C) resulting in the optimum turquoise of the Cr$_{0.4}$Bi$_{0.6}$VO$_4$ sample (L*a*b* = 53.8/−5.8/−8.9) that maintains NIR reflectance at approximately 67% (band gap 2.13 eV). The use of mineralizer 3 (2 wt% CaCO$_3$·KCl +1 wt% NaF·Na$_4$B$_2$O$_7$) with a near thermal activity (580 °C) to the employed firing temperature (600 °C) produces a significant improvement in the turquoise shade of the glazed sample (L*a*b* = 49.6/−9.4/−9.4) with similar NIR reflectance (66%) and band gap (2.09 eV).

Cr-doped samples prepared by gel methods (CO and citrate MOD routes) produce yellow–green powders that darken with the amount of chromium, decreasing NIR reflectance and increasing the band gap. The corresponding glazed samples produce black shades with high NIR reflectance (69–71%) associated with a shift in the absorbance bands of Cr probably in dodecadeltahedral site to lower wavelengths relative to the ceramic sample and a significant increase in absorption at approximately 500 nm associated with the darkening of glazes.

The Cr-doped ceramic samples show an interesting degradation half time of 85 min that increases to 150 min for the citrate sample (lower than the 497 min for the simple photolysis of the CONTROL test).

To the best of our knowledge, Ca- or Cr-doped polyfunctional BiVO$_4$ pigments have been prepared for the first time for paints, and for glazes in the case of chromium, producing colors from turquoise to black depending on whether the synthesis is a conventional ceramic route or through citrate gels, showing a high NIR reflectance that makes them suitable as cool pigments, to refresh the walls and roofs of buildings colored with them, and gives them photocatalytic activity.

**Author Contributions:** Conceptualization, G.M. and M.L.; methodology, G.M. and J.A.B.; validation, G.M., M.L. and J.A.B.; investigation, G.M., M.L. and J.A.B.; writing—original draft preparation, G.M.; writing—review and editing, G.M., M.L. and J.A.B.; supervision, G.M.; project administration, G.M. and M.L.; funding acquisition, All authors have read and agreed to the published version of the manuscript.

**Funding:** This research was funded by Universitat Jaume I (UJI B2021-73 Project).

**Institutional Review Board Statement:** Not applicable.

**Informed Consent Statement:** Not applicable.

**Data Availability Statement:** Data sharing is not applicable to this article.

**Conflicts of Interest:** The authors declare no conflict of interest.

## References

1. Malathia, A.; Madhavana, J.; Ashokkumarb, M.; Arunachalam, P.A. Review on BiVO$_4$ photocatalyst: Activity enhancement methods for solar photocatalytic applications. *Appl. Catal. A Gen.* **2018**, *555*, 47–74. [CrossRef]
2. Schiavello, M. *Photoelectrochemistry, Photocatalysis, and Photoreactors: Fundamentals and Developments*; Reidel: Dordrecht, The Netherlands, 1985; ISBN 978-94-015-7725-0.
3. Monrós, G. *Scheelite and Zircon: Brightness, Color and NIR Reflectance in Ceramics*; Nova Science Publishers: New York, NY, USA, 2021; ISBN 978-1-53619-332-9.
4. Jovanovića, S.D.; Babićc, K.S.; Porobića, M.M.; Dramićanin, M.D. A comparative study of photocatalytically active nanocrystalline tetragonal T zircon type and monoclinic scheelite-type bismuth vanadate. *Ceram. Int.* **2018**, *44*, 17953–17961. [CrossRef]
5. Luo, W.; Yang, Z.; Li, Z.; Zhang, J.; Liu, J.; Zhao, Z.; Wang, Z.; Yan, S.; Yu, T.; Zou, Z. Solar hydrogen generation from seawater with a modified BiVO$_4$ photoanode. *Energy Environ. Sci.* **2011**, *4*, 4046–4051. [CrossRef]
6. Telpande, D.V.P.H.D. Stability Testing of Non-Toxic Bismuth Vanadate According to Pharmaceutical Criteria. *Int. J. Innov. Res. Sci. Eng. Technol.* **2015**, *4*, 5350–5354. [CrossRef]
7. Parker, S.P. *McGraw-Hill Dictionary of Scientific & Technical Terms*, 6th ed.; The McGraw-Hill Companies, Inc.: New York, NY, USA, 2003.
8. Gorbach, S.L. Bismuth therapy in gastrointestinal diseases. *Gastroenterology* **2003**, *99*, 863–875. [CrossRef] [PubMed]
9. Laraib, I.; Alves-Carneiro, M.; Janotti, A. Effects of Doping on the Crystal Structure of BiVO$_4$. *J. Phys. Chem. C* **2019**, *123*, 26752–26757. [CrossRef]
10. Abdi, F.F.; Starr, D.E.; Ahmet, I.Y.; Van De Krol, R. Photocurrent Enhancement by Spontaneous Formation of a p–n Junction in Calcium-Doped Bismuth Vanadate Photoelectrodes. *ChemPlusChem* **2018**, *83*, 941–946. [CrossRef] [PubMed]
11. Li, S.L.; Bychkov, K.L.; Butenko, D.S.; Terebilenko, K.V.; Zhu, Y.; Han, W.; Baumer, V.N.; Slobodyanik, M.S.; Ji, H.; Klyui, N.I. Scheelite-related M$^{II}_x$Bi$_{1-x}$V$_{1-x}$Mo$_x$O$_4$ (M$^{II}$–Ca, Sr) solid solution-based photoanodes for enhanced photoelectrochemical water oxidation. *Dalton Trans.* **2020**, *49*, 2345–2355. [CrossRef] [PubMed]
12. Krysiak, O.A.; Junqueira, J.R.C.; Conzuelo, F.; Bobrowski, T.; Wilde, P.; Wysmolek, A.; Schuhmann, W. Tuning Light-Driven Water Oxidation Efficiency of Molybdenum-Doped BiVO$_4$ by Means of Multicomposite Catalysts Containing Nickel, Iron, and Chromium Oxides. *ChemPlusChem* **2020**, *85*, 327–333. [CrossRef] [PubMed]
13. Okuno, K.; Kato, H.; Vequizo, J.J.M.; Yamakata, A.; Kobayashi, H.; Kobayashi, M.; Kakihana, M. Expansion of the photoresponse window of a BiVO$_4$ photocatalyst by doping with chromium(VI). *RSC Adv.* **2018**, *8*, 38140–38145. [CrossRef] [PubMed]
14. Commission International de L'Eclairage (CIE). *Recommendations on Uniform Colour Spaces, Colour-Difference Equations, Psychometrics Colour Terms*; CIE: Paris, France, 1978.
15. Tauc, J.; Grigorovici, R.; Vancu, A. Optical Properties and Electronic Structure of Amorphous Germanium. *Phys. Status Solidi* **1996**, *15*, 627–637. [CrossRef]
16. Kumar, K.V.; Porkodi, K.; Rocha, F. Langmuir–Hinshelwood kinetics—A theoretical study. *Catal. Commun.* **2008**, *9*, 82–84. [CrossRef]
17. Shannon, R.D. Revised effective ionic radii and systematic studies of interatomic distances in halides and chalcogenides. *Acta Cryst.* **1976**, *A32*, 751–766. [CrossRef]
18. Adeli, P.; Bazak, J.D.; Huq, A.; Goward, G.R.; Nazar, L.F. Influence of aliovalent cation substitution and mechanical compression on Li-ion conductivity and diffusivity in argyrodite solid electrolytes. *Chem. Mater.* **2020**, *33*, 146–157. [CrossRef]
19. Burdett, J.K.; Hoffmann, R.; Fay, R.C. Eight-Coordination. *Inorg. Chem.* **1978**, *17*, 2553–2567. [CrossRef]
20. Gargori, C.; Cerro, S.; Galindo, R.; García, A.; Llusar, M.; Monrós, G. Iron and chromium doped perovskite (CaMO$_3$ M = Ti, Zr) ceramic pigments, effect of mineralizer. *Ceram. Int.* **2012**, *38*, 4453–4460. [CrossRef]
21. Enríquez, E.; Reinosa, J.J.; Fuertes, V.; Fernández, J.F. Advanced and challenges of ceramic pigments for inkjet printing. *Ceram. Int.* **2022**, *48*, 31080–31101. [CrossRef]

**Disclaimer/Publisher's Note:** The statements, opinions and data contained in all publications are solely those of the individual author(s) and contributor(s) and not of MDPI and/or the editor(s). MDPI and/or the editor(s) disclaim responsibility for any injury to people or property resulting from any ideas, methods, instructions or products referred to in the content.

*Article*

# Zinc Oxide Films Fabricated via Sol-Gel Method and Dip-Coating Technique–Effect of Sol Aging on Optical Properties, Morphology and Photocatalytic Activity

Katarzyna Wojtasik [1,*], Magdalena Zięba [1], Cuma Tyszkiewicz [1], Wojciech Pakieła [2], Grażyna Żak [3], Olgierd Jeremiasz [4,5], Ewa Gondek [6], Kazimierz Drabczyk [4] and Paweł Karasiński [1,*]

[1] Department of Optoelectronics, Silesian University of Technology, B. Krzywoustego 2, 44-100 Gliwice, Poland
[2] Department of Engineering Materials and Biomaterials, Silesian University of Technology, Konarskiego 18a, 44-100 Gliwice, Poland
[3] Oil and Gas Institute–National Research Institute, Lubicz 25A, 31-503 Kraków, Poland
[4] Institute of Metallurgy and Materials Science, Polish Academy of Sciences, Reymonta 25, 30-059 Kraków, Poland
[5] Helioenergia Sp. z o.o., Rybnicka 68, 44-238 Czerwionka-Leszczyny, Poland
[6] Department of Physics, Cracow University of Technology, Podchorążych 1, 31-084 Kraków, Poland
* Correspondence: katarzyna.wojtasik@polsl.pl or katarzyna.wojtasik@pk.edu.pl (K.W.); pawel.karasinski@polsl.pl (P.K.)

**Citation:** Wojtasik, K.; Zięba, M.; Tyszkiewicz, C.; Pakieła, W.; Żak, G.; Jeremiasz, O.; Gondek, E.; Drabczyk, K.; Karasiński, P. Zinc Oxide Films Fabricated via Sol-Gel Method and Dip-Coating Technique–Effect of Sol Aging on Optical Properties, Morphology and Photocatalytic Activity. *Materials* **2023**, *16*, 1898. https://doi.org/10.3390/ma16051898

Academic Editor: Aleksej Zarkov

Received: 1 February 2023
Revised: 21 February 2023
Accepted: 22 February 2023
Published: 24 February 2023

**Copyright:** © 2023 by the authors. Licensee MDPI, Basel, Switzerland. This article is an open access article distributed under the terms and conditions of the Creative Commons Attribution (CC BY) license (https://creativecommons.org/licenses/by/4.0/).

**Abstract:** Zinc oxide layers on soda-lime glass substrates were fabricated using the sol-gel method and the dip-coating technique. Zinc acetate dihydrate was applied as the precursor, while diethanolamine as the stabilizing agent. This study aimed to determine what effect has the duration of the sol aging process on the properties of fabricated ZnO films. Investigations were carried out with the sol that was aged during the period from 2 to 64 days. The sol was studied using the dynamic light scattering method to determine its distribution of molecule size. The properties of ZnO layers were studied using the following methods: scanning electron microscopy, atomic force microscopy, transmission and reflection spectroscopy in the UV-Vis range, and the goniometric method for determination of the water contact angle. Furthermore, photocatalytic properties of ZnO layers were studied by the observation and quantification of the methylene blue dye degradation in an aqueous solution under UV illumination. Our studies showed that ZnO layers have grain structure, and their physical–chemical properties depend on the duration of aging. The strongest photocatalytic activity was observed for layers produced from the sol that was aged over 30 days. These layers have also the greatest porosity (37.1%) and the largest water contact angle (68.53°). Our studies have also shown that there are two absorption bands in studied ZnO layers, and values of optical energy band gaps determined from positions of maxima in reflectance characteristics are equal to those determined using the Tauc method. Optical energy band gaps of the ZnO layer fabricated from the sol aged over 30 days are $E_g^I$ = 4.485 eV and $E_g^{II}$ = 3.300 eV for the first and second bands, respectively. This layer also showed the highest photocatalytic activity, causing the pollution to degrade 79.5% after 120 min of UV irradiation. We believe that ZnO layers presented here, thanks to their attractive photocatalytic properties, may find application in environmental protection for the degradation of organic pollutants.

**Keywords:** sol-gel; dip-coating; zinc oxide; thin layers; energy band gap; hydrophilicity; photocatalytic activity

## 1. Introduction

Zinc oxide (ZnO) has attractive physicochemical properties for many applications in modern technology, which determines its high application potential. It is a wide-band gap oxide semiconductor with a direct energy gap. The wide energy gap of bulk zinc oxide ($Eg_{dir}$ = 3.37 eV) causes it to have good transmission properties in the visible range of electromagnetic radiation. It has high exciton binding energy (~60 meV) at room temperature [1]. It is a non-toxic, chemically stable material with sensor properties and high photocatalytic activity [2]. Depending on manufacturing technology, fabricated ZnO samples can have different morphology. Zero-dimensional, 1D, 2D, and 3D dimensional structures can be produced [3–6]. ZnO films are currently being intensively developed for applications in optoelectronics [7], in chemical sensors as sensor layers [8], and for use as protective coatings [9]. Considering optoelectronics, photovoltaics is currently the primary area of ZnO application. ZnO layers are used as photoanodes of DSSCs (Dye-Sensitized Solar Cells) [10]. Moreover, ZnO layers in solar cells, thanks to their good transmission properties and appropriate electrical properties, are used as TCO (Transparent Conductive Oxide) layers [11]. They are successfully replacing the previously used ITO layers. Finally, zinc oxide layers are used as ETL (Electron Transport Layer) in perovskite solar cells [12]. This article focuses on the fabrication and characterization of zinc oxide thin layers, with particular emphasis put on investigations of the effect sol aging has on the morphology, optical, and photocatalytic properties of ZnO layers.

Many methods are used to produce ZnO thin layers, including atomic layer deposition (ALD) [13], sputtering [14], spray pyrolysis [8], electrodeposition [15], microwave assisted synthesis [16], and sol-gel [17]. Innovative methods are also used, e.g., simultaneous etching and Al doping of $H_2O$-oxidized ZnO nanorods to generate Al-doped ZnO nanotubes [18]. The applied method affects the morphology and properties of the ZnO layers produced [19,20]. The sol-gel method offers the greatest opportunities in terms of shaping the structure of the produced material. From a technical point of view, the sol-gel method may seem like a simple method, however, the development of layers with the desired properties requires an understanding of the phenomena occurring in the processes of sol synthesis and layering [21]. It is necessary to carry out many time-consuming studies on the influence of the composition of the sols, the parameters of the hydrolysis and condensation processes, the conditions of layer deposition (e.g., the withdrawal speed, which is the fundamental parameter of the dip-coating method), and the conditions of their annealing on the final properties of the layers. The subjects of the research are the type of precursor and its concentration, chelating and stabilizing agents, the temperature of sol synthesis, annealing of sol layers, the nature of the solvent, and the conditions of sol aging [22–25]. All these factors affect the structure, properties, and morphology of materials produced by the sol-gel method. This paper presents the results of our research on the production of ZnO layers by the sol-gel method and the dip-coating technique, with particular emphasis on the influence of the sol aging time on their properties.

Numerous research groups have reported results of studies on the influence of the duration of sol aging processes on properties of ZnO layers fabricated from sols [17,23,26–30]. As one can notice, studies were carried out for sols aged from several days [17] to several dozen days [23]. Aghkonbad et al. [29] studied the influence of the duration of sol aging processes, lasting in the range of 0–36 h, on the properties of layers produced from aged sols. They reported that ZnO layers, deposited on glass substrates, fabricated from the sol aged in this range of time, have high quality and were characterized by the highest transmission in the UV-Vis range, while their optical bandgap was broadened. Marouf et al. in Ref. [27] reported the synthesis of a sol without the addition of a stabilizing agent. They examined the sol during the period of 13 days. They found that the optimal aging time is 7 days, after which the sol reached stability, and the layers from that day were characterized by high transmittance (exceeding 70% in the visible range) and smooth surface. Toubane et al. reported in Ref. [23] studies in which sols were aged over the longest period, 30 days. They concluded that, as a result of increasing the duration of the sol aging process, fabricated ZnO layers have better transmittance in the UV-Vis range (up to 96%), increased optical band gap $Eg$, and improved smoothness. That group also reported investigations of the influence of the duration of sol aging on the photocatalytic activity, as well as the results of their visual assessment during their aging. Considering sols, they were synthesized from zinc acetate dihydrate (precursor), both without a stabilizing agent [27] and with the addition of DEA [23].

Zinc oxide has very good photocatalytic properties. It is frequently used by researchers for the photodegradation of pollutants because it has very good stability and availability, and the mobility of electrons in ZnO is high [19]. The latest literature reports indicate a wide interest in modified ZnO materials due to the possibility of shifting the spectral range of absorbed electromagnetic radiation in the presence of a certain photocatalyst. This is achieved by doping ZnO with metals or non-metals [31], or by producing hybrid structures with other metal oxides [32] or nanocomposites [33–35]. Excellent photocatalytic and antibacterial activity was also observed for hybrid systems of zinc oxide nanoparticles with reduced graphene oxide and gold nanoparticles [36]. However, the knowledge of the dependence of photocatalytic activity on the structure of pure ZnO is crucial to conduct more complex experiments.

This work presents the results of research on the influence of the duration of the sol aging process on the properties of ZnO layers. The sol was aged for 64 days, and ZnO layers were fabricated during this period. The sol was synthesized using zinc acetate dihydrate as the precursor, ethanol as the solvent, and DEA as a stabilizing agent. Glass substrates were coated with the single sol layer using the dip-coating technique. The subjects of research presented in this paper are the temporal stability of the sol, the optical properties of pure ZnO layers, produced from that sol, as well as their morphology and photocatalytic activity. The latter was determined from observations of the degradation of methylene blue (MB) using the Langmuir–Hinshelwood method. The influence of sol aging time on transmittance, optical band gap, Urbach energy, porosity, morphology, and surface roughness of produced ZnO layers was determined. The diameters of the nanocrystals were estimated from the transmission spectra in the absorption region by the analysis of the quantum size effect. The hydrophilic properties of the produced ZnO films were also studied. So far, investigations of the aforementioned properties of ZnO layers carried out for so lengthy duration of the sol aging process have not been published. We found that parameters of the ZnO layer fabricated from the sol aged for 30 days, take extreme values.

## 2. Materials and Methods

### 2.1. Materials

Zinc oxide layers were produced by the sol-gel method and the dip-coating technique. The sol was prepared by a procedure similar to that reported by Liu et al. in Ref. [37]. Zinc acetate dihydrate ($Zn(CH_3COO)_2 \cdot 2H_2O$, ZAD, purchased from Sigma-Aldrich, Steinheim-Germany) as a precursor of zinc oxide, nonaqueous ethanol (EtOH, 99.8%, Avantor Performance Materials, Gliwice-Poland) as a solvent, and diethanolamine (DEA, Sigma-Aldrich, Steinheim-Germany) as a chelating agent. Deionized water was used directly from the deionizer (Polwater DL2-100S613TUV, Labopol Solution&Technologies, Kraków, Poland). All reagents were used without prior purification. The sol obtained in this way was filtered through a 0.2 μm PTFE syringe filter (Puradisc 25 TF, Whatman, Maidstone, UK).

### 2.2. Methods

If zinc acetate dihydrate (ZAD) is used as the precursor of ZnO, sols can be prepared in two ways. One can form them in (i) an acid medium in the system ethanol-acetic acid-water-ZAD [38], or (ii) a mild alkaline medium ethanol–DEA–ZAD [39]. The choice of the reaction method (method (i) or (ii)) affects the structure of the material produced, the surface morphology, and the surface properties of the produced layers [1]. In the studies reported here, we used method (ii) due to the slower gelation time, which probably results, among others, from the complexation of Zn(II) ions by DEA [23]. A strong, stable bond is formed between DEA and $Zn^{2+}$, so the gelation process is slower.

The scheme of the ZnO layer fabrication process is shown in Figure 1. A portion of ZAD was added to 50 mL of ethanol and stirred at (50 °C) for 30 min, resulting in a

0.6 mol/L solution, to which DEA and $H_2O$ were added. The molar ratios of ZAD:DEA and ZAD:$H_2O$ were 1:0.6 and 1:2, respectively. The solution was then stirred in an ultrasonic bath for 2.5 h at 50 °C. A clear solution was obtained. It was subsequently filtered through a 0.2 µm PTFE syringe filter. The sol prepared in this way, closed in a vessel, was stored at a temperature of 18 °C. Soda-lime glass microscope slides (Menzel Gläser, Thermo Scientific), of dimensions 76 × 26 × 1 mm$^3$, were coated with sol films after a certain elapsed time. Glass substrates were cleaned according to the procedure described in Ref. [30]. All ZnO layers were produced at the same withdrawal speed of the substrates from sol of 4.7 cm/min. Each substrate after deposition of the layer was annealed at 485 °C for 60 min.

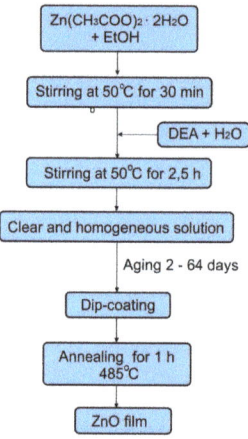

**Figure 1.** The course of the procedure for producing thin ZnO films.

ZnO powder was prepared for FTIR measurements. For this purpose, 3 mL of the sol was annealed at 485 °C for 60 min. The resulting powder had a gray color.

The size distribution of particles in sol was determined using the dynamic light scattering (DLS) method. The DLS was measured at a fixed angle of $\theta$ = 173° using a Zetasizer Nano (Malvern, Worcestershire, UK) with a laser beam of wavelength $\lambda_0$ = 633 nm. The IR spectra of ZnO powder were recorded in the frequency range of 400–4000 cm$^{-1}$ using the Thermo Nicolet i-5 apparatus (Thermo Fisher Scientific, Waltham, MA, USA) with an ATR attachment. The thicknesses and refractive indices of ZnO layers were measured by monochromatic ellipsometry using the SE400 ellipsometer (Sentech, model 2003, $\lambda$ = 632.8 nm, Berlin, Germany). Transmittance and reflectance spectra were recorded using the UV-Vis AvaSpec-ULS2048LTEC Spectrophotometer (Avantes, Apeldoorn, The Netherlands) and Lab-grade Reflection Probes QR400-7-SR (Ocean Optics, Orlando, FL, USA). The AvaLight-DH-S-BAL (Avantes, Apeldoorn, The Netherlands) was used as a light source. The spectra were recorded in the wavelength range of 200–1100 nm at room temperature. The surface morphology of the ZnO layers was studied by atomic force microscopy (AFM), using the N_TEGRA Prima platform (NT-MDT, Moscow, Russia) and NOVA SPM software license number 1.1.120108. In addition, the layer surfaces and their cross-sections were imaged by scanning electron microscopy (SEM) using SEM Supra 35 (Zeiss, Oberkochen, Germany). The layers were observed at magnifications of 100 × 10$^3$, 200 × 10$^3$, and 300 × 10$^3$ at an accelerating voltage of 2–10 kV. Surface morphology observations and layer thickness measurements were performed in the InLens mode (objective lens). A 10 µm aperture was used for observation. The chemical composition of the layers was investigated using energy dispersive spectroscopy (EDS) UltraDry EDS Detector (Thermo Fisher Scientific EDS, Waltham, MA, USA). Measurements of the chemical composition were carried out with accelerating voltage varying between 3 and 15 kV. The working distance WD for secondary electron SE imaging was 14 mm according to EDS calibration. Apertures of 20 and 30 µm were used for the analysis. The results of the chemical composition were corrected using the Phi-Rho-Z method. The water contact angles of the surfaces of the produced ZnO layers were determined by the goniometric method using the goniometer Ossila (L2004A1, Ossila Ltd., Sheffield, UK). The contact angles were measured in five repetitions for each layer. The obtained values were averaged. To determine the photocatalytic activity of produced ZnO layers, their influence on the decomposition of an organic dye imitating organic impurities in an aqueous solution was examined. Following the Langmuir–Hinshelwood procedure, the tests were carried out on layers corresponding to different durations of the sol aging process. Methylene blue (Merck, Darmstadt-Germany) at a concentration of 5 mg/L was used as a dye. A UV diode M365LP1 operating at a wavelength of $\lambda$ = 365 nm (THORLABS Industriemontagen GmbH & Co. KG, Erfurt, Germany) was used as the light source.

## 3. Results and Discussion
### 3.1. Sol Properties
#### 3.1.1. DLS

The results of DLS measurements, carried out after 2, 14, 30, 41, and 64 days counted from the day the sol was synthesized, are shown in Figure 2. Samples were marked with labels ZnO-2, ZnO-14, ZnO-30, ZnO-41, and ZnO-64, respectively. Distributions of hydrodynamic diameters of solid particles by the intensity and by volume, respectively, were recorded. The measurement results for the ZnO-2 sol are shown in Figure 2a,b. In both cases, high peaks are visible. Maximum values are reached for the hydrodynamic diameter equal to 2 nm. One can observe in Figure 2a, a low-intensity peak with a maximum for hydrodynamic diameter of 300 nm. The presence of this peak is associated with the formation of agglomerates composed of nanocrystals. The analogous peak is not visible in Figure 2b, which proves the negligible concentration of these agglomerates in the sol.

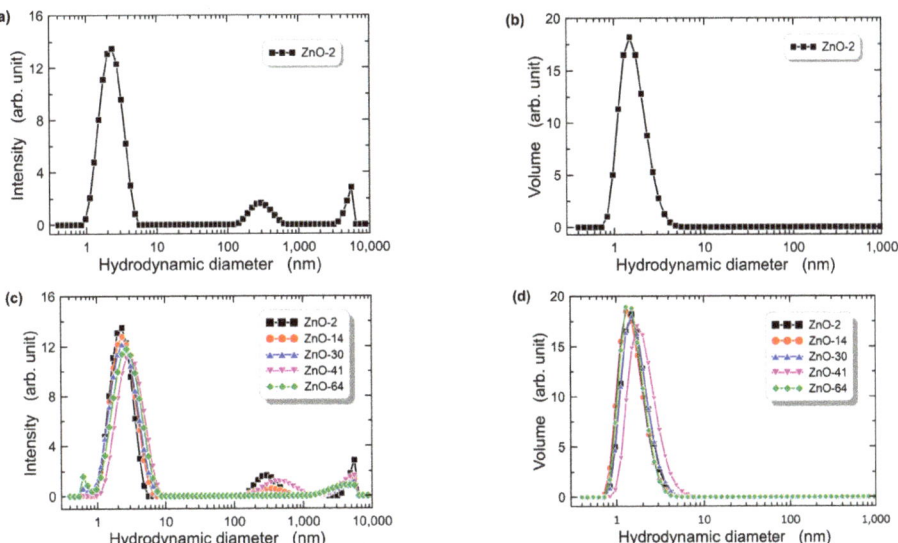

Figure 2. Hydrodynamic diameters of particles in ZnO sol; size distribution by intensity (a,c), and size distribution by volume (b,d).

The results of measurements for longer durations of the sol aging process are shown in Figure 2c,d, respectively. One can see that similar hydrodynamic diameter distributions were recorded, both in terms of intensity and volume. The position of the main peak in the particle size distribution after the intensity is slightly shifted by ca. 1 nm towards larger diameters. In some distributions, one can observe additional peaks above 300 nm, resulting from particle agglomerates, similar to Figure 2a. The particle size distribution by volume is monomodal in each case, which indicates that, despite an increasing duration of aging, the share of agglomerates is negligible. The presented results show that there was a slight increase in particle diameters during the process of sol aging, which indicates slow hydrolysis and condensation processes. The sol was stored at the temperature of 18 °C during the aging. In addition, the resulting particles do not show a clear tendency to form aggregates.

3.1.2. FTIR

The FTIR spectrum recorded for the ZnO powder is shown in Figure 3. The interpretation of the visible absorption bands was carried out based on the FTIR spectrum of the same material as presented in Refs [24,40–42].

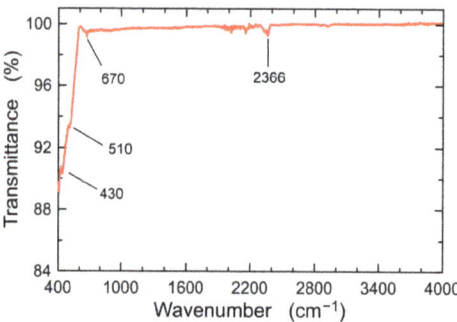

**Figure 3.** FTIR spectrum for the ZnO powder in the range of 400–4000 cm$^{-1}$.

The characteristics show absorption peaks at frequencies of 430, 510, and 670 cm$^{-1}$, which is the effect of Zn-O bond vibrations [42]. Whereas, the absorption band at 2366 cm$^{-1}$ is related to the absorption of atmospheric $CO_2$. The spectrum does not reveal the presence of bands resulting from vibrations of bonds in the water molecule, nor the characteristic C=O bands. This means that the applied annealing temperature (485 °C) is suitable for removing organic residues and water from the sol and, consequently, from the powder [40,42].

3.2. ZnO Layers

3.2.1. Optical Properties

Thin ZnO layers were produced after the following periods: 2, 14, 30, 41, and 64 days. These periods were measured from the day on which the sol was produced. Glass substrates were coated with sol layers using the dip-coating technique. The substrate withdrawal speed from sol was equal to 4.7 cm/min. Subsequently, the layers were annealed at 485 °C for 1 h. In the following part of the manuscript, these layers will be referred to as ZnO-$t_a$, where $t_a$ is the aging time. Table 1 summarizes the determined parameters of the produced films.

One can observe from the data in Col. 3, that thickness of ZnO layers varies non-monotonously with the sol aging time. For aging times up to 30 days, an increase in layer thickness is observed, while for longer times, thickness is decreasing. The refractive index (Col. 4) shows exactly the opposite dependence on the aging time of the sol. One can see from the comparison of the data in Col. 3, 4, and 5 that the increase in thickness is accompanied by an increase in porosity and a decrease in refractive index. Subsequent aging results in the opposite direction of changes. Namely, thickness and porosity are decreasing, while the refractive index is increasing.

Table 1. Parameters of the films corresponding to different sol aging times (d—thickness of layers; n—refractive index of layers; P—porosity; $E_g^{II}$—second of the energy band gap; $E_g^{I}$—first of the energy band gap; $\Delta E_g^{I} = E_g^{I} - E_g^{bulk}$; D—diameter of nanocrystals; k—degradation rate).

| Sol Aging Time (Day) | Name | d (nm) | n (λ = 632.8 nm) | P (%) | $E_g^{II}$ (eV) | $E_g^{I}$ (eV) | $\Delta E_g^{I}$ | D (nm) | Contact Angle (°) | k (min$^{-1}$) × 10$^{-3}$ |
|---|---|---|---|---|---|---|---|---|---|---|
| 1 | 2 | 3 | 4 | 5 | 6 | 7 | 8 | 9 | 10 | 11 |
| 2 | ZnO-2 | 42 | 1.631 | 28.2 | 3.293 | 4.455 | 1.085 | 0.64 | 65.27 | 7.5 |
| 14 | ZnO-14 | 45 | 1.605 | 30.6 | 3.290 | 4.431 | 1.061 | 0.65 | 59.99 | 11.1 |
| 30 | ZnO-30 | 52 | 1.537 | 37.1 | 3.300 | 4.485 | 1.115 | 0.61 | 68.53 | 12.3 |
| 41 | ZnO-41 | 46.5 | 1.572 | 33.7 | 3.282 | 4.428 | 1.058 | 0.65 | 63.35 | 5.4 |
| 64 | ZnO-64 | 45.5 | 1.570 | 33.9 | 3.290 | 4.414 | 1.044 | 0.68 | 49.61 | 4.4 |

Transmittance characteristics in the UV-Vis spectral range of all ZnO-$t_a$ structures are presented in Figure 4. The transmittance of a soda-lime glass substrate is also plotted for comparison. In a wide spectral range (λ > 400 nm), all structures show slightly lower transmittance than soda-lime glass substrate. A slight decrease in transmittance is the result of the difference between the refractive indices of ZnO and soda-lime glass substrate. The ZnO has a higher refractive index than soda-lime glasses.

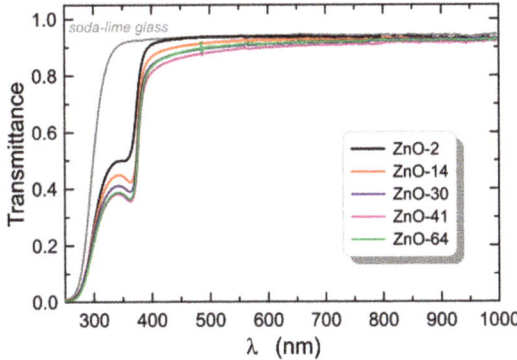

Figure 4. Transmittance characteristics of the layers formed in subsequent days of aging of the ZnO sol.

Within this spectral range, the transmittance decreases with increasing duration of sol aging time despite the observed decrease in the refractive index (Table 1). The thickness of layers is also initially increasing with the duration of aging and, as will be shown later, their roughness also increases. Thus, as the aging time increases, despite the decrease in refractive index, both thickness and surface roughness have a decisive influence on the reduction in transmission in the spectral range above 400 nm.

One can observe two sharp drops of transmittance in the spectral range below 400 nm. They are the result of strong light absorption in ZnO films. The widths of the optical band gaps were determined using the Tauc relationship [43]:

$$\alpha \cdot h\nu = B(h\nu - E_g)^r, \qquad (1)$$

where $\alpha$ is the absorption coefficient, $h\nu$ is the photon energy, $B$ is the constant, $E_g$ is the determined width of the optical band gap, and $r = 1/2$ results from the direct optical band gap. The absorption coefficient $\alpha$ was determined taking into account the light absorption of $A_{gs}$ in the soda-lime glass substrate, the reflectance of the tested structure $R$, and the fact that, in transmission measurements, the light passes through two layers of the same thickness. In the dip-coating technique, two identical layers are deposited on both sides of the substrate. Hence, the absorption coefficient was calculated using the formula [44]:

$$\alpha = -\frac{1}{2d_{eff}} \ln \frac{T}{(1-R)(1-A_{gs})},\qquad(2)$$

where $d_{eff} = d \times (1-P)$. The porosity $P$ was determined from the Lorentz-Lorenz formula [45]:

$$\frac{n^2-1}{n^2+2} = \left(1 - \frac{P}{100\%}\right)\frac{n_d^2-1}{n_d^2+2},\qquad(3)$$

where $n_d = 1.9888$ is the refractive index of wurtzite at $\lambda = 632.8$ nm. Wurtzite is one of the three ZnO crystallographic systems [46]. The wurtzite structure is the only one that is stable at room temperature and atmospheric pressure [47]. The calculated porosities are presented in Table 1 in Col. 5.

The plot of $(\alpha \cdot h\nu)^2$ versus photon energy for the ZnO-64 is presented in Figure 5. One can find on this characteristic two ranges where the dependency on photon energy is linear. Zeros of the lines determined by the linear approximation in these ranges are values of optical band gaps. As one can observe, therefore, there are two values of energy. Both of them are different from the value of the optical band gap for the bulk ZnO crystal which equals ca. 3.37 eV. This fact indicates the presence of two types of absorption in our ZnO films. The first one, for photon energy equal to energy band gap, $E_g^I = 4.414$ eV $> 3.37$ eV is connected with the charge transfer excitations in ZnO nanocrystals, and the blue shift of the energy band gap is caused by the quantum size effect. The second absorption type for photon energies below the band gap of bulk ZnO crystal ($E_g^{II} = 3.290$ eV $< 3.37$ eV) is caused by free exciton. This exciton absorption is responsible for the occurrence of minima in the transmittance characteristics shown in Figure 4 for the wavelength of about 365 nm. Due to the high absorption of photons at this wavelength, a large fraction of the excitations of free excitons may be bound to surface defects. Excitation of free excitons is observed in bulk ZnO, as well as in polycrystalline layers [48–51].

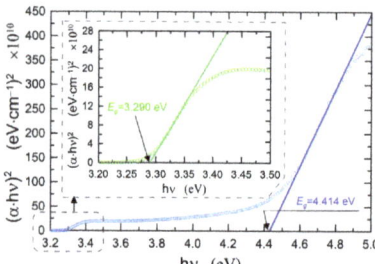

Figure 5. Plots of $(\alpha h\nu)^2$ versus photon energy for the structure ZnO-64.

The determined values of optical band gaps for individual ZnO films are listed in Table 1, Col. 6, and 7, respectively. For each absorption case, the band gaps determined have similar values, and the differences are at the level of $10^{-2}$ eV. For both absorption bands, the highest $E_g$ value was determined for ZnO-30. In the calculations of the absorption coefficient $\alpha$ (Equation (2)), the influence of the soda-lime glass substrate absorption was taken into account. However, since the determined band gap $E_g^I = 4.414$ eV lies in the spectral range of the strong absorption of the substrate, we fabricated ZnO films on silica substrates (SiO$_2$) to confirm the existence of this absorption band. In Figure 6, the solid blue line shows the transmittance, and the solid red line shows the reflectance. The reflectance characteristics show a clear maximum at the wavelength of $\lambda = 377$ nm, which corresponds to the excitation wavelength of free excitons ($E_g^{II} = 3.290$ eV). A more detailed analysis also showed slight disturbances in the reflectance characteristics near the wavelength $\lambda$ = 290 nm. An enlarged fragment of the reflectance characteristics is presented in the insert chart. In this spectral range, the reflectance characteristics were differentiated with respect to the wavelength. The $dR/d\lambda$ plot has a clear minimum with a value slightly greater than zero at the wavelength of $\lambda \sim$290 nm (~4.28 eV). By compensating the slope of the transmittance characteristic, we determined the location of this maximum at the wavelength of $\lambda$ = 282 nm (4.40 eV). The obtained results confirm for the tested ZnO layers the presence of the absorption band at the wavelength of $\lambda \sim$281 nm and the optical band gap of $E_g$ = 4.41 eV. In addition, the analysis of the reflectance characteristics showed that its local maxima occur at wavelengths corresponding to the positions of the absorption bands, and that the energies calculated from them exactly correspond to the values of the band gaps determined by the Tauc method.

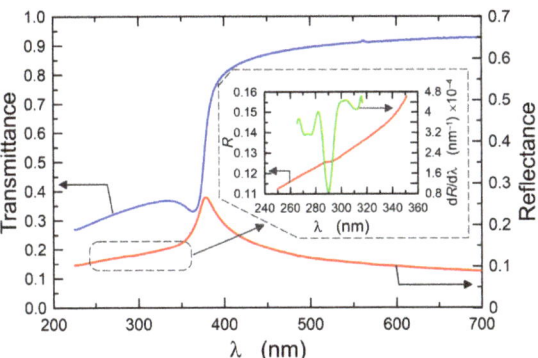

**Figure 6.** Transmittance and reflectance of ZnO-64 structure.

### 3.2.2. Quantum Size Effect

The quantity $\Delta E$ of blue-shift of the energy band gap is in a relationship with a diameter $D$ of nanocrystals. For the parabolic energy band near the band gap, the average value of $D$ may be estimated from the formula [44,52,53]:

$$\Delta E = \frac{2\pi^2 \hbar^2}{m^* D^2} - \frac{0.893 e^2}{\pi \varepsilon \varepsilon_0 D} - 0.248 E^*_{Ry}, \qquad (4)$$

where $\hbar$ is the reduced Planck constant, $\hbar = 1.0545 \cdot 10^{-34}$ J·s; $m^*$ is the reduced effective mass of the electron–hole pair, $m^* = m_e m_h/(m_e+m_h)$; $m_e$ and $m_h$ are the effective mass of the electron and hole, respectively; $\varepsilon_0$ is the electric permittivity in a vacuum; $\varepsilon$ is the dielectric constant for zinc oxide; $E^*_{Ry}$ is the effective Rydberg energy, $E^*_{Ry} = (m^*/(2\hbar^2)) \cdot (e^2/4\pi\varepsilon_0\varepsilon)^2$. Calculations were carried out by accepting values of $m_e$, $m_h$, and $\varepsilon$ the same as in Ref. [52]: $m_e = 0.24 m_0$, $m_h = 0.45 m_0$, and $\varepsilon = 3.7$ ($m_0 = 9.11 \cdot 10^{-31}$ kg). Dimensions of crystallites in the ZnO layers were determined by comparing obtained values of $E_g$ with the literature values for the bulk material $E_g^{bulk}$ = 3.37 eV [1]. They are presented in Table 1, Col. 9. The diameters of ZnO nanocrystals determined in this way are consistent with the results of DLS measurements.

### 3.2.3. Urbach Energy

The data presented in Table 1 (Col. 9) shows that the size of the nanocrystals in ZnO films practically does not change as sol aging time increases. The Urbach energy was determined to track changes in the structure of the layer material. This way, variations in the material structure can be assessed. [54]. The Urbach energy was determined by plotting the relation in the spectral range close to the first absorption band [43]:

$$\alpha(h\nu) = \alpha_0 \cdot exp^{\left(\frac{h\nu}{E_U}\right)},\qquad(5)$$

where $\alpha_0$ is a constant and $E_U$ is the Urbach energy. The high value of the Urbach energy proves that the structure of the material is varying in terms of the size or shape of nanocrystals. The highest Urbach energy values are obtained in amorphous materials or materials with a large number of defects [55,56]. The determined Urbach energy values for individual ZnO films are shown in Figure 7. The presented data show that, since the beginning of the sol aging process during the first 30 days (ZnO-2, ZnO-14 and ZnO-30), the value of the Urbach energy increases. These structures have lower porosity and presumably fewer defects. The Urbach energy for the ZnO-30 structure is $E_U$ = 16.7 meV, and is the highest value among the tested layers. This is probably related to the fact that the ZnO-30 has the highest porosity, and the nanocrystals have the smallest diameters. After this time, there is a clear decrease in the Urbach energy for the ZnO-41 and ZnO-64 layers, which may be related to the lower porosity of the layer material and greater compactness.

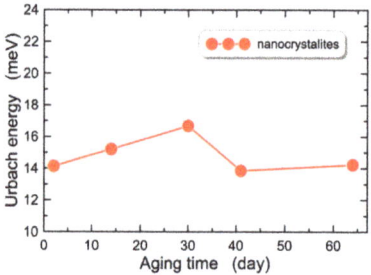

**Figure 7.** Urbach energy of the films formed in successive days of aging of the ZnO sol.

3.2.4. SEM and EDS

SEM images of the upper surface of ZnO films, recorded in the in-lens mode, are presented in Figure 8. On the other hand, Figure 10 presents SEM images of cross-sections of these layers. Based on these images, we can conclude that zinc oxide layers produced on soda-lime glass substrate have a granular structure. We can also conclude that the ZnO layer is built of grains with a diameter of <70 nm, while from the quantum size effect, it was found that these grains are built of nanocrystals with a diameter less than 0.65 nm. One can observe in Figure 8a that the ZnO-2 layer does not cover uniformly the glass substrate. The SEM image shows the incoherence of the layer and areas of the substrate covered with single ZnO grains.

A similar inhomogeneity of the substrate coverage is presented in Ref. [57]. Authors reported similar behavior for the layer fabricated after 4 days since the production of their sol. Authors of Ref. [29] also reported such a case for the layer fabricated after 2 days of production of their sol. Obtained EDS spectra are presented in Figure 9. The two peaks at 1.01 keV and 0.53 keV validate the presence of zinc and oxygen, respectively. Besides peaks from the soda-lime glass substrate, the EDS spectra reveal the presence of Zn and O elements, which confirms that the ZnO layer has a grain structure.

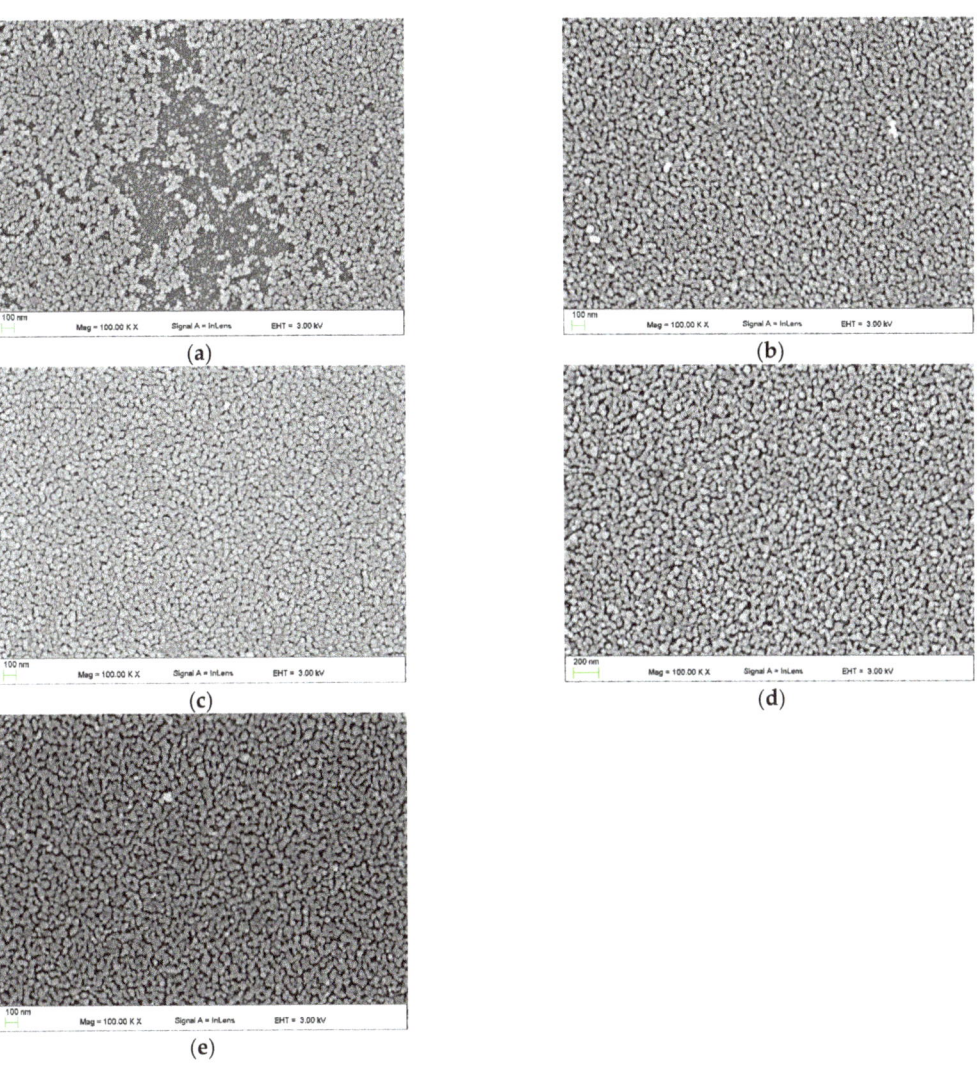

**Figure 8.** Top view SEM images of the ZnO films, respectively: ZnO-2 (**a**), ZnO-14 (**b**), ZnO-30 (**c**), ZnO-41 (**d**), and ZnO-64 (**e**).

One can observe from SEM images (Figure 8b–e) that ZnO layers are uniformly covering the substrates if they are fabricated from the sol which was aged for longer periods. As can be seen, these layers are crack-free and have other structural defects. Moreover, the diameter of the ZnO grains does not change as the duration of the sol aging process increases. Presumably, the sol obtained according to the tested procedure has, immediately after its synthesis, too high surface tension, which in turn lowers the wettability of the substrate. As a result, some areas of the soda-lime glass substrate are not covered with the ZnO layer [58,59]. The uniformity of substrate coverage with the ZnO layer would be improved if the sol was enriched with a surfactant, or some kind of substrate conditioning procedure was applied before the coating process. The images of cross-sections of the ZnO-2, ZnO-14, and ZnO-30 layers (Figure 10a–c) show their grain structure with spherical grain shapes with clearly outlined grain boundaries. In addition, it is noteworthy that the value of film thickness obtained from the analysis of the SEM cross-section image is in good agreement with the one measured using monochromatic ellipsometry. On the other hand, in the SEM images of the ZnO-41 and ZnO-64 layers (Figure 10d,e), grain boundaries are less clear, and the layers are more compact. These observations correlate with the determined decrease in their porosity (Table 1, Col. 5).

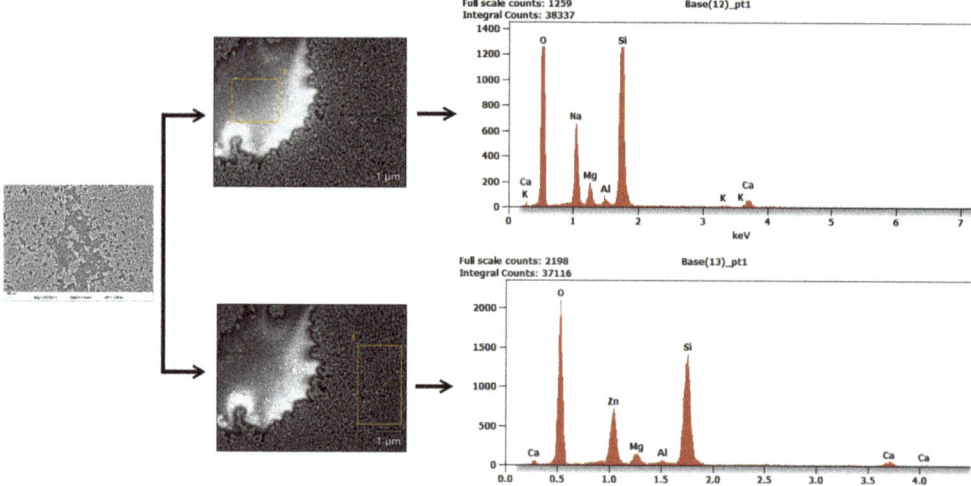

**Figure 9.** EDS analysis of a thin film of ZnO-2.

3.2.5. AFM

Figure 11 shows the AFM images of the surface of the ZnO-2 (Figure 11a), ZnO-30 (Figure 11c), and ZnO-64 (Figure 11e) layers and their profiles along the selected scans. These images confirm the grainy structure of ZnO films. Moreover, they illustrate the influence of the duration of the sol aging process on the magnitude of the ZnO layer's surface roughness. The analysis of the ZnO layer's surface profiles showed that the largest difference in height between the highest and lowest point on the surfaces of $1 \times 1$ μm$^2$ occurs for the ZnO-64 layer and equals 50.3 nm. Calculated values of root mean square ($rms$) surface roughness, over the area of $1 \times 1$ μm$^2$, for the ZnO-2, ZnO-30, and ZnO-64 layers, are 5.02 nm, 3.36 nm, and 8.91 nm, respectively. Similar $rms$ values, not so clearly correlated with the sol aging time, were reported in Ref. [23].

Based on results obtained by SEM and AFM methods, we can conclude that the duration of the sol aging process affects the structure of ZnO layers fabricated from that sol. Layers produced from the sol after the shortest duration of the aging process are characterized by discontinuities, while for longer duration, produced layers (ZnO-41, ZnO-64) are characterized by greater compactness. ZnO layers fabricated on every stage of the sol aging process are grainy.

3.2.6. Contact Angles

The measured water contact angles for ZnO layers are presented in Col. 10 of Table 1. One can observe that all layers, regardless of the duration of sol aging, are hydrophilic in nature, as the contact angle in each case is less than 90°. Depending on the time of sol aging, the hydrophilic character of ZnO layers changes slightly, except for the layer made of the longest aged sol. The highest value of the contact angle was obtained for the ZnO-30 layer. It equals 68.53°. Presumably, this is related to the porosity of the layer material, which is also the highest for this layer [60]. The lowest value of the water contact angle, 49.61°, was obtained for the ZnO-64 layer.

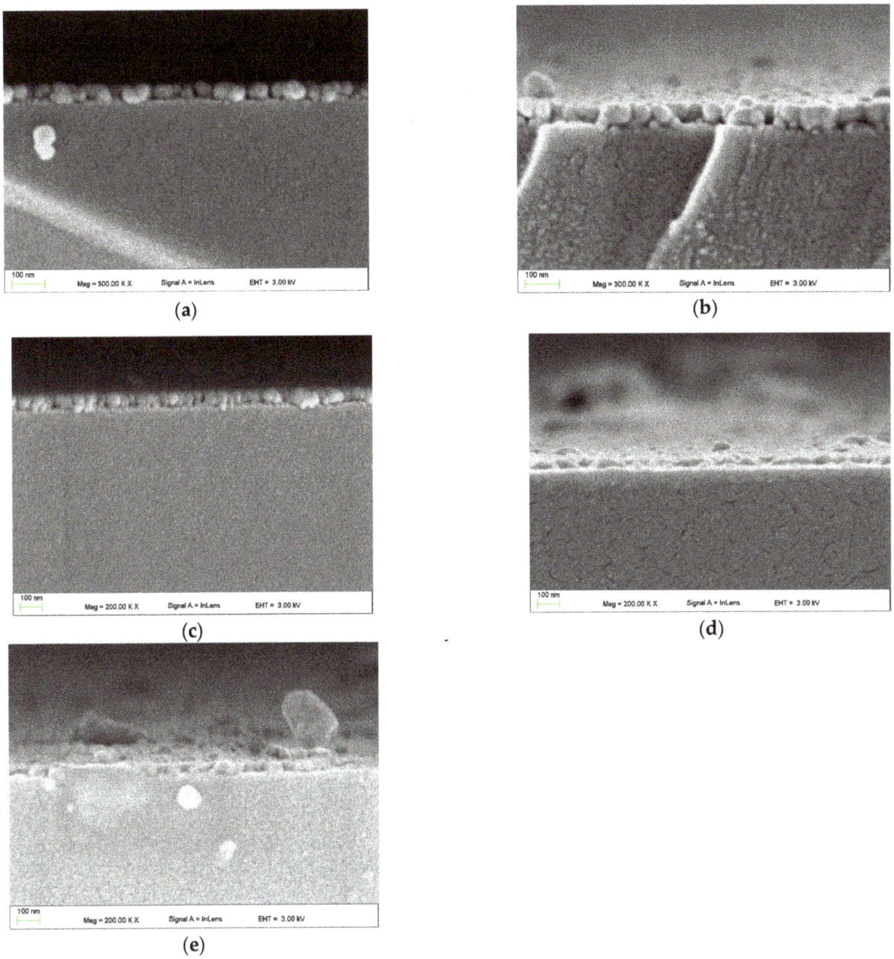

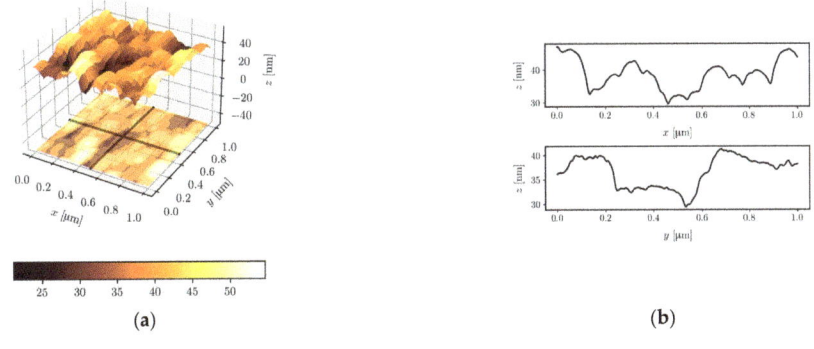

**Figure 10.** Cross-section SEM images of the ZnO films, respectively: ZnO-2 (**a**), ZnO-14 (**b**), ZnO-30 (**c**), ZnO-41 (**d**), and ZnO-64 (**e**).

**Figure 11.** *Cont.*

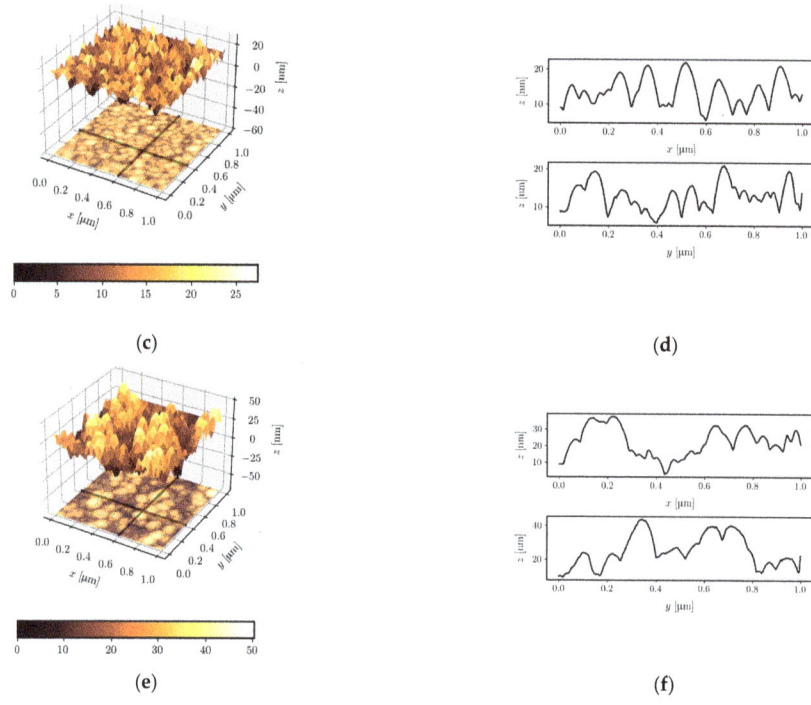

**Figure 11.** AFM images and corresponding line profiles for ZnO-2 (**a,b**), ZnO-30 (**c,d**), and ZnO-64 (**e,f**) films.

### 3.2.7. Photocatalytic Activity

The photocatalytic activity of the fabricated ZnO layers was investigated by the decomposition of methylene blue (MB) in an aqueous solution. MB was used as a representative impurity. The concentration of the MB solution used was 5 mg/L. ZnO layers, whose photocatalytic activity was investigated, were fabricated on glass substrates of dimensions 20 × 20 × 1 mm$^3$. These structures were placed horizontally in a 30 mm diameter beaker filled with 45 mL of MB solution. The solution was agitated magnetically for 30 min in the dark, to achieve equilibrium between the adsorption and desorption processes. Subsequently, the examined structures were irradiated using the LED operating at a central wavelength of $\lambda$ = 365 nm, driven by an electric current of intensity 700 mA. The distance between the ZnO layer and the light source was 6 cm. The total exposure time for each sample was 120 min. The MB degradation process started due to the exposure of submerged ZnO layers to UV light. Subsequently, dye samples of 1 mL were collected from the solution, at regular time intervals of 30 min, and analyzed using UV-Vis spectroscopy. The intensity of the MB absorption band for the collected portion of the dye was measured at the wavelength of $\lambda$ = 664 nm. The intensity of the absorption band for the MB solution recorded at the wavelength of $\lambda$ = 664 nm decreased with the time of its exposure to UV radiation. This is evidence of the degradation of MB dye molecules under the influence of UV irradiation. The processes of dye degradation under the influence of UV radiation are described by the photocatalytic efficiency:

$$\eta(\%) = \left(1 - \frac{c}{c_0}\right) \cdot 100\%, \qquad (6)$$

and the degradation rate $k$, which, according to the Langmuir–Hinshelwood [61] model, is defined by the equation:

$$\ln\left(\frac{c}{c_0}\right) = -kt. \qquad (7)$$

The degradation rate $k$ is also called the first-order reaction rate constant. In the equations above, $c_0$ denotes the concentration of the dye before irradiation, and $c$ denotes its concentration after irradiation. Results of the experiment on MB dye degradation are shown in Figures 12 and 13. The characteristic of photocatalytic efficiency as a function of UV irradiation time is presented in Figure 12, while linearized characteristics of the rate of change of the relative MB concentration are presented in Figure 13. As can be seen from the presented data, the presence of the ZnO film during MB dye exposure in each case contributes to the increase in photocatalytic performance. The determined MB dye degradation rates $k$ are shown in Col. 11 of Table 1. The MB dye degradation rate without the presence of ZnO is $k = 1.4 \times 10^{-3}$ min$^{-1}$. According to Ref. [22], differences in photocatalytic activity largely depend on the thickness of the tested layers. However, our research indicates that the highest $k$ was obtained for the structure ZnO-30, which has the highest porosity. One can observe that structures ZnO-41 and ZnO-46, which are only slightly thinner than ZnO-30 and thicker than ZnO-2 and ZnO-14, have significantly lower photocatalytic activity than much thinner structures ZnO-2 and ZnO-14. Presumably, the photocatalytic activity depends mainly on the specific surface area (BET) which increases with increasing porosity and diversity of material structure reflected by a value of the Urbach energy. One can observe that the ZnO-30 structure has the highest Urbach energy value (Figure 7), while the structures ZnO-41 and ZnO-46 have lower Urbach energy values, comparable with values for ZnO-41 and ZnO-46. In addition, the ZnO-30 structure has the largest contact angle. These features determine its best photocatalytic properties.

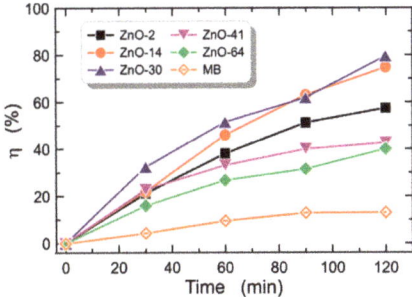

**Figure 12.** Plots of photodegradation efficiency ($\eta$) as function time for the degradation of MB dye without (orange empty diamonds) and with the presence of the catalyst.

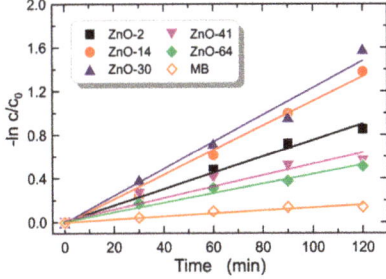

**Figure 13.** Plots of $-\ln(c/c_0)$ as a function of reaction time for the degradation of MB dye without (orange empty diamonds) and with the presence of the catalyst.

In Table 2 are presented results of the literature survey, which is covering the scope and results of investigations on the influence of the duration of the sol aging process on various physicochemical properties of ZnO thin films. The scope and results presented in this paper are juxtaposed with those obtained by other research groups.

**Table 2.** Comparison of the present article with published articles based on sol aging time and thin layers ZnO.

| Sol Aging Time | Sol Composition | Methods, T (°C), Substrates | Results | Ref. |
|---|---|---|---|---|
| 0–36 h | zinc acetate dihydrate, monoethanolamine, 2-methoxyethanol | spin-coating, 300, 500, glass substrates | the optimal aging time is 24 h, the ZnO films were of high quality and had the best properties (transmission in the UV-Vis range, bandgap broadening) | [30] |
| 0–13 days | zinc acetate dihydrate, methanol | dip-coating (multilayers), 500, glass substrates | the optimal aging time is 7 days, the layers were characterized by high transmittance (above 70% in the visible range) and a smooth surface | [27] |
| 0–30 days | zinc acetate dihydrate, diethanolamine, ethanol | dip-coating, 450, glass substrates | the longer the sol aging time, the ZnO layers having better transmittance in the UV-Vis range (up to 96%), increased optical band gap Eg, and improved smoothness | [23] |
| 2–64 days | zinc acetate dihydrate, diethanolamine, ethanol | dip-coating (monolayer), 485, glass substrates | the optimal sol aging time is 30 days, the ZnO layers are characterized by the best optical and photocatalytic properties, after this time the layers have weaker optical and photocatalytic properties | this work |

## 4. Conclusions

This paper presents the results of investigations on ZnO layers fabricated using the sol-gel method and dip-coating technique. In the course of sol synthesis, zinc acetate dihydrate was applied as the precursor, while diethanolamine was the stabilizing agent. ZnO layers were produced on soda-lime glass substrates. The layers were annealed at the temperature of 485 °C for 60 min. The maximum duration of sol aging was 64 days. The sol was kept at the temperature of 18 °C throughout the whole aging process. At the end of that process, the gelation process had not ended, and sol was still suitable for coating processes.

Presented studies aimed at the determination of the effect which has the duration of the sol aging process on the properties of produced ZnO layers. They have a thickness varying in the range from 42 nm to 52 nm. The DLS investigations showed that synthesized sol was very stable during the whole aging process. FTIR studies of ZnO powder showed that there are no organic remnants in ZnO layers after the annealing process. The surface morphology of ZnO layers was investigated using SEM and AFM microscopy. Both of these methods revealed that ZnO layers have a granular structure. The spectrophotometric studies showed that the absorption in the UV range has two components: first, below the band gap of bulk ZnO crystal ($E_g^{II}$ = 3.290 eV), which is caused by excitation of free exciton and second, connected with the charge transfer excitations in ZnO nanocrystals. The blue energy shift is observed in the second case as a result of the quantum size effect. Moreover, we showed that the optical energy band gaps determined from positions of maxima in reflectance characteristics are equal to those determined using the Tauc method.

The obtained results indicate that there is an optimum duration of the sol-gel aging process that results in the fabrication of layers having extreme values of optical and morphological parameters. All produced ZnO layers showed photocatalytic activity against the aqueous solution of methylene blue dye, however, the most efficient dye degradation was observed in the solution characterized by $k = 12.3 \cdot 10^{-3}$ min$^{-1}$ and $\eta = 79.5\%$ for ZnO layers aged for 30 days. It is to be noted that ZnO layers aged for this period have the greatest porosity ($p = 37.1\%$), the weakest hydrophilicity (water contact angle 68.53°), and the greatest optical energy band gaps ($E_g^I = 4.485$ eV and $E_g^{II} = 3.300$ eV). The high photocatalytic activity of produced ZnO layers allows us to conclude that they may find their applications in environmental applications where the degradation of organic pollutants is required.

**Author Contributions:** Conceptualization, K.W. and P.K.; methodology, K.W., M.Z., C.T., W.P., G.Ż. and O.J.; software, C.T., W.P. and O.J.; validation, P.K., K.D. and E.G.; formal analysis, K.W., M.Z. and P.K.; investigation, K.W., P.K., E.G. and K.D.; resources, K.W., M.Z., C.T. and P.K.; data curation, K.W., M.Z., C.T., W.P., G.Ż. and O.J.; writing—original draft preparation, K.W., M.Z., C.T. and P.K.; writing—review and editing, K.W., M.Z. and P.K.; visualization, K.W. and M.Z.; supervision, P.K., K.D. and E.G.; project administration, K.W. and M.Z.; funding acquisition, K.W. and M.Z. All authors have read and agreed to the published version of the manuscript.

**Funding:** This research was funded by Statutory Research for the Department of Optoelectronics, Silesian University of Technology BKM-761/RE4/2022 and BKM-762/RE4/2022.

**Institutional Review Board Statement:** Not applicable.

**Informed Consent Statement:** Not applicable.

**Data Availability Statement:** Data sharing is not applicable to this article.

**Conflicts of Interest:** The authors declare no conflict of interest.

## References

1. Znaidi, L. Sol-gel-deposited ZnO thin films: A review. In *Materials Science and Engineering B: Solid-State Materials for Advanced Technology*; Elsevier: Amsterdam, The Netherlands, 2010; Volume 174, pp. 18–30. [CrossRef]
2. Mirzaei, A.; Chen, Z.; Haghighat, F.; Yerushalmi, L. Removal of pharmaceuticals and endocrine disrupting compounds from water by zinc oxide-based photocatalytic degradation: A review. *Sustain. Cities Soc.* **2016**, *27*, 407–418. [CrossRef]
3. Li, Z.; Yu, X.; Zhu, Y.; Liu, S.; Wen, X.; Lu, H.; Wang, C.; Li, X.; Li, M.-Y.; Yang, Y. High performance ZnO quantum dot (QD)/magnetron sputtered ZnO homojunction ultraviolet photodetectors. *Appl. Surf. Sci.* **2022**, *582*, 152352. [CrossRef]
4. Gowthaman, N.; Chen, J.-W.; Dee, C.F.; Chai, S.-P.; Chang, W.S. Nanostructural dimension and oxygen vacancy synergistically induced photoactivity across high surface area monodispersed AuNPs/ZnO nanorods heterojunction. *J. Alloy Compd.* **2022**, *920*, 165836. [CrossRef]
5. Yu, L.; Cai, W.; Wang, L.; Lee, C.-Y.; Sun, S.; Xiao, Y.; Shen, K.-C.; Chen, C. Three-Dimensional crystallized ZnO crosslinked nanowire structure. *Inorg. Chem. Commun.* **2022**, *140*, 109413. [CrossRef]
6. Foumani, I.R.; Pat, S. Optical and surface properties of Gd-doped ZnO thin films deposited by thermionic vacuum arc deposition technology. *Inorg. Chem. Commun.* **2022**, *144*, 109831. [CrossRef]
7. Malik, G.; Mourya, S.; Jaiswal, J.; Chandra, R. Effect of annealing parameters on optoelectronic properties of highly ordered ZnO thin films. *Mater. Sci. Semicond. Process* **2019**, *100*, 200–213. [CrossRef]
8. Kaur, M.; Shaheera, M.; Pathak, A.; Gadkari, S.C.; Debnath, A.K. Highly sensitive NO$_2$ sensor based on ZnO nanostructured thin film prepared by SILAR technique. *Sens. Actuators B Chem.* **2021**, *335*, 129678. [CrossRef]
9. Ennaceri, H.; Erfurt, D.; Wang, L.; Köhler, T.; Taleb, A.; Khaldoun, A.; El Kenz, A.; Benyoussef, A.; Ennaoui, A. Deposition of multifunctional TiO2 and ZnO top-protective coatings for CSP application. *Surf. Coatings Technol.* **2016**, *298*, 103–113. [CrossRef]
10. Jiang, C.Y.; Sun, X.; Lo, G.Q.; Kwong, D.L.; Wang, J.X. Improved dye-sensitized solar cells with a ZnO-nanoflower photoanode. *Appl. Phys. Lett.* **2007**, *90*, 263501. [CrossRef]
11. Tiwari, A.; Sahay, P. Modification in the physical properties of nanocrystalline ZnO thin films by Sn/Ni co-doping for transparent conductive oxide applications. *Phys. B Condens. Matter* **2022**, *629*, 413638. [CrossRef]
12. Ahmadi, S.H.; Ghaffarkani, M.; Ameri, M.; Safari, N.; Mohajerani, E. Solvent selection for fabrication of low temperature ZnO electron transport layer in perovskite solar cells. *Opt. Mater.* **2020**, *106*, 109977. [CrossRef]
13. Szindler, M.; Szindler, M.; Basiaga, M.; Łoński, W.; Kaim, P. Application of ALD Thin Films on the Surface of the Surgical Scalpel Blade. *Coatings* **2021**, *11*, 1096. [CrossRef]
14. Kaim, P.; Lukaszkowicz, K.; Szindler, M.; Szindler, M.M.; Basiaga, M.; Hajduk, B. The Influence of Magnetron Sputtering Process Temperature on ZnO Thin-Film Properties. *Coatings* **2021**, *11*, 1507. [CrossRef]
15. Pelicano, C.M.; Yanagi, H. Effect of rubrene:P3HT bilayer on photovoltaic performance of perovskite solar cells with electrodeposited ZnO nanorods. *J. Energy Chem.* **2018**, *27*, 455–462. [CrossRef]
16. Luo, L.; Rossell, M.D.; Xie, D.; Erni, R.; Niederberger, M. Microwave-Assisted Nonaqueous Sol–Gel Synthesis: From Al:ZnO Nanoparticles to Transparent Conducting Films. *ACS Sustain. Chem. Eng.* **2013**, *1*, 152–160. [CrossRef]

17. Amakali, T.; Daniel, L.; Uahengo, V.; Dzade, N.; de Leeuw, N.H. Structural and Optical Properties of ZnO Thin Films Sol–Gel Methods. *Crystals* **2020**, *10*, 132. [CrossRef]
18. Pelicano, C.M.; Yanagi, H. Enhanced charge transport in Al-doped ZnO nanotubes designed via simultaneous etching and Al doping of $H_2O$-oxidized ZnO nanorods for solar cell applications. *J. Mater. Chem. C* **2019**, *7*, 4653–4661. [CrossRef]
19. Laurenti, M.; Cauda, V. Porous Zinc Oxide Thin Films: Synthesis Approaches and Applications. *Coatings* **2018**, *8*, 67. [CrossRef]
20. Darvishi, M.; Jamali-Paghaleh, F.; Jamali-Paghaleh, M.; Seyed-Yazdi, J. Facile synthesis of ZnO/rGO hybrid by microwave irradiation method with improved photoactivity. *Surfaces Interfaces* **2017**, *9*, 167–172. [CrossRef]
21. Brinker, C.J.; Scherer, G.W. *Sol-Gel Science: The Physics and Chemistry of Sol-Gel Processing*; Academic Press: Cambridge, MA, USA, 2013. [CrossRef]
22. Chaitra, U.; Kekuda, D.; Rao, K.M. Effect of annealing temperature on the evolution of structural, microstructural, and optical properties of spin coated ZnO thin films. *Ceram. Int.* **2017**, *43*, 7115–7122. [CrossRef]
23. Toubane, M.; Tala-Ighil, R.; Bensouici, F.; Bououdina, M.; Cai, W.; Liu, S.; Souier, M.; Iratni, A. Structural, optical and photocatalytic properties of ZnO nanorods: Effect of aging time and number of layers. *Ceram. Int.* **2016**, *42*, 9673–9685. [CrossRef]
24. Sheikhi, S.; Aliannezhadi, M.; Tehrani, F.S. Effect of precursor material, pH, and aging on ZnO nanoparticles synthesized by one-step sol–gel method for photodynamic and photocatalytic applications. *Eur. Phys. J. Plus* **2022**, *137*, 60. [CrossRef]
25. Benramache, S.; Rahal, A.; Benhaoua, B. The effects of solvent nature on spray-deposited ZnO thin film prepared from Zn $(CH_3COO)_2$, $2H_2O$. *Optik* **2014**, *125*, 663–666. [CrossRef]
26. Varol, S.F.; Babür, G.; Çankaya, G.; Kölemen, U. Synthesis of sol–gel derived nano-crystalline ZnO thin films as TCO window layer: Effect of sol aging and boron. *RSC Adv.* **2014**, *4*, 56645–56653. [CrossRef]
27. Marouf, S.; Beniaiche, A.; Guessas, H.; Azizi, A. Morphological, Structural and Optical Properties of ZnO Thin Films Deposited by Dip Coating Method. *Mater. Res.* **2017**, *20*, 88–95. [CrossRef]
28. Ibrahim, N.; Al-Shomar, S.; Ahmad, S.H. Effect of aging time on the optical, structural and photoluminescence properties of nanocrystalline ZnO films prepared by a sol–gel method. *Appl. Surf. Sci.* **2013**, *283*, 599–602. [CrossRef]
29. Pérez-González, M.; Tomás, S.; Morales-Luna, M.; Arvizu, M.; Tellez-Cruz, M. Optical, structural, and morphological properties of photocatalytic $TiO_2$–ZnO thin films synthesized by the sol–gel process. *Thin Solid Films* **2015**, *594*, 304–309. [CrossRef]
30. Aghkonbad, E.M.; Aghgonbad, M.M.; Sedghi, H. A Study on Effect of Sol Aging Time on Optical Properties of ZnO Thin Films: Spectroscopic Ellipsometry Method. *Micro Nanosyst.* **2019**, *11*, 100–108. [CrossRef]
31. Mancuso, A.; Sacco, O.; Mottola, S.; Pragliola, S.; Moretta, A.; Vaiano, V.; De Marco, I. Synthesis of Fe-doped ZnO by supercritical antisolvent precipitation for the degradation of azo dyes under visible light. *Inorganica Chim. Acta* **2023**, *549*, 121407. [CrossRef]
32. Rini, N.P.; Istiqomah, N.I.; Sunarta; Suharyadi, E. Enhancing photodegradation of methylene blue and reusability using CoO/ZnO composite nanoparticles. *Case Stud. Chem. Environ. Eng.* **2023**, *7*, 100301. [CrossRef]
33. Sher, M.; Javed, M.; Shahid, S.; Hakami, O.; Qamar, M.A.; Iqbal, S.; Al-Anazy, M.M.; Baghdadi, H.B. Designing of highly active g-C3N4/Sn doped ZnO heterostructure as a photocatalyst for the disinfection and degradation of the organic pollutants under visible light irradiation. *J. Photochem. Photobiol. A Chem.* **2021**, *418*, 113393. [CrossRef]
34. Qamar, M.A.; Javed, M.; Shahid, S.; Iqbal, S.; Abubshait, S.A.; Abubshait, H.A.; Ramay, S.M.; Mahmood, A.; Ghaithan, H.M. Designing of highly active g-C3N4/Co@ZnO ternary nanocomposites for the disinfection of pathogens and degradation of the organic pollutants from wastewater under visible light. *J. Environ. Chem. Eng.* **2021**, *9*, 105534. [CrossRef]
35. Naghani, M.E.; Neghabi, M.; Zadsar, M.; Ahangar, H.A. Synthesis and characterization of linear/nonlinear optical properties of graphene oxide and reduced graphene oxide-based zinc oxide nanocomposite. *Sci. Rep.* **2023**, *13*, 1496. [CrossRef]
36. Maiti, M.; Sarkar, M.; Maiti, S.; Liu, D. Gold decorated shape-tailored zinc oxide-rGO nanohybrids: Candidate for pathogenic microbe destruction and hazardous dye degradation. *Colloids Surfaces A Physicochem. Eng. Asp.* **2022**, *641*, 128465. [CrossRef]
37. Liu, Z.; Jin, Z.; Li, W.; Qiu, J. Preparation of ZnO porous thin films by sol–gel method using PEG template. *Mater. Lett.* **2005**, *59*, 3620–3625. [CrossRef]
38. Wang, X.-H.; Shi, J.; Dai, S.; Yang, Y. A sol-gel method to prepare pure and gold colloid doped ZnO films. *Thin Solid Films* **2003**, *429*, 102–107. [CrossRef]
39. Znaidi, L.; Illia, G.S.; Le Guennic, R.; Sanchez, C.; Kanaev, A. Elaboration of ZnO Thin Films with Preferential Orientation by a Soft Chemistry Route. *J. Sol-Gel Sci. Technol.* **2003**, *26*, 817–821. [CrossRef]
40. Khan, Z.R.; Khan, M.S.; Zulfequar, M. Optical and Structural Properties of ZnO Thin Films Fabricated by Sol-Gel Method. *Mater. Sci. Appl.* **2011**, *02*, 340–345. [CrossRef]
41. Bindu, P.; Thomas, S. Estimation of lattice strain in ZnO nanoparticles: X-ray peak profile analysis. *J. Theor. Appl. Phys.* **2014**, *8*, 123–134. [CrossRef]
42. Kayani, Z.N.; Iqbal, M.; Riaz, S.; Zia, R.; Naseem, S. Fabrication and properties of zinc oxide thin film prepared by sol-gel dip coating method. *Mater. Sci.* **2015**, *33*, 515–520. [CrossRef]
43. Tauc, J. *Amorphous and Liquid Semiconductors*; Springer Science & Business Media: Berlin/Heidelberg, Germany, 2012.
44. Karasiński, P.; Gondek, E.; Drewniak, S.; Kajzer, A.; Waczyńska-Niemiec, N.; Basiaga, M.; Izydorczyk, W.; Kouari, Y.E. Porous titania films fabricated via sol gel rout—Optical and AFM characterization. *Opt. Mater.* **2016**, *56*, 64–70. [CrossRef]
45. Millán, C.; Santonja, C.; Domingo, M.; Luna, R.; Satorre, M.Á. An experimental test for effective medium approximations (EMAs). *Astron. Astrophys.* **2019**, *628*, A63. [CrossRef]

46. Ozgür, Ü.; Alivov, Y.I.; Liu, C.; Teke, A.; Reshchikov, M.A.; Doğan, S.; Avrutin, V.; Cho, S.-J.; Morkoç, H. A comprehensive review of ZnO materials and devices. *J. Appl. Phys.* **2005**, *98*, 041301. [CrossRef]
47. Sokolov, P.; Baranov, A.; Bell, A.; Solozhenko, V. Low-temperature thermal expansion of rock-salt ZnO. *Solid State Commun.* **2014**, *177*, 65–67. [CrossRef]
48. Hamby, D.W.; Lucca, D.A.; Klopfstein, M.J.; Cantwell, G. Temperature dependent exciton photoluminescence of bulk ZnO. *J. Appl. Phys.* **2003**, *93*, 3214. [CrossRef]
49. Travnikov, V.; Freiberg, A.; Savikhin, S. Surface excitons in ZnO crystals. *J. Lumin* **1990**, *47*, 107–112. [CrossRef]
50. Rai, R.C.; Guminiak, M.; Wilser, S.; Cai, B.; Nakarmi, M.L. Elevated temperature dependence of energy band gap of ZnO thin films grown by e-beam deposition. *J. Appl. Phys.* **2012**, *111*, 073511. [CrossRef]
51. Davis, K.; Yarbrough, R.; Froeschle, M.; White, J.; Rathnayake, H. Band gap engineered zinc oxide nanostructures *via* a sol–gel synthesis of solvent driven shape-controlled crystal growth. *RSC Adv.* **2019**, *9*, 14638–14648. [CrossRef]
52. Brus, L.E. Electron–electron and electron-hole interactions in small semiconductor crystallites: The size dependence of the lowest excited electronic state. *J. Chem. Phys.* **1984**, *80*, 4403–4409. [CrossRef]
53. Tan, S.T.; Chen, B.J.; Sun, X.W.; Fan, W.J.; Kwok, H.S.; Zhang, X.H.; Chua, S.J. Blueshift of optical band gap in ZnO thin films grown by metal-organic chemical-vapor deposition. *J. Appl. Phys.* **2005**, *98*, 013505. [CrossRef]
54. Üzar, N. Investigation of detailed physical properties and solar cell performances of various type rare earth elements doped ZnO thin films. *J. Mater. Sci. Mater. Electron.* **2018**, *29*, 10471–10479. [CrossRef]
55. Speaks, D.T. Effect of concentration, aging, and annealing on sol gel ZnO and Al-doped ZnO thin films. *Int. J. Mech. Mater. Eng.* **2020**, *15*, 2. [CrossRef]
56. Chala, S.; Bdirina, M.; Elbar, M.; Naoui, Y.; Benbouzid, Y.; Taouririt, T.E.; Labed, M.; Boumaraf, R.; Bouhdjar, A.F.; Sengouga, N.; et al. Dependence of Structural and Optical Properties of ZnO Thin Films Grown by Sol–Gel Spin-Coating Technique on Solution Molarity. *Trans. Electr. Electron. Mater.* **2022**, *1*, 3. [CrossRef]
57. Bhujel, K.; Ningthoujam, S.S.; Singh, L.R.; Rai, S. Effect of solution aging on properties of spin coated zinc oxide thin films. *Mater. Today Proc.* **2019**, *46*, 6419–6422. [CrossRef]
58. El Hallani, G.; Fazouan, N.; Liba, A.; Khuili, M. The effect of sol aging time on Structural and Optical properties of sol gel ZnO doped Al. *J. Physics Conf. Ser.* **2016**, *758*, 012021. [CrossRef]
59. Gao, D.-G.; Chen, C.; Ma, J.-Z.; Lv, B.; Jia, X.-L. Preparation, characterization and application of ZnO sol containing quaternary ammonium salts. *J. Sol-Gel Sci. Technol.* **2013**, *65*, 336–343. [CrossRef]
60. Poddighe, M.; Innocenzi, P. Hydrophobic Thin Films from Sol–Gel Processing: A Critical Review. *Materials* **2021**, *14*, 6799. [CrossRef]
61. Islam, M.R.; Rahman, M.; Farhad, S.; Podder, J. Structural, optical and photocatalysis properties of sol–gel deposited Al-doped ZnO thin films. *Surfaces Interfaces* **2019**, *16*, 120–126. [CrossRef]

**Disclaimer/Publisher's Note:** The statements, opinions and data contained in all publications are solely those of the individual author(s) and contributor(s) and not of MDPI and/or the editor(s). MDPI and/or the editor(s) disclaim responsibility for any injury to people or property resulting from any ideas, methods, instructions or products referred to in the content.

Article

# Sol-Gel Synthesis and Characterization of Yttrium-Doped MgFe$_2$O$_4$ Spinel

Dovydas Karoblis [1], Kestutis Mazeika [2], Rimantas Raudonis [1], Aleksej Zarkov [1] and Aivaras Kareiva [1,*]

[1] Institute of Chemistry, Vilnius University, Naugarduko 24, LT-03225 Vilnius, Lithuania
[2] Center of Physical Sciences and Technology, LT-02300 Vilnius, Lithuania
* Correspondence: aivaras.kareiva@chgf.vu.lt; Tel.: +370-5219-3110

**Abstract:** In this study, an environmentally friendly sol-gel synthetic approach was used for the preparation of yttrium-doped MgFe$_2$O$_4$. Two series of compounds with different iron content were synthesized and A-site substitution effects were investigated. In the first series, the iron content was fixed and the charge balance was suggested to be compensated by a partial reduction of Fe$^{3+}$ to Fe$^{2+}$ or formation of interstitial O$^{2-}$ ions. For the second series of samples, the iron content was reduced in accordance with the substitution level to compensate for the excess of positive charge, which accumulates due to replacing divalent Mg$^{2+}$ with trivalent Y$^{3+}$ ions. Structural, morphological and magnetic properties were inspected. It was observed that single-phase compounds can only form when the substitution level reaches 20 mol% of Y$^{3+}$ ions and iron content is reduced. The coercivity as well as saturation magnetization decreased with the increase in yttrium content. Mössbauer spectroscopy was used to investigate the iron content in both tetrahedral and octahedral positions.

**Keywords:** magnesium spinel; sol-gel synthesis; ferrite spinel; solid solutions; magnetic properties

**Citation:** Karoblis, D.; Mazeika, K.; Raudonis, R.; Zarkov, A.; Kareiva, A. Sol-Gel Synthesis and Characterization of Yttrium-Doped MgFe$_2$O$_4$ Spinel. *Materials* **2022**, *15*, 7547. https://doi.org/10.3390/ma15217547

Academic Editor: Christian Müller

Received: 28 September 2022
Accepted: 25 October 2022
Published: 27 October 2022

**Publisher's Note:** MDPI stays neutral with regard to jurisdictional claims in published maps and institutional affiliations.

**Copyright:** © 2022 by the authors. Licensee MDPI, Basel, Switzerland. This article is an open access article distributed under the terms and conditions of the Creative Commons Attribution (CC BY) license (https://creativecommons.org/licenses/by/4.0/).

## 1. Introduction

Spinel ferrites with general formula MFe$_2$O$_4$ (where M = Fe, Co, Mn, Ni, Mg, Cu, Zn) are considered to be an important class of inorganic materials displaying a large variety of properties, including mechanical hardness, chemical stability, high electrical resistivity and good thermal stability [1]. In addition, these compounds are soft magnetic materials with low coercivity, high magnetization saturation and remnant magnetization at room temperature [2]. A variety of physical, electrical, dielectric, magnetic, optical and catalytic properties make ferrite spinels applicable in medicine [3], water and wastewater treatment [4], nonvolatile memory devices [5], catalysis [6], gas sensing [7], microwave absorption [8] etc.

Amongst all the spinel ferrites, magnesium ferrite (MgFe$_2$O$_4$) attracted considerable attention from the scientific community due to its high adsorption capacity, suitable bandgap and non-toxicity [9,10]. This compound has a partially inverse spinel structure, where divalent Mg$^{2+}$ ions partly fill the tetrahedral site. It is known to be ferrimagnetic below Neel temperature; however, in the case of nanoscale particles depending on preparation conditions it can also be superparamagnetic [11,12]. Various synthesis approaches, such as co-precipitation [13], solvothermal [14], one-pot solution combustion [15], hydrothermal [16] and microwave-assisted ball milling [17] were utilized for the preparation of magnesium ferrite. Another widely applied synthesis technique is the aqueous sol-gel method [18–20]. Using different complexing agents allows the preparation of phase-pure magnesium ferrite spinel at a low processing temperature in a relatively short time. Moreover, the mixing of starting materials at the atomic level leads to the ease of doping with other elements [20].

One of the ways to tune the physical properties of spinel ferrites is doping with different monovalent, divalent or trivalent cations. For example, Mg$^{2+}$ substitution by Zn$^{2+}$

led to an increase in magnetization and magnetic moments [21]; the introduction of 80% of $Co^{2+}$ ions into the $MgFe_2O_4$ structure resulted in the smallest core losses [22]. In particular, the effects of substitution by $Y^{3+}$ ions were investigated for many ferrite spinels, including $MnFe_2O_4$ [23], $Mn_{0.5}Zn_{0.5}Fe_2O_4$ [24], $CoFe_2O_4$ [25], $CdFe_2O_4$ [26] and $ZnFe_2O_4$ [27]. In most cases, trivalent $Fe^{3+}$ ions were substituted, while the amount of A-site cation remained the same. While yttrium has a larger ionic radius than other cations in the spinel crystal lattice, previous works [23–27] have shown that up to 30% can be successfully introduced into the ferrite spinel structure without the formation of any impurity phases.

To the best of our knowledge, there is only one work regarding $MgFe_2O_4$ spinel substituted with yttrium, where a double sintering technique was applied [28]. In this study, the introduction of $Y^{3+}$ in favor of $Fe^{3+}$ ions resulted in increased resistivity, decreased dielectric constant and reduced particle size. Additionally, it was shown that only 2 mol% of iron can be substituted since a higher yttrium amount resulted in the formation of the $YFeO_3$ impurity phase.

In this work, we investigated $Mg^{2+}$ substitution with $Y^{3+}$ ions by preparing two different series of $Mg_{1-x}Y_xFe_{2-\delta}O_4$ powders applying an environmentally friendly sol-gel method. Since divalent ion is substituted by trivalent, in the first series (with $\delta = 0$) the amount of Fe amount was fixed, while in the second ($\delta = x/3$) the charge was compensated by an appropriately reduced amount of Fe. The structural, morphological and magnetic properties of the obtained spinels were evaluated. Moreover, the maximal yttrium substitution level, when monophasic compounds form, was also inspected.

## 2. Materials and Methods

For the preparation of yttrium-doped $MgFe_2O_4$ spinels, magnesium (II) nitrate hexahydrate ($Mg(NO_3)_2 \cdot 6H_2O$, ≥98%, Chempur, Karlsruhe, Germany), iron (III) nitrate nonahydrate ($Fe(NO_3)_3 \cdot 9H_2O$, 99.9%, Alfa Aesar, Haverhill, MA, USA) and yttrium (III) nitrate hexahydrate ($Y(NO_3)_3 \cdot 6H_2O$, 99.9%, Sigma-Aldrich, St. Louis, MO, USA,) were used as starting materials. During the first step, the appropriate amounts of nitrates required for the synthesis of 1 g of material were dissolved in 20 mL of deionized water. After that, citric acid monohydrate ($C_6H_8O_7 \cdot H_2O$, 99.9%, Chempur) and ethylene glycol ($C_2H_6O_2$, ≥99.5%, Sigma-Aldrich) were added to the mixture (the molar ratio between total metal ions, citric acid and ethylene glycol was 1:1:2, respectively). The temperature of magnetic stirrer was set at 90 °C and the above solution was homogenized for 1 h under constant mixing. After that, the temperature was increased up to 120 °C for complete solvent evaporation and the gel was obtained. The acquired gel was left to dry overnight at 130 °C in the oven, carefully ground in agate mortar and annealed in air at 800 °C for 5 h with a heating rate of 5°/min.

PerkinElmer STA 6000 Simultaneous Thermal Analyzer was used to perform thermogravimetric and differential scanning calorimetric (TG/DTG-DSC) analysis. A small amount of dried gel (5–10 mg) was heated at a 10 °C/min heating rate from 30 to 850 °C in dry flowing air (20 mL/min). Rigaku Miniflex II diffractometer using a primary beam Cu K$\alpha$ radiation ($\lambda$ = 1.541838 Å) was used for X-ray diffraction (XRD) analysis. The 2$\Theta$ angle of the diffractometer was selected in 20°–80° range while moving 5°/min. To calculate the crystallite size the Scherrer's equation ($D = \frac{K\lambda}{\beta cos\Theta}$, where $K$—shape factor 0.89, $\lambda$—X-ray wavelength, $\beta$—full width at half maximum in radian, $\Theta$—Bragg diffraction angle) was used. To determine the instrumental broadening, the $\beta$ was measured for corundum standard. Fourier-transform infrared spectroscopy (FT-IR) was performed using an Alpha FT-IR spectrophotometer (Bruker, Ettlingen, Germany) in the range of 4000–400 cm$^{-1}$. The morphology of solid solutions was examined using the Hitachi SU-70 (Tokyo, Japan) scanning electron microscope (SEM). Magnetometer consisting of the lock-in amplifier SR510 (Stanford Research Systems, Sunnyvale, USA), the gauss/teslameter FH-54 (Magnet Physics, Cologne, Germany) and the laboratory magnet supplied by the power source SM 330-AR-22 (Delta Elektronika, Zierikzee, The Netherlands) were applied to record magnetization dependences on the applied magnetic field. Mössbauer spectra were measured using $^{57}Co(Rh)$ source and Mössbauer spectrometer (Wissenschaftliche Elektronik

GmbH, Starnberg, Germany). For low temperature measurements, closed cycle He cryostat (Advanced Research Systems, Macungie, USA) was applied. One or two hyperfine field distributions, separate sextet and singlet/doublet were used to fit to Mössbauer spectra applying WinNormos Dist software. Isomer shift is given relative to α-Fe.

## 3. Results

The thermal decomposition behavior of the obtained gels as well as possible minimal annealing temperature were investigated by thermogravimetric analysis. TG/DTG/DSC curves of two xerogels with different compositions (Mg-Fe-O and Mg-Y-Fe-O) are depicted in Figures 1 and 2. The first degradation stage for both samples takes place in 50–130 °C range, where negligible weight loss (around 2%) can be seen. At these temperatures, the removal of adsorbed water occurs. Two degradation steps can be witnessed for Mg-Fe-O xerogel at 200–330 °C, while (0.8)Mg-(0.2)Y-(1.933)Fe-O xerogel has only one in this temperature range. These decomposition steps (where around 50% of the initial weight was lost) could be related to the decomposition of metal complexes with ethylene glycol and citric acid. A small exothermic peak found in DSC curve at around 250 °C supports the combustion reaction. Moreover, the next step centered around 370 °C in the DTG curve for both xerogels could be attributed to the thermal decomposition of metal nitrates and organic species. The last degradation step, in the 440–500 °C range for Mg-Fe-O xerogel and 380–470 °C range for (0.8)Mg-(0.2)Y-(1.933)Fe-O is related to pyrolysis as well as combustion of intermediates species formed during gelation and residual organic parts. Both xerogels lose around 80 °C of total weight according to TG curves. Interestingly, the sample containing 20 mol% of yttrium has a lower decomposition temperature compared to the sample with only magnesium and iron ions.

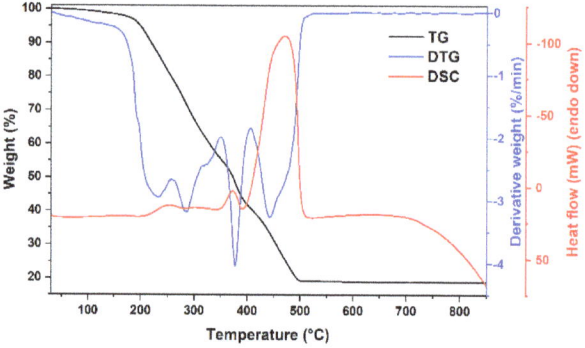

**Figure 1.** TG/DTG/DSC curves of Mg-(2)Fe-O xerogel.

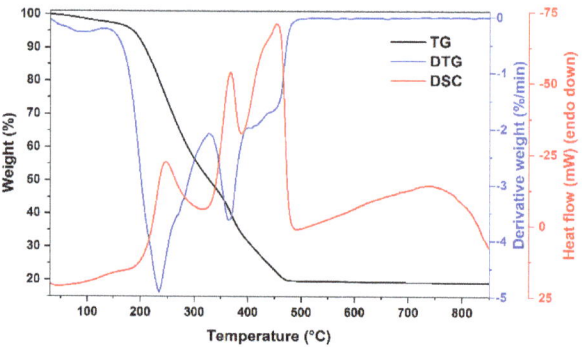

**Figure 2.** TG/DTG/DSC curves of (0.8)Mg-(0.2)Y-(1.933)Fe-O xerogel.

While 460 °C was determined to be the lowest possible annealing temperature for the synthesis of $Mg_{0.8}Y_{0.2}Fe_{1.933}O_4$ spinel, only 800 °C temperature was sufficient for the formation of the spinel phase. Two different compositions of yttrium-doped spinels were prepared at this temperature and the results of the X-ray diffraction analysis are presented in Figure 3.

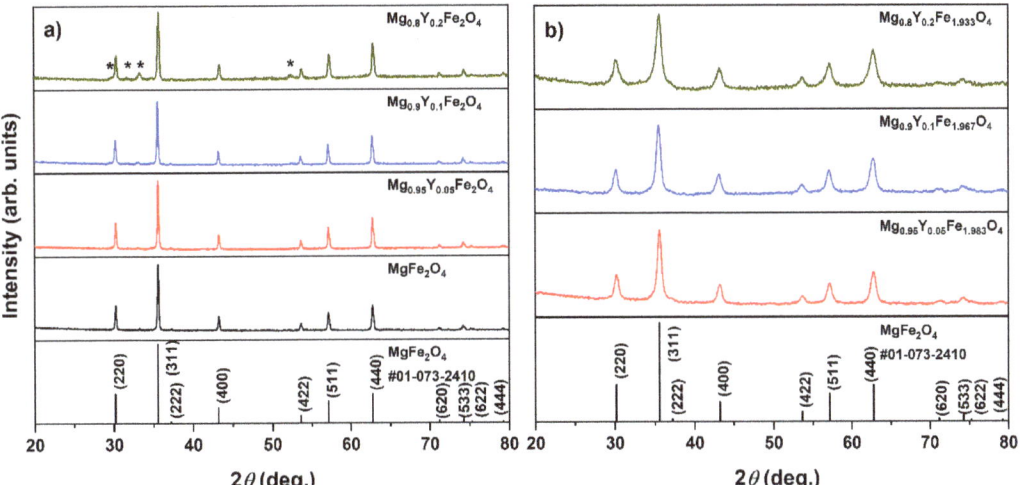

**Figure 3.** XRD patterns of $Mg_{1-x}Y_xFe_{(2-\delta)}O_4$, where $\delta = 0$ (**a**) and $\delta = x/3$ (**b**). An asterisk represents the diffraction peaks ascribed to $YFeO_3$ impurity phase.

The undoped $MgFe_2O_4$ sample seems to be nearly monophasic with only a negligible amount of $Fe_2O_3$ as a neighboring phase. The phase purity of solid solutions depends on the iron content. In the first case, when $Mg^{2+}$ was substituted with $Y^{3+}$ ions and the iron content remained fixed throughout the whole series (Figure 3a), the formation of the hexagonal $YFeO_3$ (#00-048-0529) impurity phase occurred. The amount of neighboring yttrium ferrite phase increased with the increase of $Y^{3+}$ content. For the second series of samples (Figure 3b), since $Mg^{2+}$ and $Y^{3+}$ ions have different valencies, the excess of positive charge was compensated with appropriate reducing of the $Fe^{3+}$ content. $Mg_{1-x}Y_xFe_{(2-\delta)}O_4$ ($\delta = x/3$) solid solutions were monophasic until $x = 0.2$. With the increase of yttrium content, these materials demonstrated considerably broader diffraction peaks, indicating the formation of smaller particles. The crystallite size decreased from ca. 40 nm for $MgFe_2O_4$ to 10–13 nm for yttrium-containing solid solutions. This effect can be assumed as evidence of the introduction of yttrium ions into the crystal lattice. Moreover, only a slight shift of diffraction peaks to lower $2\Theta$ can be seen, due to the difference in ionic radii between magnesium and yttrium ions (0.72 Å vs. 0.9 Å in VI-fold coordination) [29]. It should be noted that yttrium ions may also be located in Fe-sites at the octahedral position, since the iron amount in $Mg_{1-x}Y_xFe_{(2-\delta)}O_4$ ($\delta = x/3$) solid solutions was also reduced. While the orthorhombic $YFeO_3$ phase is considered to be a thermodynamically stable one, the hexagonal perovskite phase can be prepared with a similar sol-gel methodology at lower temperatures [30,31]. It can be summarized that the compensation of excess of positive charge by reducing iron content was crucial for the preparation of single-phase spinel ferrites.

FT-IR spectroscopy was additionally performed to investigate the structural changes caused by the introduction of $Y^{3+}$ ions into the spinel structure; the spectra of $Mg_{1-x}Y_xFe_{(2-\delta)}O_4$ ($\delta = x/3$) solid solutions are demonstrated in Figure 4. Since a relatively high annealing temperature was used for the preparation of ferrite spinels, no absorption bands were observed in the 4000–800 $cm^{-1}$ range, which could be related to hydroxide, carbonate or residual organic species. According to the previous study of various spinel ferrites [32], the FT-IR spectrum of $MgFe_2O_4$ contains two bands assigned to the Fe-O bond centered at 565 and 406 cm$^{-1}$. The first band is associated with the intrinsic vibrations in the tetrahedral site, while the latter is attributed to octahedral groups. In our case, the position of absorption bands is slightly different, which could indicate a redistribution of cations between A and B sites. A similar inversion of cations between both sites was previously observed for nanosized $MgFe_2O_4$ spinel prepared via sol-gel synthesis technique [33]. The absence of change in the position of the lower intensity band and monotonous, but a non-significant shift of the most intense absorption band with increasing yttrium content could suggest that $Y^{3+}$ ions occupy tetrahedral positions in the lattice.

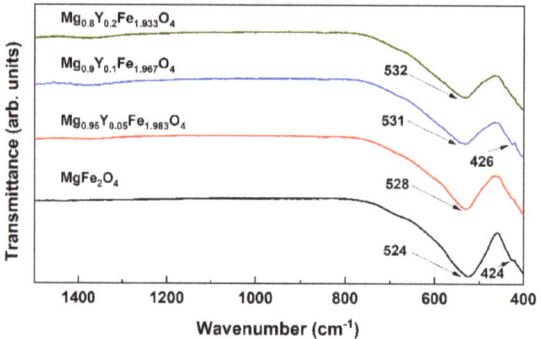

Figure 4. FT-IR spectra of $Mg_{1-x}Y_xFe_{(2-\delta)}O_4$ ($\delta = x/3$) specimens.

Scanning electron microscopy (SEM) was employed to evaluate surface morphology as well as the particle size of two different yttrium-containing solid solutions and the results are presented in Figure 5. In both cases, samples consist of smaller particles, which are connected to each other forming larger aggregates. For $Mg_{0.95}Y_{0.05}Fe_{1.983}O_4$ sample (Figure 5a) the size of these aggregates varied in the 200–400 nm range, while for the sample containing 10 mol% of yttrium (Figure 5b), assemblies of particles were larger (200–600 nm). Interestingly, while the increase in $Y^{3+}$ content resulted in the formation of larger aggregates, for individual particles the opposite effect can be seen. ImageJ was used to estimate the size of separate particles for both solid solutions, and most particles of $Mg_{0.9}Y_{0.1}Fe_{1.967}O_4$ spinel lay in the range of 30–100 nm, while $Mg_{0.95}Y_{0.05}Fe_{1.983}O_4$ sample was comprised of slightly larger particles varying from 50 to 150 nm.

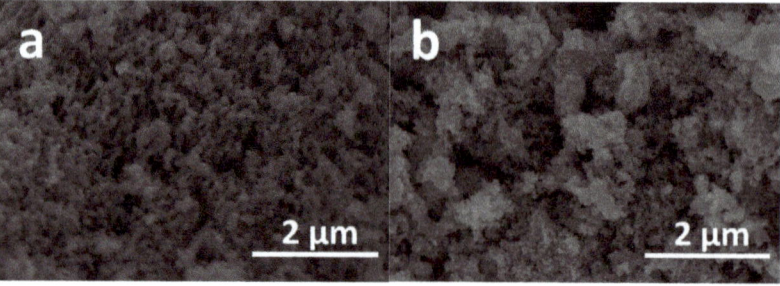

Figure 5. SEM images of $Mg_{0.95}Y_{0.05}Fe_{1.983}O_4$ (a) and $Mg_{0.9}Y_{0.1}Fe_{1.967}O_4$ (b) solid solutions.

With an increase of Y amount the saturation magnetization (at maximal applied field) of hysteresis loops of $Mg_{1-x}Y_xFe_{2-\delta}O_4$ (Figure 6) decreased from 26 emu/g to 16 emu/g ($\delta = 0$) or 20 emu/g ($\delta = x/3$), while the coercivity decreased from 70 to 17 Oe

($\delta = 0$) or 5 Oe ($\delta = x/3$). Cations in Mg ferrite are distributed between tetrahedral A and octahedral B sublattices which are denoted in formula $(Fe_\alpha Mg_{1-\alpha})[Fe_{2-\alpha}Mg_\alpha]O_4$ by round and square brackets, respectively. The cation redistribution is also known for other compounds with spinel structures, such as $MgAl_2O_4$ [34]. The Mg ferrite inversion degree is high ($\alpha \approx 0.9$) with Mg occupying predominantly octahedral sublattice [35–37]. The magnetization of Mg ferrite is determined by the difference in magnetic moments of Fe in tetrahedral and octahedral sublattices. By changing the chemical composition, Fe cations may redistribute between A and B sublattices of $Mg_{1-x}Y_xFe_{2-\delta}O_4$ in a way causing a change in magnetization. However, the major factor causing a decrease in magnetization could be a decrease in grain (nanoparticle) size and an increase in the contribution of a magnetically disordered, magnetically dead intergranular layer. The decrease in coercivity can also be explained by the formation of smaller superparamagnetic nanograins [38].

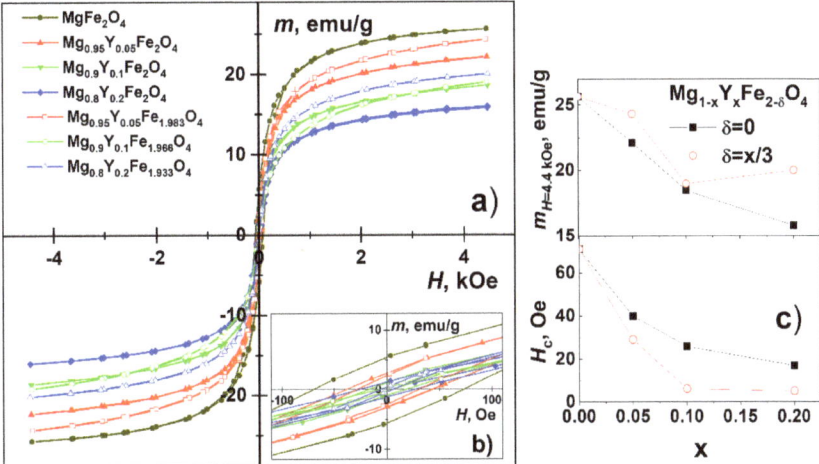

**Figure 6.** Magnetization hysteresis loops of $Mg_{1-x}Y_xFe_{2-\delta}O_4$ (**a**,**b**) and their parameters (**c**): saturation magnetization $m_{H = 4.4\ Oe}$ and coercivity $H_c$.

Mössbauer spectra of the $MgFe_2O_4$ sample (Figure 7a) showed broader spectral lines as compared to those of previously studied polycrystalline Mg ferrite [35,36]. The broadening of room temperature Mössbauer spectra and the relative area of superparamagnetic doublet increased with an increase of $Y^{3+}$ content (Table 1). Two broad overlapping subspectra expressed by hyperfine field distribution P(B) attributable to octahedral A and tetrahedral B sublattices were distinguished by different isomer shifts: $\delta_A \approx 0.15$ mm/s and $\delta_B \approx 0.4$ mm/s for $Mg_{1-x}Y_xFe_{2-\delta}O_4$ samples with $x \leq 0.1$. However, at larger spectral broadening, when $x \geq 0.1$, only one P(B) distribution having an isomer shift of $\approx 0.30$ mm/s was used, which was an average isomer shift of two A and B subspectra. The isomer shift $\approx 0.30$ mm/s was characteristic of the superparamagnetic doublet. The decrease in average hyperfine field and increase of doublet area can be explained by the increase of the superparamagnetic relaxation rate of Fe spins in case of a decrease in grain size. The average hyperfine field of the whole spectrum $<B_{all}>$ decreased up to 50% (Table 1) with an increase of x. It was the smallest for $Mg_{0.9}Y_{0.1}Fe_{1.967}O_4$ having the largest contribution of doublet in the spectrum. It can be noted that Mössbauer spectra do not indicate that $Fe^{2+}$ ions may form as there were no characteristic shifts of spectral shape while increasing the substitution of $Mg^{2+}$ by $Y^{3+}$.

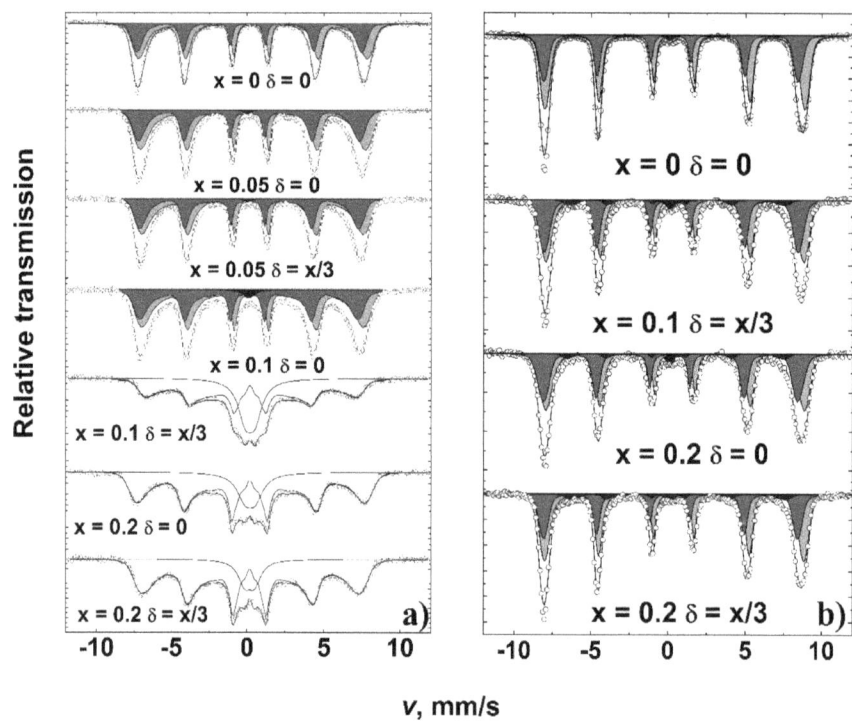

**Figure 7.** Mössbauer spectra of $Mg_{1-x}Y_xFe_{2-\delta}O_4$ at 296 K (**a**) and 10 K (**b**).

**Table 1.** $Mg_{1-x}Y_xFe_{2-\delta}O_4$ Mössbauer spectra parameters at 293 K: average hyperfine field <B>, subspectra A and B area ratio $I_A/I_B$, singlet relative area $I_s$, isomer shifts δ. Indexes A, B and s indicate tetrahedral, octahedral sublattices and singlet, respectively.

| x | δ | $<B_A>$, T | $<B_B>$, T | $I_A/I_B$ | $I_s$, % | $δ_A$, mm/s | $δ_B$, mm/s |
|---|---|---|---|---|---|---|---|
| 0 | 0 | 45.0 | 45.0 | 0.62 | 0 | 0.16 ± 0.01 | 0.39 ± 0.01 |
| 0.05 | 0 | 41.6 | 43.4 | 0.7 | 1 | 0.14 ± 0.01 | 0.40 ± 0.01 |
| 0.05 | 0.017 | 41.2 | 42.3 | 0.7 | 2 | 0.16 ± 0.01 | 0.40 ± 0.01 |
| 0.1 | 0 | 39.6 | 40.8 | 0.74 | 1 | 0.15 ± 0.01 | 0.40 ± 0.01 |
| | | $<B_{AB}>$, T | $<B_{all}>$, T | | | $δ_{AB}$, mm/s | $δ_s$, mm/s |
| 0.1 | 0.034 | 31.4 | 21.8 | - | 31 | 0.30 ± 0.01 | 0.31 ± 0.01 |
| 0.2 | 0 | 35.3 | 28.4 | - | 20 | 0.29 ± 0.01 | 0.31 ± 0.01 |
| 0.2 | 0.067 | 34.1 | 29.4 | - | 14 | 0.29 ± 0.01 | 0.29 ± 0.01 |

At 10 K the width of hyperfine distributions of two major subspectra shrank (Figure 7b). The additional subspectrum of 5% spectral area (Table 2) was distinguished for yttrium-containing samples. The additional subspectrum with an average hyperfine field of 42–43 T can be attributed to Fe-disordered sites because of Y presence in the neighborhood of Fe sites or the formation of hexagonal $YFeO_3$. We were not able to find previously published low-temperature Mössbauer data for hexagonal $YFeO_3$. However, for Fe in hexagonal $YMnO_3$, the positions of the lines in the Mössbauer spectrum (hyperfine field of 40–44.5 T) at 12 K [39] are in rather good agreement with those of additional subspectra.

Table 2. $Mg_{1-x}Y_xFe_{2-\delta}O_4$ Mössbauer spectra parameters at 10 K: subspectra relative intensity I, isomer shift δ, quadrupole shift 2ε and average hyperfine field <B>.

| x | δ | I, % | δ, mm/s | 2ε, mm/s | <B>, T | |
|---|---|---|---|---|---|---|
| 0 | 0 | 37 | 0.27 ± 0.01 | 0.01 ± 0.01 | 51.3 | Tetrahedral A |
|   |   | 63 | 0.51 ± 0.01 | 0.05 ± 0.01 | 52.3 | Octahedral B |
| 0.1 | 0.034 | 34 | 0.27 * | 0.10 ± 0.01 | 51.0 | Tetrahedral A |
|   |   | 61 | 0.51 * | −0.01 ± 0.01 | 51.7 | Octahedral B |
|   |   | 5 | 0.39 ± 0.02 | 0.54 ± 0.05 | 42.7 | Disordered/h-$YFeO_3$ |
| 0.2 | 0.067 | 36 | 0.27 ± 0.01 | 0.03 ± 0.01 | 51.2 | Tetrahedral A |
|   |   | 59 | 0.51 ± 0.01 | 0.03 ± 0.01 | 51.9 | Octahedral B |
|   |   | 5 | 0.45 ± 0.02 | 0.63 ± 0.05 | 42.6 | Disordered/$YFeO_3$ |
| 0.2 | 0 | 35 | 0.27 * | 0.12 ± 0.01 | 51.2 | Tetrahedral A |
|   |   | 60 | 0.51 * | 0.01 ± 0.01 | 51.7 | Octahedral B |
|   |   | 5 | 0.33 ± 0.03 | 0.55 ± 0.07 | 43.4 | Disordered/$YFeO_3$ |

* Fixed.

## 4. Conclusions

A sol-gel synthetic approach using ethylene glycol and citric acid was successfully utilized for the preparation of $Mg_{1-x}Y_xFe_{2-\delta}O_4$ (δ = x/3) solid solutions. The iron content played a key role in the phase purity of the final products and only when δ = x/3 monophasic spinels could be obtained. At fixed iron content (δ = 0), the formation of a secondary $YFeO_3$ phase occurred. While the size of individual particles was smaller with the increase in yttrium amount, the size of aggregates was larger. Intercalation of $Y^{3+}$ ions caused a decrease in the saturation of magnetization and coercivity. According to the Mössbauer spectroscopy studies, with the increase in yttrium amount for $Mg_{1-x}Y_xFe_{2-\delta}O_4$ solid solutions the amount of iron located in the tetrahedral position increased. Low-temperature Mössbauer measurements revealed the formation of hexagonal $YFeO_3$ or disordered phase.

**Author Contributions:** Conceptualization, A.K. and A.Z.; methodology, D.K. and R.R.; validation, D.K. and A.Z.; formal analysis, K.M.; investigation, D.K.; resources, A.Z.; data curation, K.M. and R.R.; writing—original draft preparation, D.K.; writing—review and editing, A.Z.; visualization, K.M.; supervision, A.K. All authors have read and agreed to the published version of the manuscript.

**Funding:** This research received no external funding.

**Conflicts of Interest:** The authors declare no conflict of interest.

## References

1. Amiri, M.; Salavati-Niasari, M.; Akbari, A. Magnetic nanocarriers: Evolution of spinel ferrites for medical applications. *Adv. Colloid Interface Sci.* **2019**, *265*, 29–44. [CrossRef] [PubMed]
2. Vinnik, D.A.; Sherstyuk, D.P.; Zhivulin, V.E.; Zhivulin, D.E.; Starikov, A.Y.; Gudkova, S.A.; Zherebtsov, D.A.; Pankratov, D.A.; Alekhina, Y.A.; Perov, N.S. Impact of the Zn–Co content on structural and magnetic characteristics of the Ni spinel ferrites. *Ceram. Int.* **2022**, *48*, 18124–18133. [CrossRef]
3. Kefeni, K.K.; Msagati, T.A.M.; Nkambule, T.T.I.; Mamba, B.B. Spinel ferrite nanoparticles and nanocomposites for biomedical applications and their toxicity. *Mater. Sci. Eng. C* **2020**, *107*, 110314. [CrossRef] [PubMed]
4. Ivanets, A.; Prozorovich, V.; Roshchina, M.; Grigoraviciute-Puroniene, I.; Zarkov, A.; Kareiva, A.; Wang, Z.; Srivastava, V.; Sillanpää, M. Heterogeneous Fenton oxidation using magnesium ferrite nanoparticles for ibuprofen removal from wastewater: Optimization and kinetics studies. *J. Nanomater.* **2020**, *2020*, 8159628. [CrossRef]
5. Hu, W.; Qin, N.; Wu, G.; Lin, Y.; Li, S.; Bao, D. Opportunity of Spinel Ferrite Materials in Nonvolatile Memory Device Applications Based on Their Resistive Switching Performances. *J. Am. Chem. Soc.* **2012**, *134*, 14658–14661. [CrossRef]
6. Ivanets, A.; Prozorovich, V.; Sarkisov, V.; Roshchina, M.; Grigoraviciute-Puroniene, I.; Zarkov, A.; Kareiva, A.; Masindi, V.; Wang, C.; Srivastava, V. Effect of magnesium ferrite doping with lanthanide ions on dark-, visible-and UV-driven methylene blue degradation on heterogeneous Fenton-like catalysts. *Ceram. Int.* **2021**, *47*, 29786–29794. [CrossRef]
7. Šutka, A.; Gross, K.A. Spinel ferrite oxide semiconductor gas sensors. *Sens. Actuators B Chem.* **2016**, *222*, 95–105. [CrossRef]
8. Xie, X.; Wang, B.; Wang, Y.; Ni, C.; Sun, X.; Du, W. Spinel structured $MFe2O4$ (M = Fe, Co, Ni, Mn, Zn) and their composites for microwave absorption: A review. *Chem. Eng. J.* **2022**, *428*, 131160. [CrossRef]

9. Jiang, Z.; Chen, K.; Zhang, Y.; Wang, Y.; Wang, F.; Zhang, G.; Dionysiou, D.D. Magnetically recoverable MgFe$_2$O$_4$/conjugated polyvinyl chloride derivative nanocomposite with higher visible-light photocatalytic activity for treating Cr (VI)-polluted water. *Sep. Purif. Technol.* **2020**, *236*, 116272. [CrossRef]
10. Sharma, L.; Kakkar, R. Magnetically retrievable one-pot fabrication of mesoporous magnesium ferrite (MgFe$_2$O$_4$) for the re-mediation of chlorpyrifos and real pesticide wastewater. *J. Environ. Chem. Eng.* **2018**, *6*, 6891–6903. [CrossRef]
11. Shahjuee, T.; Masoudpanah, S.M.; Mirkazemi, S.M. Thermal Decomposition Synthesis of MgFe$_2$O$_4$ Nanoparticles for Magnetic Hyperthermia. *J. Supercond. Nov. Magn.* **2019**, *32*, 1347–1352. [CrossRef]
12. Reza Barati, M.; Selomulya, C.; Suzuki, K. Particle size dependence of heating power in MgFe$_2$O$_4$ nanoparticles for hyper-thermia therapy application. *J. Appl. Phys.* **2014**, *115*, 17B522. [CrossRef]
13. Ajeesha, T.; Ashwini, A.; George, M.; Manikandan, A.; Mary, J.A.; Slimani, Y.; Almessiere, M.A.; Baykal, A. Nickel substituted MgFe$_2$O$_4$ nanoparticles via co-precipitation method for photocatalytic applications. *Phys. B: Condens. Matter* **2021**, *606*, 412660. [CrossRef]
14. Shen, Y.; Wu, Y.; Li, X.; Zhao, Q.; Hou, Y. One-pot synthesis of MgFe$_2$O$_4$ nanospheres by solvothermal method. *Mater. Lett.* **2013**, *96*, 85–88. [CrossRef]
15. Heidari, P.; Masoudpanah, S.M. A facial synthesis of MgFe$_2$O$_4$/RGO nanocomposite powders as a high performance micro-wave absorber. *J. Alloys. Compd.* **2020**, *834*, 155166. [CrossRef]
16. Ali, N.A.; Idris, N.H.; Din, M.F.M.; Mustafa, N.S.; Sazelee, N.A.; Yap, F.A.H.; Sulaiman, N.N.; Yahya, M.S.; Ismail, M. Nanolayer-like-shaped MgFe$_2$O$_4$ synthesised via a simple hydrothermal method and its catalytic effect on the hydrogen storage properties of MgH$_2$. *RSC Adv.* **2018**, *8*, 15667–15674. [CrossRef]
17. Chen, D.; Zhang, Y.; Tu, C. Preparation of high saturation magnetic MgFe$_2$O$_4$ nanoparticles by microwave-assisted ball milling. *Mater. Lett.* **2012**, *82*, 10–12. [CrossRef]
18. Feng, Y.; Li, S.; Zheng, Y.; Yi, Z.; He, Y.; Xu, Y. Preparation and characterization of MgFe$_2$O$_4$ nanocrystallites via PVA sol-gel route. *J. Alloys. Compd.* **2017**, *699*, 521–525. [CrossRef]
19. Araújo, J.C.R.; Araujo-Barbosa, S.; Souza, A.L.R.; Iglesias, C.A.M.; Xavier, J.; Souza, P.B.; Pla Cid, C.C.; Azevedo, S.; da Silva, R.B.; Correa, M.A.; et al. Tuning structural, magnetic, electrical, and dielectric properties of MgFe$_2$O$_4$ synthesized by sol-gel followed by heat treatment. *J. Phys. Chem. Solids* **2021**, *154*, 110051. [CrossRef]
20. Uke, S.J.; Mardikar, S.P.; Bambole, D.R.; Kumar, Y.; Chaudhari, G.N. Sol-gel citrate synthesized Zn doped MgFe$_2$O$_4$ nanocrystals: A promising supercapacitor electrode material. *Mater. Sci. Energy Technol.* **2020**, *3*, 446–455. [CrossRef]
21. Phor, L.; Chahal, S.; Kumar, V. Zn$^{2+}$ substituted superparamagnetic MgFe2O4 spinel-ferrites: Investigations on structural and spin-interactions. *J. Adv. Ceram.* **2020**, *9*, 576–587. [CrossRef]
22. Wang, X.; Kan, X.; Liu, X.; Feng, S.; Zheng, G.; Cheng, Z.; Wang, W.; Chen, Z.; Liu, C. Characterization of microstructure and magnetic properties for Co$^{2+}$ ions doped MgFe$_2$O$_4$ spinel ferrites. *Mater. Today Commun.* **2020**, *25*, 101414. [CrossRef]
23. Ahmad, Y.; Raina, B.; Thakur, S.; Bamzai, K.K. Magnesium and yttrium doped superparamagnetic manganese ferrite nanoparticles for magnetic and microwave applications. *J. Magn. Magn. Mater.* **2022**, *552*, 169178. [CrossRef]
24. Almessiere, M.A.; Güner, S.; Slimani, Y.; Baykal, A.; Shirsath, S.E.; Korkmaz, A.D.; Badar, R.; Manikandan, A. Investigation on the structural, optical, and magnetic features of D$^{3+}$ and Y$^{3+}$ co-doped Mn$_{0.5}$Zn$_{0.5}$Fe$_2$O$_4$ spinel ferrite nanoparticles. *J. Mol. Struct.* **2022**, *1248*, 131412. [CrossRef]
25. Sharma, R.; Kumar, V.; Bansal, S.; Singhal, S. Boosting the catalytic performance of pristine CoFe$_2$O$_4$ with yttrium (Y$^{3+}$) in-clusion in the spinel structure. *Mater. Res. Bull.* **2017**, *90*, 94–103. [CrossRef]
26. Amin, N.; Hasan, M.S.U.; Majeed, Z.; Latif, Z.; un Nabi, M.A.; Mahmood, K.; Ali, A.; Mehmood, K.; Fatima, M.; Akhtar, M. Structural, electrical, optical and dielectric properties of yttrium substituted cadmium ferrites prepared by Co-Precipitation method. *Ceram. Int.* **2020**, *46*, 20798–20809. [CrossRef]
27. Cvejić, Ž.; Rakić, S.; Jankov, S.; Skuban, S.; Kapor, A. Dielectric properties and conductivity of zinc ferrite and zinc ferrite doped with yttrium. *J. Alloys. Compd.* **2009**, *480*, 241–245. [CrossRef]
28. Ishaque, M.; Khan, M.A.; Ali, I.; Khan, H.M.; Iqbal, M.A.; Islam, M.U.; Warsi, M.F. Investigations on structural, electrical and dielectric properties of yttrium substituted Mg-ferrites. *Ceram. Int.* **2015**, *41*, 4028–4034. [CrossRef]
29. Shannon, R.D. Revised effective ionic radii and systematic studies of interatomic distances in halides and chalcogenides. *Acta Crystallogr. Sect. A* **1976**, *32*, 751–767. [CrossRef]
30. Zhang, Y.; Yang, J.; Xu, J.; Gao, Q.; Hong, Z. Controllable synthesis of hexagonal and orthorhombic YFeO$_3$ and their visi-ble-light photocatalytic activities. *Mater. Lett.* **2012**, *81*, 1–4. [CrossRef]
31. Zhang, R.-L.; Chen, C.-L.; Jin, K.-X.; Niu, L.-W.; Xing, H.; Luo, B. Dielectric behavior of hexagonal and orthorhombic YFeO$_3$ prepared by modified sol-gel method. *J. Electroceramics* **2014**, *32*, 187–191. [CrossRef]
32. Waldron, R.D. Infrared Spectra of Ferrites. *Phys. Rev.* **1955**, *99*, 1727–1735. [CrossRef]
33. Pradeep, A.; Priyadharsini, P.; Chandrasekaran, G. Sol–gel route of synthesis of nanoparticles of MgFe$_2$O$_4$ and XRD, FTIR and VSM study. *J. Magn. Magn. Mater.* **2008**, *320*, 2774–2779. [CrossRef]
34. Seeman, V.; Feldbach, E.; Kärner, T.; Maaroos, A.; Mironova-Ulmane, N.; Popov, A.I.; Shablonin, E.; Vasil'chenko, E.; Lush-chik, A. Fast-neutron-induced and as-grown structural defects in magnesium aluminate spinel crystal with different morphology. *Opt. Mater.* **2019**, *91*, 42–49. [CrossRef]

35. Šepelák, V.; Baabe, D.; Mienert, D.; Litterst, F.J.; Becker, K.D. Enhanced magnetisation in nanocrystalline high-energy milled MgFe$_2$O$_4$. *Scr. Mater.* **2003**, *48*, 961–966. [CrossRef]
36. De Grave, E.; Govaert, A.; Chambaere, D.; Robbrecht, G. A Mössbauer effect study of MgFe$_2$O$_4$. *Phys. B C* **1979**, *96*, 103–110. [CrossRef]
37. Franco Jr, A.; Silva, M.S. High temperature magnetic properties of magnesium ferrite nanoparticles. *J. Appl. Phys.* **2011**, *109*, 07B505. [CrossRef]
38. Majetich, S.A.; Sachan, M. Magnetostatic interactions in magnetic nanoparticle assemblies: Energy, time and length scales. *J. Phys. D Appl. Phys.* **2006**, *39*, R407–R422. [CrossRef]
39. Karoblis, D.; Zarkov, A.; Garskaite, E.; Mazeika, K.; Baltrunas, D.; Niaura, G.; Beganskiene, A.; Kareiva, A. Study of gadolinium substitution effects in hexagonal yttrium manganite YMnO$_3$. *Sci. Rep.* **2021**, *11*, 2875. [CrossRef]

Article

# Reaching Visible Light Photocatalysts with Pt Nanoparticles Supported in $TiO_2$-$CeO_2$

Ixchel Alejandra Mejia-Estrella [1], Alejandro Pérez Larios [2], Belkis Sulbarán-Rangel [1,*] and Carlos Alberto Guzmán González [3,*]

[1] Department of Water and Energy, University of Guadalajara, Campus Tonalá, Tonalá 45425, Mexico
[2] Department of Engineering, University of Guadalajara, Campus Altos, Tepatitlán de Morelos 47635, Mexico
[3] Department of Applied Basic Sciences, University of Guadalajara, Campus Tonalá, Tonalá 45425, Mexico
* Correspondence: belkis.sulbaran@academicos.udg.mx (B.S.-R.); cguzman09@hotmail.com (C.A.G.G.)

**Abstract:** Nanostructured catalysts of platinum (Pt) supported on commercial $TiO_2$, as well as $TiO_2$-$CeO_2$ (1, 5 and 10 wt% $CeO_2$), were synthesized through the Sol-Gel and impregnation method doped to 1 wt% of Platinum, in order to obtain a viable photocatalytic material able to oxidate organic pollutants under the visible light spectrum. The materials were characterized by different spectroscopy and surface techniques such as Specific surface area (BET), X-ray photoelectron spectroscopy (XPS), XRD, and TEM. The results showed an increase in the diameter of the pore as well as the superficial area of the supports as a function of the $CeO_2$ content. TEM images showed Pt nanoparticles ranking from 2–7 nm, a decrease in the particle size due to the increase of $CeO_2$. The XPS showed oxidized $Pt^{2+}$ and reduced $Pt^0$ species; also, the relative abundance of the elements $Ce^{3+}/Ce^{4-}$ and $Ti^{4+}$ on the catalysts. Additionally, a shift in the Eg band gap energy (3.02–2.82 eV) was observed by UV–vis, proving the facticity of applying these materials in a photocatalytic reaction using visible light. Finally, all the synthesized materials were tested on their photocatalytic oxidation activity on a herbicide used worldwide; 2,4-Dichlorophenoxyacetic acid, frequently use in the agriculture in the state of Jalisco. The kinetics activity of each material was measured during 6 h of reaction at UV–Vis 190–400 nm, reaching a removal efficiency of 98% of the initial concentration of the pollutant in 6 h, compared to 32% using unmodified $TiO_2$ in 6 h.

**Keywords:** nanocatalysts; photocatalysts; band gap energy; sol-gel and impregnation method

**Citation:** Mejia-Estrella, I.A.; Pérez Larios, A.; Sulbarán-Rangel, B.; Guzmán González, C.A. Reaching Visible Light Photocatalysts with Pt Nanoparticles Supported in $TiO_2$-$CeO_2$. Materials 2022, 15, 6784. https://doi.org/10.3390/ma15196784

Academic Editor: Aleksej Zarkov

Received: 28 July 2022
Accepted: 23 September 2022
Published: 30 September 2022

**Publisher's Note:** MDPI stays neutral with regard to jurisdictional claims in published maps and institutional affiliations.

**Copyright:** © 2022 by the authors. Licensee MDPI, Basel, Switzerland. This article is an open access article distributed under the terms and conditions of the Creative Commons Attribution (CC BY) license (https://creativecommons.org/licenses/by/4.0/).

## 1. Introduction

As the population continues to grow, pollution has increased in all water resources, producing an urgent need to create solutions to remediate it. Studying the literature regarding a reliable technology for water treatment able to oxidate persistent organic molecules heterogeneous catalysts, it has been proven to be able to remove a wide range of contaminants [1–3]. By doping the reaction with metallic and nanometric catalysts helps to increase the surface energy of the individual particles, increasing the probability of aggregation, which can reduce the specific surface area of the catalyst and its efficiency since they are widely used materials due to their relatively low cost, and they can be reused [4,5]. To avoid aggregation, it is important to immobilize the active metal nanoparticles on mesoporous solid support, as well as to transform the metal into its oxide form [6].

The morphological structure of the catalyst support determines the dispersion of the nanoparticles and the surface area of the catalytic active sites. Various semiconductor materials including titanium dioxide ($TiO_2$), zinc oxide (ZnO), vanadium pentoxide ($V_2O_5$), cerium oxide ($CeO_2$) and tungsten trioxide ($WO_3$) have been extensively studied by photo catalysis reaction [1,7,8]. One of the most researched compositions for a support system is $TiO_2$ because it is an effective, inexpensive, and stable photocatalyst used for the decomposition of organics [1,4]. However, $TiO_2$ can only absorb the ultraviolet portion and can only take advantage of about 4% of the intensity of the sunlight spectrum due to its high bandgap (3.0–3.2 eV) [1,9]. This represents a major limitation since its photocatalytic properties are not fully used; even so, an alternative that has been explored to extend its photo response range to the region is to dope the surface with metallic nanoparticles or combine it with another support [4]. Selected methods like doping and composites have been attempted to achieve photo-initiation into the visible spectrum, therefore decreasing cost and increasing efficiency [6]. Another support that has gained importance recently and that can be used as a photocatalyst is $CeO_2$. This is due to its unique redox properties that consist of reversibly creating and eliminating oxygen vacancies on the surface [10].

The relation between the photocatalytic activity of $TiO_2$ and $CeO_2$ composite under UV and visible light has been studied [8,11–13]. Liu et al. in 2005 found that $TiO_2$-$CeO_2$ under visible illumination exhibits more photocatalytic activity than pure $TiO_2$ and $CeO_2$ films. Another study reported by Tian et al. in 2013, who prepared heterostructures of $CeO_2/TiO_2$ nanobelts using a hydrothermal method. These authors found that both UV and visible photocatalytic activities of $CeO_2/TiO_2$ nanobelt heterostructures were enhanced compared to $TiO_2$ nanobelts and $CeO_2$ nanoparticles. More recently, Henych et al. (2021) found the strong interaction of Ti with Ce within the composites led to the formation of $Ce^{3+}$ and $Ti^{<4+}$ states, reduction of titania crystallite size, change of acid-base and surface properties, and synergetic effects that are all responsible for highly improved degradation efficiency of organophosphorus compounds. In addition, the thermocatalytic, photocatalytic, and photothermocatalytic oxidation of some volatile organic compounds, 2-propanol, ethanol, and toluene, were investigated over brookite $TiO_2$-$CeO_2$ composites [11]. Other studies have focused on improving catalyst synthesis methods using green methods [14] or adding doping of $TiO_2$-$CeO_2$ supports to improve photocatalytic properties [15,16].

In order to improve the combination of $TiO_2$ and $CeO_2$ supports, the incorporation of platinum nanoparticles was studied in this research. Platinum nanoparticles are advantageous in biological, biosensor, electro-analytical, analytical, and catalytic applications [17]. They are unique because of their large surface area and their numerous catalytic applications, such as their use as automotive catalytic converters and as petrochemical cracking catalysts [18]. As mentioned above, the present work takes advantage of the combined photoactivity properties of two semiconductors $TiO_2$-$CeO_2$ supports and the platinum nanoparticles forming Pt/$TiO_2$-$CeO_2$ photocatalyst. The $TiO_2$-$CeO_2$ supports were prepared by sol-gel method at different contents of the cerium oxide (2–10 wt%) and the platinum nanoparticles (1.0 wt%) were prepared by the impregnation method to obtain Pt/$TiO_2$-$CeO_2$ photocatalysts. This project is innovative because it will use the semiconductor, $TiO_2$ mixed with $CeO_2$, mechanically alloyed, to shift photo-initiation into the visible range.

## 2. Materials and Methods

### 2.1. Materials

Cerium (IV) oxide reagent grade, 97%, was used as a support for the catalyst in the sol-gel method. The $TiO_2$ P25 reagent grade 99.5%, commercial salt of hexachloroplatinic for a precursor of nanoparticles of Pt at 37.5% purity ($H_2PtCL_6 * 6H_2O$), the reactive used to reach pH in the preparation methods was nitric acid ($HNO_3$) reagent at 65%, ethanol ($C_2H_5OH$) at 99.8% hydrochloric acid (HCl) at 99.9%. All reagents were obtained from Sigma-Aldrich (Toluca, México).

### 2.2. Support Preparation

#### 2.2.1. Impregnation Synthesis

The $TiO_2$ supports was prepared using $TiO_2$ Degussa P25 (Aldrich, 99.5%) it was first placed in a thermal treatment at 500 °C for 2 h with an airflow of 50 mL/min. The $CeO_2$ was incorporated into the $TiO_2$ with different contents (1, 5, and 10% by weight of $CeO_2$) with an aqueous solution of Ce $(NO_3)_3 * 6H_2O$. This solution was added to the $TiO_2$ that was placed in a ball flask. The mixture was stirred for 3 h on a rotary evaporator. Afterwards, the samples were dried under vacuum in a 60 °C water bath. Subsequently they were dried in an oven at 120 °C for 12 h and calcined at 500 °C for 4 h with an air flow (50 mL/min) and a heating ramp of 2 °C/min. This process was performed in duplicate.

2.2.2. Sol-Gel Synthesis

To prepare the $TiO_2$ support material by sol-gel, titanium IV butoxide (Aldrich, 97%) was used as a $TiO_2$ precursor with water, ethanol, and a few drops of $HNO_3$ to fix the pH in the solution to 3. The preparation of the supports was made in a three-necked flask, mixing dropwise the n-Butoxide into the mix in the water/alkoxide solution (8:1 molar ratio). Then the mixture was placed to reflux and stirred vigorously for 24 h. The temperature of the preparation was maintained in a range between 75–80 °C. The samples were dried under vacuum on a rotary evaporator with a 75 °C water bath. Finally, the supports were calcined at 500 °C for 4 h with an airflow (50 mL/min) and a heating ramp of 2 °C/min [19].

The process to add the $CeO_2$ into the support web was the same as described previous for the $TiO_2$ sol-gel, with the difference of adding the reagent Ce $(NO_3)_3$ * $6H_2O$ by previously preparing a solution in order to obtain the desired percentages in the support web. This process was performed in duplicate and in parallel.

### 2.3. Pt Catalysts Preparation

The catalysts were prepared by wet impregnation, using the support material previously synthesized with $TiO_2$ and $CeO_2$. Prior to catalyst preparation, both $TiO_2$ and $TiO_2$-$CeO_2$ supports were previously air-dried at 100 °C for 24 h. Subsequently, the supports were added to a ball flask to which a 0.001 M hydrochloric acid solution was used to adjust the pH. Then the mixed solution with the support was left stirring until it became homogenized, after which the Pt solution was added, and commercial salt of $H_2PtCl_6$ * $6H_2O$ at 37.5% purity was employed as a precursor for Pt in order to obtain a semiconductor material with a 1% weight metal content. The suspended solution was heated to 60 °C with vigorous stirring for 4 h. The leftover solids were dried in the oven at 120 °C for 24 h and calcined at 500 °C with an airflow of 1 $cm^3$/s and heating rate of 2 °C/min. Finally, the samples were placed in a vacuum desiccator in amber glass vials wrapped in foil paper to mitigate the exposure to light [20].

### 2.4. Characterization Techniques

The $TiO_2$-$CeO_2$ supports and the platinum nanoparticles that form the Pt/$TiO_2$-$CeO_2$ photocatalyst synthesized by sol-gel and impregnation at different concentrations were characterized to determine which method improved their photocatalytic properties. The determination of the specific surface area was carried out using the standard Brunauer, Emmett and Teller (BET) method using nitrogen physisorption in Micromeritics ASAP 2020. The X-ray diffraction (XRD) was used to determine their phases and crystallinity. This was carried out using an Empyrean by Malvern Panalytical, Almelo, equipped with Cu-Kα radiation (λ = 0.154 nm). The phase content of anatase and rutile were calculated with the XRD intensity of the characteristic peaks of the phases [21,22], as shown in Equation (1).

$$W_A = \frac{K_A I_A}{(K_A I_A + I_R)} \tag{1}$$

where $W_A$ is the mole fractions of anatase, $I_A$ and $I_R$ are the X-ray integrated intensities of the anatase and the rutile, respectively, and $K_A$ = 0.886.

The presence of elements in the catalyst and the percentage of each element were determined using X-ray photoelectron spectrometry (XPS) Phoipos 150 (ESCALAB 210, VG Scientific Ltd., East Grinstead, UK) and Raman spectroscopy (Cora 5500 Anton Paar, Anton Paar, Germany). Transmission Electron Microscopy (TEM) has been used to explicate the innermost structure, morphology, and exact particle size of the composite system (FEI TITAN G2 80–300, Hillsborough, OR, USA) operated at 300 keV. The UV–Vis (UV–Visible) spectrophotometer (Shimadzu UV-2600, Kyoto, Japan) was used to determine the energy level of the band gap for all composites of Pt/$TiO_2$-$CeO_2$ photocatalyst synthesized.

## 2.5. Photocatalytic Reaction

The photodegradation experiments of 2,4-D was carried out at room temperature, using a slurry reactor, a glass beaker of 200 mL capacity with 150 mL of a mother solution with 200 ppm of 2,4-Dimethylamine salt, and 200 mg of catalyst mixed with air (BOYUS air pump 4000 B, with pressure of 0.012 MPa and an output of 3.2 L per minute). The reaction was kept in agitation for 30 min in complete darkness until the adsorption desorption equilibrium had reached light striking the reactor with a UV lamp (4 watts). The run time for adsorption tests was 6 h under darkness, at the natural pH of the slurry. An example was obtained every 30 min, using a filtrating syringe to extract 4 mL of the slurry and using a membrane to separate the suspension material. Every example was measured in a UV–Vis (UV-2600 Shimadzu) at 190–400 nm. To obtain the kinetics activity of each material. The concentration of the reaction was calculated from the absorption band at 282 nm, applying the equation of Beer–Lambert. The conversion percentage was calculated using the Equation (2).

$$X_{2,4D} = \frac{2,4D_0 - 2,4D_f}{2,4D_0} \times 100\% \qquad (2)$$

where $X_{2,4D}$ is the percentage of the 2,4-D conversion, $2,4D_0$ is the concentration of the pollutant at the beginning of the reaction, and $2,4D_f$ is the concentration of the pollutant at the end of the reaction.

The heterogenous photocatalysts were carried out by employing the following catalyst: Aeroxide P-25® Commercial $TiO_2$, $TiO_2$-$CeO_2$ (5, 10 wt%), Pt-$TiO_2$-$CeO_2$ (5, 10 wt%) synthesis by impregnation and $TiO_2$-$CeO_2$ (1, 5, 10 wt%), Pt-$TiO_2$-$CeO_2$ (1, 5, 10 wt%) synthesis by sol-gel.

## 3. Results and Discussion

### 3.1. Specific Area by the BET Method

In order to investigate the effect that was created in the surface of the support with the addition of $CeO_2$ in different concentrations, several material characterization techniques were made. Starting with the specific area determined by the BET method, as well as the average pore diameter of the $TiO_2$ and $TiO_2$-$CeO_2$ supports, we can observe that the effect of adding $CeO_2$ to $TiO_2$ increased the pore diameter and the specific area decreased. The impregnation method generated a higher diameter of pores and low specific surface area than the sol-gel method, probably because the sol-gel method had better dispersion of $CeO_2$ in the $TiO_2$. Similar effects have been reported by other authors [23]. Table 1 shows the results of the specific surface area, and the pore diameter of the catalyst support.

Table 1. Specific surface area, and pore diameter for $TiO_2$-$CeO_2$ catalysts support.

| Support | Method | Diameter Pore (Å) | Specific Surface Area (m²/g) |
|---|---|---|---|
| $TiO_2$ | Sol-gel | 52.54 | 185.59 |
| $TiO_2$-$CeO_2$ 1% | Sol-gel | 53.08 | 181.46 |
| $TiO_2$-$CeO_2$ 10% | Sol-gel | 77.51 | 104.36 |
| $TiO_2$-$CeO_2$ 1% | Impregnation | 295.06 | 43.56 |
| $TiO_2$-$CeO_2$ 10% | Impregnation | 304.05 | 43.61 |

### 3.2. X-ray Diffraction (XRD)

An X-ray diffraction characterization test was performed to evaluate the content of the anatase and rutile phases in the supports for both methods. The crystalline phases of $TiO_2$ and $TiO_2$-$CeO_2$ can be seen in the diffraction patterns found in Figure 1. For the Pt-$TiO_2$ catalyst and the mixed oxides $TiO_2$-$CeO_2$ synthesized by sol-gel and impregnation, the presence of the peaks $2\theta$ = 25.19, 37.60, 53.95, 54.36, 62.68, 75.04, and 82.7 are attributed to the anatase phase, corresponding to the plane (JCPDS no. 21-1272). Sol-gel presented the rutile phase whose peaks associated with the phase are $2\theta$ = 70.16 and impregnation in $2\theta$ = 35.81, 41.04, and 70, according to JCPDS with reference number 23-0278. The anatase phase is dominant in both methods due to the heat treatment to which the material was subjected. This is good since the anatase phase is a better photocatalyst than rutile because the exciton diffusion is twice as long [6]. In addition, in the sol-gel method it is possible to observe a decrease in the peaks of the anatase phase due to the increase of $CeO_2$ in the network of the support. For $CeO_2$ the peaks $2\theta = 47.80$, are associated with the cerenite phase that corresponds to a cubic packing of $CeO_2$. With the impregnation method, $CeO_2$ was observed at $2\theta = 27.29, 47.80$, and $56.54$. In $Pt/TiO_2$-$CeO_2$ catalysts, the cerenite phase was observed very little, probably because it is very dispersed within the $TiO_2$ structure, this can be observed in the shift to the right of the characteristic peaks of the $TiO_2$ (JCPDS, no. 04-0802).

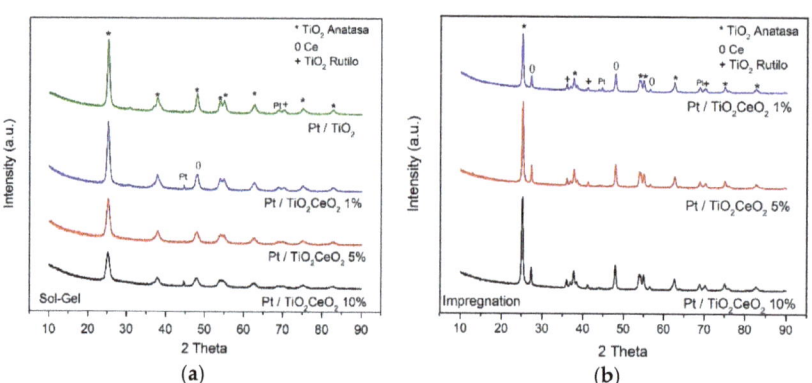

**Figure 1.** XRD of the catalysts $Pt/TiO_2$ and $Pt/TiO_2$-$CeO_2$: (**a**) Sol-gel and (**b**) Impregnation.

### 3.3. Raman Spectroscopy

On the other hand, as seen in Figure 2, in the Raman spectra of the $TiO_2$ supports and the mixed oxides $TiO_2$-$CeO_2$, which were prepared by the sol-gel and impregnation method, peaks corresponding to the anatase phase 398–400, 518–520, and 640 cm$^{-1}$ are observed. Observing a slight Raman shift, which means that $CeO_2$ has been integrated into the structure of the $TiO_2$ support. It is assumed that the $\equiv$Ti-O-Ti$\equiv$ bonds of the corresponding $TiO_2$ network of the anatase phase are disturbed by the presence of cerium oxide, which suggests some substitutions of the $Ti^{4+}$ by $Ce^{4+}$ that form $\equiv$Ti-O-Ce$\equiv$ bonds in the structure of titanium oxide.

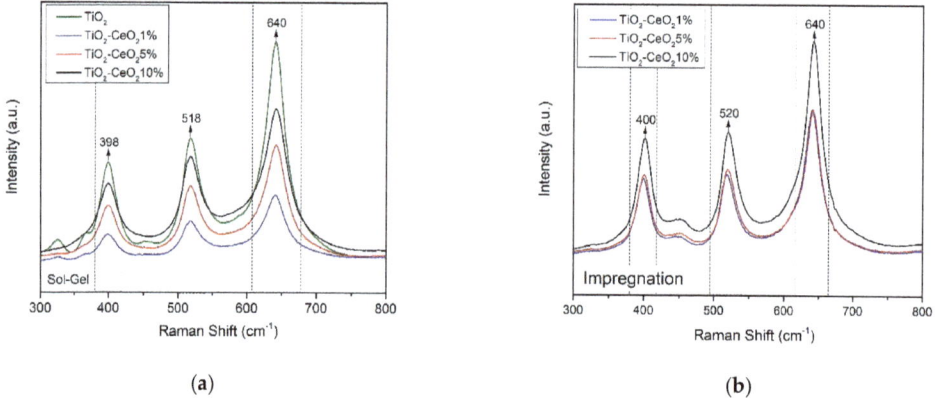

**Figure 2.** Raman spectrum for the support system $TiO_2$, $TiO_2$-$CeO_2$ (1, 5, 10 wt%): (**a**) Sol-gel and (**b**) Impregnation.

## 3.4. Transmission Electron Microscope (TEM)

In the transmission electron microscope (TEM) information obtained is the particle size by analyzing the images in software capable of measuring the diameter of the particle on a nanometric scale. Their respective alpha images of each of the elements and how it is dispersed inside and outside the support is shown in Figure 3. The particle size dispersion histograms were obtained by analyzing the series of data obtained in the TEM images, as shown in Figure 3.

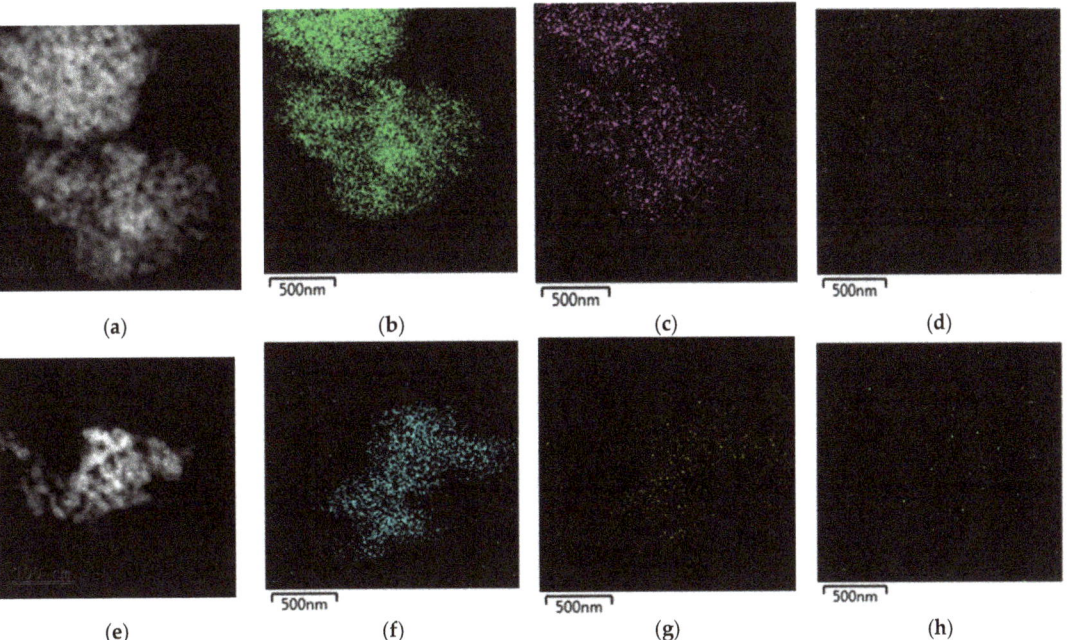

**Figure 3.** Transmission electron microscopy images of the Pt/TiO$_2$-CeO$_2$ 10%: (**a**) Sol-gel to 100 nm scale, (**b**) Sol-gel take α of the Ti, (**c**) Sol-Gel take α of the Ce, (**d**) Sol-Gel take α of the Pt, (**e**) Impregnation to 100 nm scale, (**f**) Impregnation take α of the Ti, (**g**) Impregnation take α of the Ce, and (**h**) Impregnation take α of the Pt.

Figure 4 shows how the particle size decreases, with the addition of CeO$_2$ to the network of the support, that can suggest an increment of the specific surface area of the support. The micrographs reveal that the sizes of the metallic particles for the photocatalysts ranged from 2 to 6 nm. The smallest Pt particle size was observed in Pt/TiO$_2$ CeO$_2$ 10 wt%.

## 3.5. X-ray Photoelectron Spectrometry (XPS)

The binding energy and the atomic ratios of Pt, Ti and Ce, for the Pt/TiO$_2$-CeO$_2$ catalysts (5 and 10 wt%) prepared by impregnation and sol-gel are reported in Table 2. The relative abundance of the Pt$^0$– Pt$^{2+}$ and Ce$^{3+}$ Ce$^{4+}$ species were calculated from the area under the curve of the respective peaks of the XPS spectra for the different catalysts (Figure 5). In Table 2, the corresponding binding energies for Pt 4f$_{(7/2)}$ are shown; the values of the binding energies for the Pt/TiO$_2$ and Pt/TiO$_2$-CeO$_2$ catalysts are around 73.0 to 75.9 eV corresponding to Pt$^0$ and Pt$^{2+}$. A shift in the binding energy towards higher energies can be observed with an increase the amount of cerium oxide in the Pt/TiO$_2$-CeO$_2$ catalysts (5% by weight and 10%). This is due mainly to the fact that CeO$_2$ is considered as an oxygen supplier which makes the platinum species in the reduced state Pt$^0$ transform to the oxidized species of Pt$^{2+}$ [24–26].

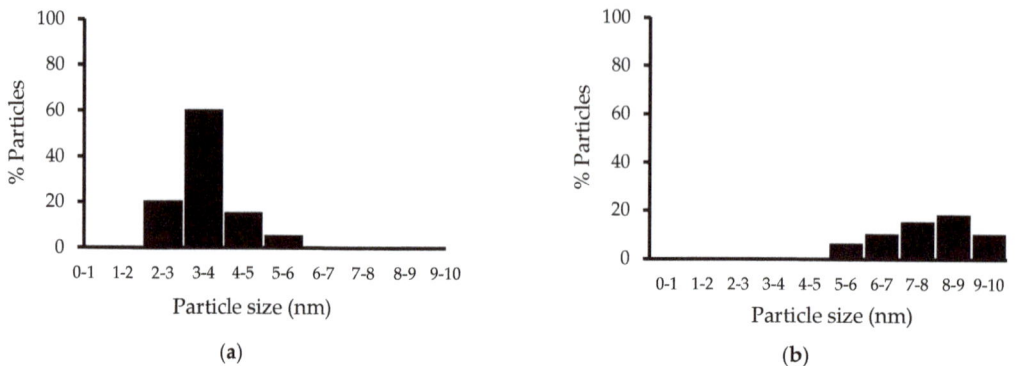

**Figure 4.** Size particle distribution of the Pt determinates by TEM for the photocatalyst Pt/TiO$_2$-CeO$_2$ 10 wt% (**a**) Sol-gel and (**b**) Impregnation.

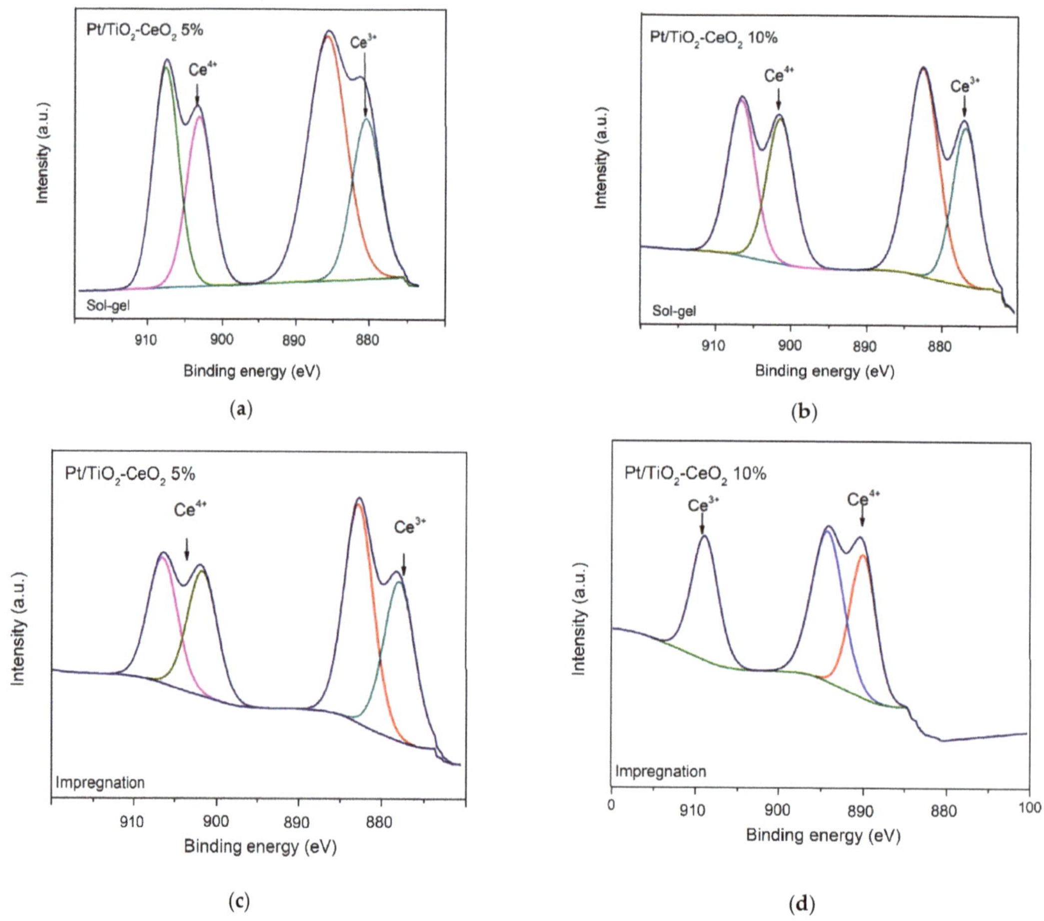

**Figure 5.** XPS spectra (Ce 3d region) for photocatalyst (**a**) Sol-gel Pt/TiO$_2$-CeO$_2$ 5%, (**b**) Sol-gel Pt/TiO$_2$-CeO$_2$ 10%, (**c**) Impregnation Pt/TiO$_2$-CeO$_2$ 5%, and (**d**) Impregnation Pt/TiO$_2$-CeO$_2$ 10%.

Table 2. Binding energy and relative abundance of the different species obtained by XPS of the catalysts.

| Support | Method | Binding Energy (eV) | | | Relative Abundance (%) | | |
|---|---|---|---|---|---|---|---|
| | | Pt (4f$_{7/2}$) | Ti (2p$_{3/2}$) | Ce (3d$_{5/2}$) | Pt$^0$– Pt$^{2+}$ | Ti$^{4+}$ | Ce$^{3+}$ Ce$^{4+}$ |
| Pt/TiO$_2$ | Sol-gel | 73.0<br>75.9 | 458.1 | - | 80–20 | 100 | - |
| Pt/TiO$_2$-CeO$_2$ 5% | Sol-gel | 75.4<br>77.09–78.09 | 458 | 880–900 | 53–47 | 100 | 56–44 |
| Pt/TiO$_2$-CeO$_2$ 10% | Sol-gel | 75.9<br>77.09–78.09 | 458.1 | 881–900.1 | 47–53 | 100 | 52–47 |
| Pt/TiO$_2$-CeO$_2$ 5% | Impregnation | - | 457.7 | 880–900 | - | 100 | 48.8–51.19 |
| Pt/TiO$_2$-CeO$_2$ 10% | Impregnation | - | 465 | 880–900 | - | 100 | 38–62 |

Table 2 reports the binding energies for TiO$_2$, which can have values ranging between 457.7–458.1 eV [27], as shown in Figure 6, which indicates that there was no modification due to the doping effect with the CeO$_2$ content, nor with the preparation method of the Pt supports. The binding energy was also determined for the Ce 3d$_{5/2}$ level (Table 2 and Figure 6); it was found in the region of 870–920 eV [28]. Relative abundance calculated from these XPS spectra showed that Ce$^{4+}$ (oxidized) species increased with increasing CeO$_2$ content relative to Ce$^{3+}$ (reduced). Coinciding with Rocha et al. (2015), it is possible to observe that at a lower concentration of CeO$_2$, a greater number of atoms in the Ce$^{3+}$ oxidation state will be obtained.

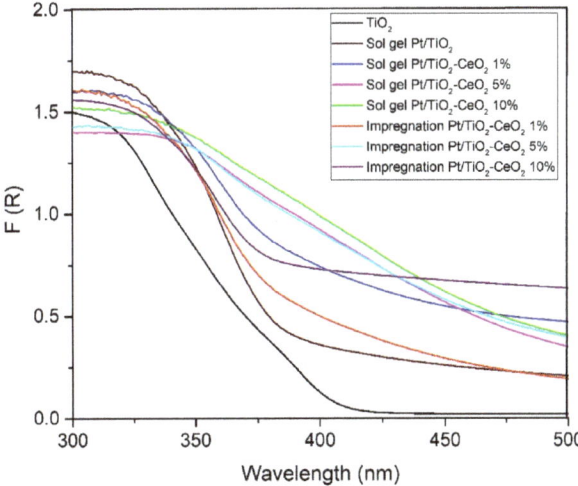

Figure 6. UV–vis spectra for the TiO$_2$-CeO$_2$ supports Sol-Gel and Impregnation.

3.6. UV–Vis for Band Gap

The photophysical properties UV–Vis absorption spectra of the catalysts were evaluated to investigate the effect of CeO$_2$ on the support network. Figure 6 shows the spectra of UV–Vis materials by diffuse reflectance for sol-gel and impregnation methods. All samples have a shift between these wavelengths, which can be attributed to the transitions of the Ti-O electrons of the TiO$_2$ and TiO$_2$-CeO$_2$ nanocrystals.

Table 3 shows the results where a change in activation energy (3.02–2.8 eV) was observed for the TiO$_2$-CeO$_2$ samples from 1% to 5% by weight of CeO$_2$, compared to the reference TiO$_2$ in anatase phase (3.4 eV). The band gap energies were calculated by a linear fit of the slope to the abscissa and are reported in Table 3. It diminished from 3.45 eV, for the bare TiO$_2$, to 2.82 eV, for the TiO$_2$-CeO$_2$ (at 5 wt%) sample. It is evident that cerium oxide modifies the bulk semiconductor properties of TiO$_2$. The shift of the Eg band gap to a lower energy can be attributed to the incorporation of $Ce^{4+}$ cations, which substitute some $Ti^{4+}$ cations.

Table 3. Band Gap Energy and Wavelengths.

| Catalyst ID Name | Band Gap (eV) | Wavelengths (nm) |
|---|---|---|
| $TiO_2$ | 3.45 | 359 |
| $Pt/TiO_2$ | 3.39 | 365 |
| $Pt/TiO_2$-$CeO_2$- 1% | 3.05 | 406 |
| $Pt/TiO_2$-$CeO_2$- 5% | 2.82 | 439 |

### 3.7. Photocatalysts Degradation of 2,4-Dichlorophenoxyacetic Acid

The photocatalytic degradation reactions of 2,4-D acid were carried out at room temperature at 298 K for 6 h, with a concentration of 200 ppm of the reagent, followed by the UV absorption band of 283 that corresponds mainly to the transition electron n → π*, which is mainly attributed to the C-Cl bond [29,30]. The percentage conversion as a function of time for both supports and catalysts impregnated and prepared by the sol-gel method at 360 min of reaction are shown in Figure 7 and Table 4.

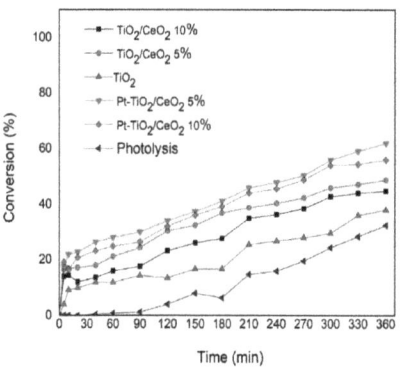

(a)

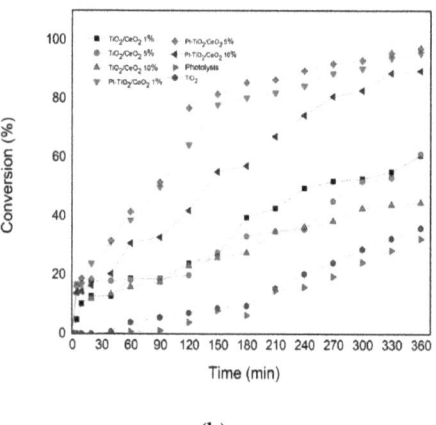

(b)

Figure 7. Photocatalysts degradation of 2,4-Dichlorophenoxyacetic acid. (a) Impregnation and (b) Sol-Gel materials.

Table 4. Photocatalysts degradation of 2,4 Dichlorophenoxiacetyc acid.

| Catalysts | Method | Pt (wt%) | X% | $C_f$ (ppm) |
|---|---|---|---|---|
| $TiO_2$ | Commercial | - | 38 | 160 |
| $TiO_2$-$CeO_2$ 5% | Impregnation | - | 49 | 128 |
| $TiO_2$-$CeO_2$ 10% | Impregnation | - | 45 | 138 |
| Pt-$TiO_2$-$CeO_2$ 5% | Impregnation | 1 | 62 | 95 |
| Pt-$TiO_2$-$CeO_2$ 10% | Impregnation | 1 | 56 | 110 |
| $TiO_2$ | Sol-Gel | - | 38 | 155 |
| $TiO_2$-$CeO_2$ 1% | Sol-Gel | - | 61 | 98 |
| $TiO_2$-$CeO_2$ 5% | Sol-Gel | - | 61 | 66 |
| $TiO_2$-$CeO_2$ 10% | Sol-Gel | - | 45 | 138 |
| Pt-$TiO_2$-$CeO_2$ 1% | Sol-Gel | 1 | 95 | 11 |
| Pt-$TiO_2$-$CeO_2$ 5% | Sol-Gel | 1 | 97 | 7 |
| Pt-$TiO_2$-$CeO_2$ 10% | Sol-Gel | 1 | 89 | 27 |
| Photolysis | - | - | 32 | 169 |

The photocatalytic degradation of 2,4-D in the absence of support or catalyst had a conversion of 32% while the maximum conversion reached was 95% and 97% for the Pt/TiO$_2$-CeO$_2$ 1% and Pt/TiO$_2$-CeO$_2$ catalysts 5% prepared by the sol-gel method (Figure 7B), and the catalysts prepared by the impregnation method reached a maximum of 62% conversion. On the other hand, the supports prepared by impregnation reached a maximum of 49% TiO$_2$ while those prepared by the sol-gel method reached up to 61% (Figure 7A). The highest yield achieved in the catalysts prepared by sol-gel Pt/TiO$_2$-CeO$_2$ 1% and Pt/TiO$_2$-CeO$_2$ 5% could be attributed to an optimal concentration of CeO$_2$, which allows the insertion within the CeO$_2$ of the TiO$_2$ and leads to the deformation of the lattice, modifying the mobility of the oxygen atoms and favoring the oxidation-reduction process [26]. In contrast, the results of XPS in the Pt/TiO$_2$-CeO$_2$ 1% and Pt/TiO$_2$-CeO$_2$ 5% catalysts showed that the proportion of oxidized species of Pt$^{2+}$ and Ce$^{4+}$ are essential to function as oxygen scavengers, which are important in oxidation-reduction processes. Additionally, the smallest particle size in the catalysts prepared with the supports by sol-gel was in a range of 2 to 6 nm. This is due to a greater specific area due to a good integration of CeO$_2$, which favors a better dispersion of the metallic nanoparticles on the surface of the supports, favoring the catalytic activity in the degradation of 2,4-D.

## 4. Conclusions

In the present work, the TiO$_2$ and TiO$_2$-CeO$_2$ supports, prepared by the sol-gel method and increasing the CeO$_2$ concentration in a 1–10 ratio in the TiO$_2$ support network, significantly increased the pore diameter, affecting the specific surface area for the catalyst. On the other hand, in the supports prepared by impregnation, no important modification was observed, either in the area or in the pore diameter due to the addition of CeO$_2$, since these remained constant. However, when comparing the results of both materials we can conclude that sol-gel supports can obtain pore diameters four times smaller than those obtained with impregnation. By having less exposed area, the Pt catalyst particles will be larger because they tend to agglomerate, as they do not have enough space to disperse efficiently. Affecting the catalytic activity of the material, the Pt particles, being well dispersed, favored the catalytic activity of the material. Another important fact is that it was possible to obtain Pt nanoparticles on the sol-gel supports in the order of 2 and 6 nm, dependent of the CeO$_2$ content in the support. A cerium oxide shift in the energy band gap was observed in the Pt/TiO$_2$-CeO$_2$ photocatalysts. It is proposed that the high activity showed by the Pt/TiO$_2$-CeO$_2$ photo-catalysts can be due to a synergetic effect between the cerium oxide and the platinum of oxidizing agent.

**Author Contributions:** Methodology, I.A.M.E. and C.A.G.G.; formal analysis, A.P.L., I.A.M.E. and C.A.G.G.; writing—original draft preparation, I.A.M.E. and B.S.-R.; writing—review and editing, B.S-R.; supervision, C.A.G.G. All authors have read and agreed to the published version of the manuscript.

**Funding:** This research received no external funding.

**Institutional Review Board Statement:** Not applicable.

**Informed Consent Statement:** Not applicable.

**Data Availability Statement:** Not applicable.

**Acknowledgments:** A graduate scholarship for Ixchel Alejandra Mejía-Estrella was provided by the National Council for Science and Technology (CONACyT) of México.

**Conflicts of Interest:** The authors declare no conflict of interest.

## References

1. Gopinath, K.; Madhav, N.; Krishnan, A.; Malolan, R.; Rangarajan, G. Present applications of titanium dioxide for the photocatalytic removal of pollutants from water: A review. *J. Environ. Manag.* **2020**, *270*, 110906. [CrossRef] [PubMed]
2. Borges, M.E.; Sierra, M.; Cuevas, E.; García, R.D.; Esparza, P. Photocatalysis with solar energy: Sunlight-responsive photocatalyst based on TiO$_2$ loaded on a natural material for wastewater treatment. *Sol. Energy* **2016**, *135*, 527–535. [CrossRef]

3. Van Deelen, T.W.; Hernández Mejía, C.; de Jong, K.P. Control of metal-support interactions in heterogeneous catalysts to enhance activity and selectivity. *Nat. Catal.* **2019**, *2*, 955–970. [CrossRef]
4. Yunarti, R.T.; Isa, I.D.; Dimonti, L.C.C.; Dwiatmoko, A.A.; Ridwan, M.; Ha, J.-M. Study of $Ag_2O/TiO_2$ nanowires synthesis and characterization for heterogeneous reduction reaction catalysis of 4-nitrophenol. *Nano-Struct. Nano-Objects* **2021**, *26*, 100719. [CrossRef]
5. Ayesha, B.; Jabeen, U.; Naeem, A.; Kasi, P.; Malghani, M.N.K.; Khan, S.U.; Akhtar, J.; Aamir, M. Synthesis of zinc stannate nanoparticles by sol-gel method for photocatalysis of commercial dyes. *Results Chem.* **2020**, *2*, 100023. [CrossRef]
6. Fonseca-Cervantes, O.; Pérez-Larios, A.; Romero Arellano, V.; Sulbaran-Rangel, B.; Guzmán González, C. Effects in Band Gap for Photocatalysis in $TiO_2$ Support by Adding Gold and Ruthenium. *Processes* **2020**, *8*, 1032. [CrossRef]
7. Pragathiswaran, C.; Smitha, C.; Mahin Abbubakkar, B.; Govindhan, P.; Anantha Krishnan, N. Synthesis and characterization of $TiO_2$/ZnO–Ag nanocomposite for photocatalytic degradation of dyes and anti-microbial activity. *Mater. Today Proc.* **2021**, *45*, 3357–3364. [CrossRef]
8. Tian, J.; Sang, Y.; Zhao, Z.; Zhou, W.; Wang, D.; Kang, X.; Liu, H.; Wang, J.; Chen, S.; Cai, H.; et al. Enhanced Photocatalytic Performances of $CeO_2/TiO_2$ Nanobelt Heterostructures. *Small* **2013**, *9*, 3864–3872. [CrossRef]
9. Chatterjee, D.; Dasgupta, S. Visible light indu ced photocatalytic degradation of organic pollutants. *J. Photochem. Photobiol. C Photochem. Rev.* **2005**, *6*, 186–205. [CrossRef]
10. Matte, L.P.; Kilian, A.S.; Luza, L.; Alves, M.C.M.; Morais, J.; Baptista, D.L.; Dupont, J.; Bernardi, F. Influence of the $CeO_2$ Support on the Reduction Properties of $Cu/CeO_2$ and $Ni/CeO_2$ Nanoparticles. *J. Phys. Chem. C* **2015**, *119*, 26459–26470. [CrossRef]
11. Bellardita, M.; Fiorenza, R.; Urso, L.; Spitaleri, L.; Gulino, A.; Compagnini, G.; Scirè, S.; Palmisano, L. Exploring the Photothermo-Catalytic Performance of Brookite $TiO_2$-$CeO_2$ Composites. *Catalysts* **2020**, *10*, 765. [CrossRef]
12. Henych, J.; Šťastný, M.; Němečková, Z.; Mazanec, K.; Tolasz, J.; Kormunda, M.; Ederer, J.; Janoš, P. Bifunctional $TiO_2/CeO_2$ reactive adsorbent/photocatalyst for degradation of bis-p-nitrophenyl phosphate and CWAs. *Chem. Eng. J.* **2021**, *414*, 128822. [CrossRef]
13. Liu, B.; Zhao, X.; Zhang, N.; Zhao, Q.; He, X.; Feng, J. Photocatalytic mechanism of $TiO_2$-$CeO_2$ films prepared by magnetron sputtering under UV and visible light. *Surf. Sci.* **2005**, *595*, 203–211. [CrossRef]
14. Gnanasekaran, L.; Rajendran, S.; Priya, A.K.; Durgalakshmi, D.; Vo, D.-V.N.; Cornejo-Ponce, L.; Gracia, F.; Soto-Moscoso, M. Photocatalytic degradation of 2,4-dichlorophenol using bio-green assisted $TiO_2$-$CeO_2$ nanocomposite system. *Environ. Res.* **2021**, *195*, 110852. [CrossRef]
15. García-Domínguez, Á.E.; Torres-Torres, G.; Arévalo-Pérez, J.C.; Silahua-Pavón, A.; Sánchez-Trinidad, C.; Godavarthi, S.; Ojeda-López, R.; Sierra-Gómez, U.A.; Cervantes-Uribe, A. Urea assisted synthesis of $TiO_2$-$CeO_2$ composites for photocatalytic acetaminophen degradation via simplex-centroid mixture design. *Results Eng.* **2022**, *14*, 100443. [CrossRef]
16. Petrović, S.; Stanković, M.; Pavlović, S.; Mojović, Z.; Radić, N.; Mojović, M.; Rožić, L. Nickel oxide on mechanochemically synthesized $TiO_2$-$CeO_2$: Photocatalytic and electrochemical activity. *React. Kinet. Mech. Catal.* **2021**, *133*, 1097–1110. [CrossRef]
17. Khan, M.A.R.; Mamun, M.S.A.; Ara, M.H. Review on platinum nanoparticles: Synthesis, characterization, and applications. *Microchem. J.* **2021**, *171*, 106840. [CrossRef]
18. Jeyaraj, M.; Gurunathan, S.; Qasim, M.; Kang, M.-H.; Kim, J.-H. A Comprehensive Review on the Synthesis, Characterization, and Biomedical Application of Platinum Nanoparticles. *Nanomaterials* **2019**, *9*, 1719. [CrossRef]
19. Guzmán, C.; del Ángel, G.; Gómez, R.; Galindo-Hernández, F.; Ángeles-Chavez, C. Degradation of the herbicide 2,4-dichlorophenoxyacetic acid over $Au/TiO_2$-$CeO_2$ photocatalysts: Effect of the $CeO_2$ content on the photoactivity. *Catal. Today* **2011**, *166*, 146–151. [CrossRef]
20. Guzmán, C.; Del Angel, G.; Fierro, J.L.G.; Bertin, V. Role of Pt Oxidation State on the Activity and Selectivity for Crotonaldehyde Hydrogenation Over Pt–Sn/$Al_2O_3$–La and Pt–Pb/$Al_2O_3$–La Catalysts. *Top. Catal.* **2010**, *53*, 1142–1144. [CrossRef]
21. Zhang, H.; Banfield, J.F. Understanding Polymorphic Phase Transformation Behavior during Growth of Nanocrystalline Aggregates: Insights from $TiO_2$. *J. Phys. Chem. B* **2000**, *104*, 3481–3487. [CrossRef]
22. He, Z.; Cai, Q.; Fang, H.; Situ, G.; Qiu, J.; Song, S.; Chen, J. Photocatalytic activity of $TiO_2$ containing anatase nanoparticles and rutile nanoflower structure consisting of nanorods. *J. Environ. Sci.* **2013**, *25*, 2460–2468. [CrossRef]
23. Rocha-Ortiz, G.; Tessensohn, M.E.; Salas-Reyes, M.; Flores-Moreno, R.; Webster, R.D.; Astudillo-Sánchez, P.D. Homogeneous electron-transfer reaction between anionic species of anthraquinone derivatives and molecular oxygen in acetonitrile solutions: Electrochemical properties of disperse red 60. *Electrochim. Acta* **2020**, *354*, 136601. [CrossRef]
24. Yang, W.-D.; Hsu, Y.-C.; Lin, W.-C.; Huang, I.-L. Characterization and photocatalytic activity of N and Pt doped titania prepared by microemulsion technique. *Adv. Mater. Sci.* **2018**, *3*, 1–5. [CrossRef]
25. Dauscher, A.; Hilaire, L.; Le Normand, F.; Müller, W.; Maire, G.; Vasquez, A. Characterization by XPS and XAS of supported $Pt/TiO_2$—$CeO_2$ catalysts. *Surf. Interface Anal.* **1990**, *16*, 341–346. [CrossRef]
26. Rocha, M.A.L.; Del Ángel, G.; Torres-Torres, G.; Cervantes, A.; Vázquez, A.; Arrieta, A.; Beltramini, J.N. Effect of the Pt oxidation state and $Ce^{3+}/Ce^{4+}$ ratio on the $Pt/TiO_2$-$CeO_2$ catalysts in the phenol degradation by catalytic wet air oxidation (CWAO). *Catal. Today* **2015**, *250*, 145–154. [CrossRef]
27. Thermo Fisher Scientific. Titanium Transition Metal Primary XPS Region: Ti2p. Available online: https://xpssimplified.com/elements/titanium.php (accessed on 24 July 2022).

28. Thermo Fisher Scientific. Platinum Transition Metal. Primary XPS Region: Pt4f. Available online: https://xpssimplified.com/elements/platinum.php#appnotes (accessed on 24 July 2022).
29. Ramos-Ramírez, E.; Gutiérrez-Ortega, N.L.; Tzompantzi-Morales, F.; Barrera-Rodríguez, A.; Castillo-Rodríguez, J.C.; Tzompantzi-Flores, C.; Santolalla-Vargas, C.E.; Guevara-Hornedo, M.d.P. Photocatalytic Degradation of 2,4-Dichlorophenol on NiAl-Mixed Oxides Derivatives of Activated Layered Double Hydroxides. *Top. Catal.* **2020**, *63*, 546–563. [CrossRef]
30. Ba-Abbad, M.M.; Kadhum, A.A.H.; Mohamad, A.B.; Takriff, M.S.; Sopian, K. Photocatalytic degradation of chlorophenols under direct solar radiation in the presence of ZnO catalyst. *Res. Chem. Intermed.* **2013**, *39*, 1981–1996. [CrossRef]

Article

# Versatile Zirconium Oxide (ZrO₂) Sol-Gel Development for the Micro-Structuring of Various Substrates (Nature and Shape) by Optical and Nano-Imprint Lithography

Nicolas Crespo-Monteiro *, Arnaud Valour, Victor Vallejo-Otero, Marie Traynar, Stéphanie Reynaud, Emilie Gamet and Yves Jourlin *

Université de Lyon, Université Jean-Monnet-Saint Etienne, Laboratoire Hubert Curien, UMR CNRS 5516, 42000 Saint-Etienne, France
* Correspondence: nicolas.crespo.monteiro@univ-st-etienne.fr (N.C.-M.); yves.jourlin@univ-st-etienne.fr (Y.J.)

**Abstract:** Zirconium oxide ($ZrO_2$) is a well-studied and promising material due to its remarkable chemical and physical properties. It is used, for example, in coatings for corrosion protection layer, wear and oxidation, in optical applications (mirror, filters), for decorative components, for anti-counterfeiting solutions and for medical applications. $ZrO_2$ can be obtained as a thin film using different deposition methods such as physical vapor deposition (PVD) or chemical vapor deposition (CVD). These techniques are mastered but they do not allow easy micro-nanostructuring of these coatings due to the intrinsic properties (high melting point, mechanical and chemical resistance). An alternative approach described in this paper is the sol-gel method, which allows direct micro-nanostructuring of the $ZrO_2$ layers without physical or chemical etching processes, using optical or nano-imprint lithography. In this paper, the authors present a complete and suitable $ZrO_2$ sol-gel method allowing to achieve complex micro-nanostructures by optical or nano-imprint lithography on substrates of different nature and shape (especially non-planar and foil-based substrates). The synthesis of the $ZrO_2$ sol-gel is presented as well as the micro-nanostructuring process by masking, colloidal lithography and nano-imprint lithography on glass and plastic substrates as well as on plane and curved substrates.

**Keywords:** zirconium oxide; sol-gel; optical lithography; nano-imprint lithography; non-planar substrates; plastic

Citation: Crespo-Monteiro, N.; Valour, A.; Vallejo-Otero, V.; Traynar, M.; Reynaud, S.; Gamet, E.; Jourlin, Y. Versatile Zirconium Oxide (ZrO₂) Sol-Gel Development for the Micro-Structuring of Various Substrates (Nature and Shape) by Optical and Nano-Imprint Lithography. *Materials* 2022, *15*, 5596. https://doi.org/10.3390/ma15165596

Academic Editor: Aleksej Zarkov

Received: 15 July 2022
Accepted: 9 August 2022
Published: 15 August 2022

**Publisher's Note:** MDPI stays neutral with regard to jurisdictional claims in published maps and institutional affiliations.

**Copyright:** © 2022 by the authors. Licensee MDPI, Basel, Switzerland. This article is an open access article distributed under the terms and conditions of the Creative Commons Attribution (CC BY) license (https://creativecommons.org/licenses/by/4.0/).

## 1. Introduction

Zirconium oxide ($ZrO_2$) is an intensively studied and used material due to its many remarkable physical and chemical properties. It has a high melting point (2700 °C), high hardness (between 11 and 18 GPa depending on the phase) [1,2], good chemical resistance [2,3], biocompatibility [4], high refractive index (2.1 at 633 nm) [5], wide band gap (5 eV) [6], high transparency in the visible and near-infrared range [7] and photoluminescence properties [8,9]. Due to its numerous properties, zirconium oxide is used in many applications such as protective coatings against corrosion, wear and oxidation [10,11], in optical applications (mirror, filters, etc.) [12–15], in anti-counterfeiting solutions [9] and in health applications such as the dental field [16,17].

There are many methods to synthesize $ZrO_2$, among which we can mention reactive sputtering [18–20], chemical vapor deposition [21,22] and atomic layer deposition [23–25]. These techniques are well known in thin film deposition processes and are widely described in the literature. However, they do not allow micro- and nano-structuring in a simple way (without etching step) of these coatings to obtain complex patterns (according to the shapes, micro-nanostructures, etc.), which limits their use as well as their properties. Another process to elaborate $ZrO_2$ thin films is the sol-gel method [26–31]. The sol-gel method has the advantage of micro-nanostructuring the films by optical lithography [29,31] or by nano-imprint lithography [32,33]. The optical lithography presents the advantage of being able to micro-nanostructure variable substrates of various shapes [29,30] and sizes [34,35] without having to resort to etching processes. Among the lithography processes, the masking lithography consists in obtaining a pattern by exposing it through a mask, letting the light pass through in the opened areas. This allows to structure quickly patterns on planar and non-planar substrates. The colloidal lithography consists in using silica nano-spheres to create photonic nano-jets underneath the layer of nano-spheres. These photonic nano-jets allow a field enhancement of the incident field, allowing the thin film micro-structuration according to the hexagonal and periodic pattern whose period is equal to the sphere diameter. If the material behaves like a positive resist, we achieve micro-holes [36] and if it behaves like a negative photoresist, we obtain nano-pillars [35,37]. Another method is nano-imprint lithography (NIL), which consists of patterning a layer by pressing a stamp (mould) on it [33]. This method has the advantage of being very fast, low cost and being able to adapt to a great number of supports and patterns. Nevertheless, it requires a sol-gel solution which can be patterned and stabilized by a thermal or UV treatment.

In this paper, we demonstrate how our $ZrO_2$ sol-gel can be used to obtain both complex patterns (shapes, micro-nanostructures, etc.) by optical lithography (mask lithography, colloidal lithography) and by nano-imprint lithography. We also show the possibility of using this versatile sol-gel and the associated structuring methods to structure complex patterns on variable substrates in their nature and geometry.

## 2. Materials and Methods

### 2.1. Sol-Gel

To elaborate photo-patternable films, a mixture of two sols with different reactivities has been prepared. The first sol (sol 1) presents low chemical reactivity due to chelation of the alkoxide groups by BzAc. The second sol (sol 2) was prepared according to the procedure described thereafter, which yielded to sols very stable over time when stored, but very reactive when used for the fabrication of metallic oxide thin films. Sol 1 was prepared by reacting Zirconium (IV) propoxide ($Zr(OPr)_4$ from Fluka) with 1-Benzoylacetone (BzAc from Aldrich) in anhydrous ethyl alcohol (EtOH from Aldrich). The $Zr(OPr)_4$/BzAc/EtOH molar composition was 1/0.9/20. Sol 2 was obtained by mixing $Zr(OPr)_4$ with deionized water, hydro-chloric acid (HCl from Roth), and butyl alcohol (BuOH from Merck) as a solvent. The $Zr(OPr)_4$ concentration in the solution was 0.4 M, and the $Zr(OPr)_4/H_2O/HCl/BuOH$ molar composition was 1/0.8/0.13/24. The sol was aged at room temperature for 2 days before being used. Finally, the photo-sensitive solution was prepared by mixing sols 1 and 2 to obtain a final sol with a $Zr(OPr)_4$ concentration of 0.6 M and a $BzAc/Zr(OPr)_4$ molar ratio of 0.6.

Then, the sol was deposited by using the spin-coating technique at a speed of 3000 rpm for 60 s, before being heated at 110 °C for 90 min, resulting in a so-called xerogel film, i.e., an inorganic polymer film made of Zr-O-Zr chains with organic chain-end groups arising from the sol formulation, mainly Zr-BzAc complexed species. The xerogel films are soluble in alcohol as far as BzAc stays complexed with $Zr(OPr)_4$. The main interest of this protocol relies on the properties of BzAc, which makes the film soluble in a solvent while being sensitive to UVA light. Indeed, under UVA illumination, the Zr-BzAc complex is partially degraded into insoluble species. Therefore, it will create a contrast of solubility, widely described in the literature [38], between illuminated and non-illuminated areas when it is selectively exposed to UVA light, allowing us to easily structure our films at different scales.

### 2.2. Optical Lithography
#### 2.2.1. Macroscopic Mask Lithography

The pattern-based masks are macroscopic patterns made from a binary black-and-white image printed on a transparent plastic sheet using an inkjet printer, allowing various macroscopic patterns. In the first step, a ZrO$_2$ xerogel layer was deposited by spin coating at 3000 rpm for 1 min. The mask was positioned above the xerogel layer during exposure to UV light. The whole area was irradiated using a UV lamp at a wavelength of 365 nm for 5 min at 200 mW/cm$^2$. Only the transparent areas of the mask let the UV light pass through, thereby allowing the pattern to be transferred onto the xerogel film after development

2.2.2. Colloidal Lithography

Silica microspheres of 1 µm diameter (in ethanol suspension (96% *v/v*) micro-mod) functionalized with a hydrophobic acrylate surface were deposited on ZrO$_2$ xerogel thin films covered with a poly(methyl methacrylate) (PMMA) layer according to the Langmuir Blodgett (LB) approach [39]. To achieve this monolayer of silica microspheres deposited on the xerogel film, an LB machine (KSV NIMA LB) (Biolin Scientific) was used. The silica microsphere monolayer was spread on the aqueous sub-phase at room temperature and left for 10 min in order to let the solvent evaporate. After compression of the silica microsphere monolayer at a barrier speed of 3 mm/min using the LB machine, the microspheres were deposited on the thin film at a surface pressure of 40 mN·m$^{-1}$ using the dipping method with a withdrawal speed of 3 mm/min. The ZrO$_2$ xerogel layer was deposited by spin coating at 3000 rpm for 1 min and the PMMA layer was deposited by spin coating at 6000 rpm for 1 min. The PMMA layer was used to protect the ZrO$_2$ thin film from water and to allow UV to pass through during film exposure. After deposition, the microspheres were illuminated at a wavelength of 365 nm for 90 s at 100 mW/cm$^2$ in order to obtain ZrO$_2$ nano-pillars with a 2D hexagonal arrangement [35].

*2.3. Nano-Imprint Lithography*

Polydimethylsiloxane (PDMS) stamps with sinusoidal micro-nanostructures of 800 nm period and 60 nm deep are used to micro-nanostructure the ZrO$_2$ xerogel films. After deposition of the ZrO$_2$ xerogel films by spin-coating at 4000 rpm for 30 s, PDMS stamp was applied to the ZrO$_2$ xerogel films in a humidity- and temperature-controlled environment (20 °C and 50% humidity) under 1 bar of pressure for 3 min. Afterwards, a UV illumination at a wavelength of 365 nm for 5 min at 200 mW/cm$^2$ is used to stabilize the patterned ZrO$_2$ films.

*2.4. Characterizations*

The film structure was analyzed using Raman micro-spectroscopy (LabRam ARAMIS from Horiba Jobin Yvon company, Kyoto, Japan) with an excitation wavelength at 633 nm (He-Ne laser) and with a He-Cd laser (325 nm). The micro-nanostructured thin films were characterized by atomic force microscopy (AFM) measurements (Dimension Icon from Bruker company, Billerica, MA, USA)) in tapping mode with a tip AppNano (ACTA) and by scanning electron microscopy (SEM) using low vacuum mode coupled and LVSED detector with a JSM-IT80 from JEOL. The film thicknesses were measured by a profilometer Veeco Dektak 3 ST.

## 3. Results and Discussion

*3.1. ZrO$_2$ Layers*

After spin-coating and drying at room temperature to evaporate solvents, the ZrO$_2$ layer was analyzed using Raman spectroscopy. Figure 1 shows the Raman spectrum of the ZrO$_2$ xerogel thin film deposited on silica substrate after annealing at 110 °C for 90 min. One can observe that the ZrO$_2$ layer has an amorphous phase. Moreover, no peaks characteristic of a ZrO$_2$ crystal phase were observed by XRD analysis (figure not shown), confirming the amorphous nature of the films. Indeed, the corresponding Raman spectrum (Figure 1) shows multiple peaks but no features of crystallized ZrO$_2$. According to Oda et al. [37], the strong peaks at 1598 and 1000 cm$^{-1}$ are assigned to 8b and 12 vibration modes of the phenyl group of BzAc and the peaks at 1297 and 1310 cm$^{-1}$ are assigned to

C=C=C symmetric vibrations in chelating rings [40]. The others small Raman peaks are supposed to be related to a BzAc or ZrO$_2$/BzAc complex.

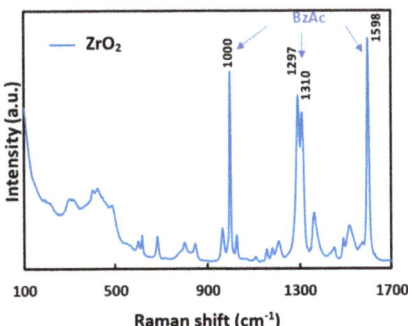

**Figure 1.** Raman spectrum of a ZrO$_2$ xerogel thin film deposited on SiO$_2$ substrate.

Apart from Raman spectroscopy, the ZrO$_2$ amorphous thin film was also analyzed with UV-Vis-NIR spectroscopy in the wavelength range of 200–2000 nm. Figure 2 shows the ZrO$_2$ xerogel thin film spectra. The black and blue curves are, respectively, the transmission and reflection spectra of ZrO$_2$ xerogel film deposited on silica substrate. The ZrO$_2$ layers are transparent and slightly yellowish (Figure 2). Transmittance analyses carried out reveal a transparent oscillating region in the visible and near infrared with a maximum transmittance higher than 85% on the one hand, and a typical absorption of ZrO$_2$ around 380 nm, where the transmittance decreases drastically, on the other hand [41]. The signal at 290 nm originates from the chelate ring (Zr(OC$_4$H$_9$)$_4$ + BzAc) [41]. The reflectance curve of the ZrO$_2$ film is in agreement with the results in transmission, with a reflectance lower than 15% and a slight purple color in reflection. The thickness of the layer was measured with a profilometer close to 300 nm.

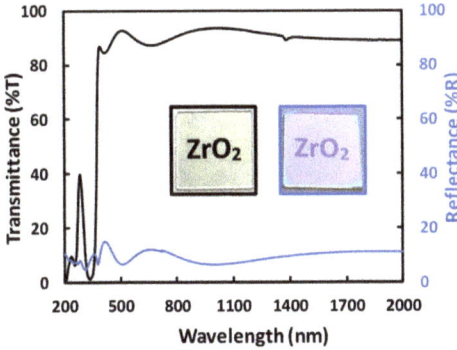

**Figure 2.** UV–Visible-NIR transmittance (in blue) and reflectance (in black) spectra of ZrO$_2$ xerogel thin film deposited on SiO$_2$ substrate. Inset: black and blue represent optical photographs of the ZrO$_2$ layers in transmission and specular reflection, respectively.

### 3.2. Micro-Nanostructuration of ZrO$_2$ Xerogel Films

#### 3.2.1. Optical Lithography

One way to obtain micro-nanostructured layers is to use the colloidal lithography technique in order to obtain amorphous ZrO$_2$ nano-pillars, according to the process described in [35] (Figure 3).

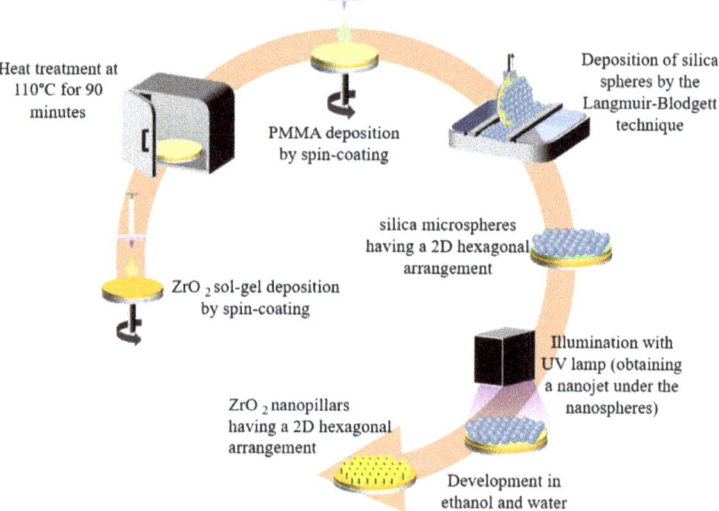

**Figure 3.** Illustration of the micro-nanostructuring process by colloidal lithography.

After homogeneous deposition of the silica microspheres following a 2D periodic hexagonal arrangement (in both $x$ and $y$ directions), the substrate was then illuminated with homogeneous UV light to create the nano-pillars (cylinder-shaped in the $z$ direction) with an arrangement also following a 2D hexagonal pattern (Figure 4C). Each microsphere behaves like a micro-lens by focusing incident UV light and creates a photonic nano-jet that emerges underneath the microsphere in the $ZrO_2$ layer [35]. After development and thermal stabilization of the $ZrO_2$ layer at 110 °C for 1 h, a nano-structuring composed of nano-pillars is obtained within the hexagonal arrangement imposed by the nano-sphere deposition.

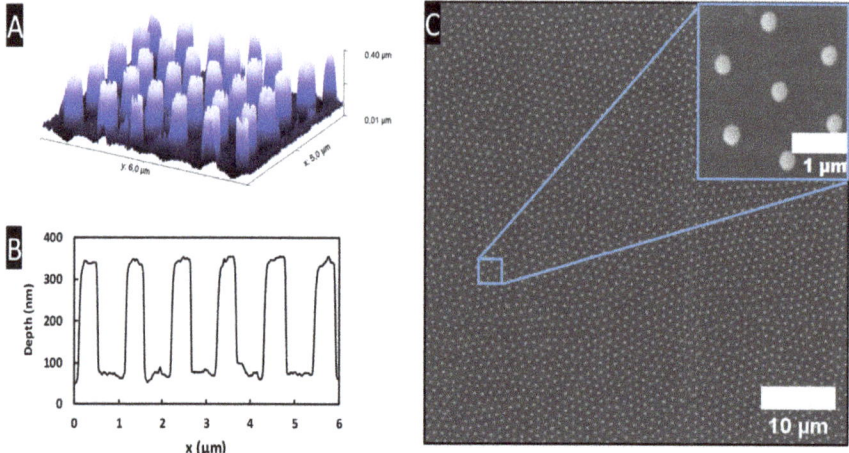

**Figure 4.** (**A,B**) Three-dimensional AFM image and profile of nano-structured $ZrO_2$ thin film. (**C**) SEM top view image of the $ZrO_2$ nano-pillars with an inset picture showing the hexagonal arrangement of the nano-pillars.

Figure 4C shows the nano-structured $ZrO_2$ xerogel thin film after UV ($\lambda$ = 365 nm) illumination and development of the xerogel layer, revealing periodically organized nano-pillars within a hexagonal arrangement. It is important to note that the 1 µm silica spheres used to make these films are not really mono-dispersed in size, which induces dislocations between small periodically well-arranged nano-sphere areas visible on the SEM image (Figure 4C). These nano-pillars appear to be fairly regular, with an average plot diameter calculated to 540 ± 30 nm with a periodicity of 1 µm corresponding to the microsphere diameter. From the 3D AFM image and profile of nano-structured $ZrO_2$ thin film (Figure 4A,B), the nano-pillars have a cylindrical shape with a height of about 280 nm.

To extend the demonstration, the process was adapted to non-conventional substrates such as non-planar substrates. The versatile macro- and micro-structuring techniques described above have been applied to standard planar substrates. However, as shown in Figure 4, the process is also suitable for use on non-conventional substrates, such as lenses with varying degrees of curvature, or on flexible and bendable plastic sheets, while retaining their optical properties. Figure 5A shows a convex lens with a multiscale structured $ZrO_2$ coating by combining macro- and micro-structuring (using colloidal lithography and UV illumination through a macroscopic mask before development and stabilization). Figure 5B presents a macroscopic $ZrO_2$ pattern on a plastic sheet.

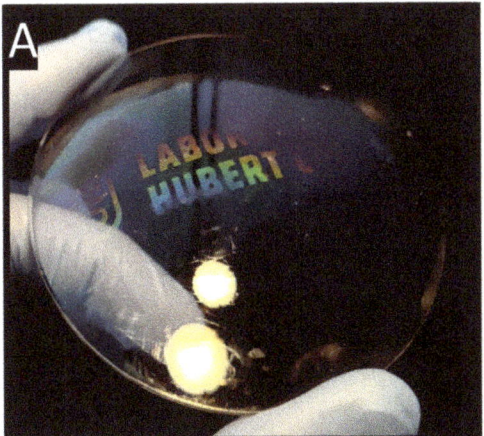
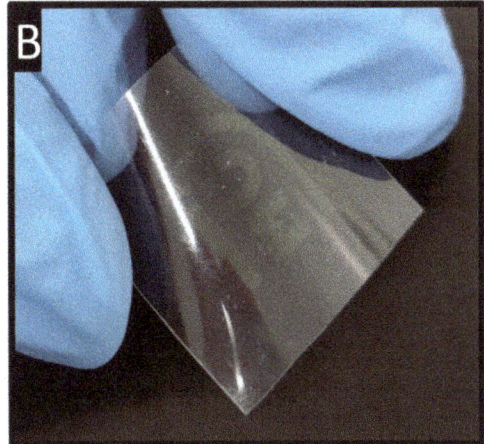

**Figure 5.** Examples of structuring on unconventional substrates: (**A**) $ZrO_2$ multiscale pattern on a convex glass optical lens and (**B**) $ZrO_2$ pattern on flexible plastic.

### 3.2.2. Nano-Imprint Lithography

Another method for structuring this $ZrO_2$ xerogel is to use the direct UV nano-imprinting process to obtain, for example, amorphous $ZrO_2$ diffraction grating. Figure 6 shows an example of structured $ZrO_2$ xerogel thin layers obtained by nano-imprint. From PDMS stamps with a sinusoidal 1D grating of 800 nm period and 60 nm deep, it is possible to obtain replicas based on $ZrO_2$ xerogel with similar characteristics to the ones of the stamp. AFM analysis illustrated in Figure 6A,B show that the $ZrO_2$ replica has a period of around 800 nm with a depth close to 50 nm. Figure 6C shows the grating pattern obtained with the nano-imprint method demonstrating the good uniformity of the $ZrO_2$ replica from both the microscopic SEM image and the good colored diffraction observed in the far field (in the −1st diffracted order direction) as shown in the inset.

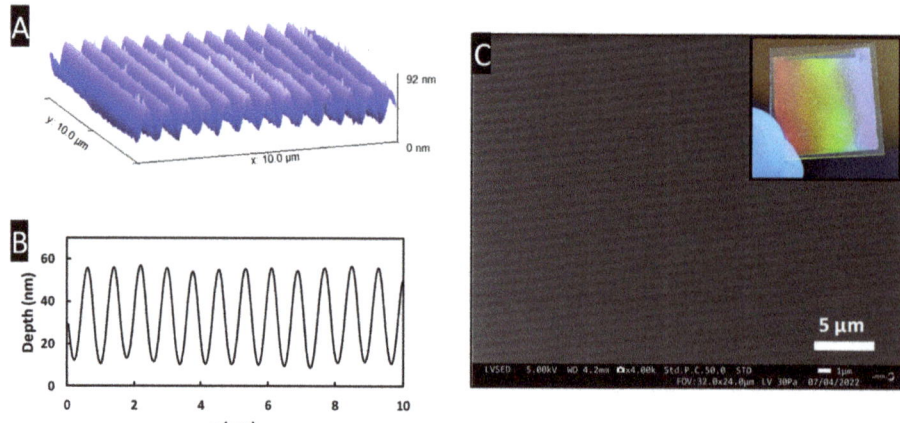

**Figure 6.** Example of structuring on planar glass substrates: ZrO$_2$ xerogel sub-micronic diffraction grating. (**A**,**B**) AFM 3D image and profile of the nano-structured ZrO$_2$ thin film. (**C**) SEM top view image of the ZrO$_2$ diffraction grating with an inset picture showing the iridescence phenomenon of the structured ZrO$_2$ layer.

Figure 7 is an illustration summarizing the different possible methods of sol-gel micro-nanostructuring.

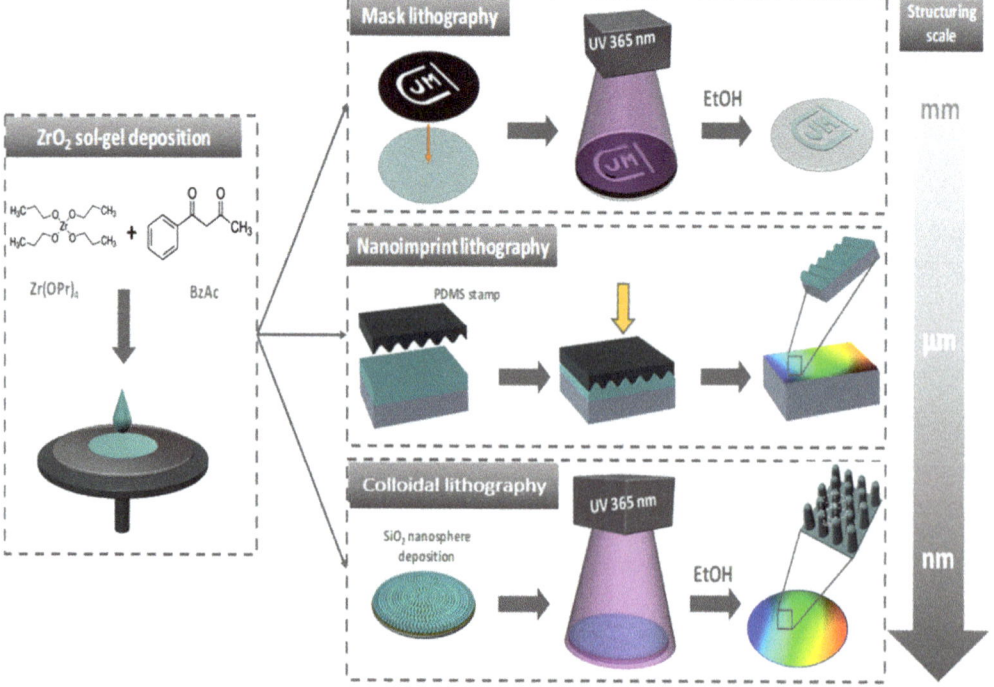

**Figure 7.** Illustration of the different methods of ZrO$_2$ sol-gel micro-nanostructuring.

## 4. Conclusions

In conclusion we firstly demonstrated that the same $ZrO_2$ sol-gel can be used both by optical and by nano-imprint lithography to realize complex patterns. This patterning by lithography is possible thanks to the presence of BzAc in the sol, which makes this xerogel layer photosensitive under UVA. The degradation of the BzAc makes the exposed areas insoluble in a solvent, thus inducing a contrast of solubility between the exposed and non-exposed areas. The xerogel then behaves as a negative photoresist, allowing its structuring at different scales by optical lithography or by nano-imprint lithography. We have also shown that after degradation of BzAc the $ZrO_2$ films are amorphous.

Secondly, we have shown the possibility to micro-nanostructure $ZrO_2$ xerogel films at different scales using optical mask or nanosphere lithography or a combination of both methods. Different patterns were realized on substrates varying in nature (plastic, glass, paper) and also in shape (planar and non-planar). Thirdly, patterns at different scales (millimeter and micrometer) were realized by nano-imprint lithography on planar substrates, opening the route to a cost-effective, fast, and direct micro-nanostructuring approach (without any etching processes) of functional coatings. This can allow an industrial development and an economic valorization of this sol-gel based process.

**Author Contributions:** N.C.-M., E.G. and Y.J. conducted the study; A.V. synthesized and deposited the films with supervision from N.C.-M., A.V., M.T., E.G. and N.C.-M. performed the micro-nanostructuring of the films by optical or nano-imprint lithography. The characterizations were performed by V.V.-O., A.V. and S.R. The paper was written by N.C.-M. and Y.J. with contributions from E.G. All authors have read and agreed to the published version of the manuscript.

**Funding:** This research was funded by French National Research Agency (ANR), grant number ANR-21-CE08-0042-01.

**Institutional Review Board Statement:** Not applicable.

**Informed Consent Statement:** Not applicable.

**Data Availability Statement:** Not applicable.

**Acknowledgments:** The authors acknowledge the French National Research Agency (ANR) for financial support in the framework of project NITRURATION (ANR-21-CE08-0042-01), and the Centre National de la Recherche Scientifique CNRS (French RENATECH+, nano-SaintEtienne plateform).

**Conflicts of Interest:** The authors declare no conflict of interest.

## References

1. Martin, P.J.; Bendavid, A. Properties of zirconium oxide films prepared by filtered cathodic vacuum arc deposition and pulsed DC substrate bias. *Thin Solid Film.* **2010**, *518*, 5078–5082. [CrossRef]
2. Gan, Z.; Yu, G.; Zhao, Z.; Tan, C.M.; Tay, B.K. Mechanical properties of zirconia thin films deposited by filtered cathodic vacuum arc. *J. Am. Ceram. Soc.* **2005**, *88*, 2227–2229. [CrossRef]
3. Sui, J.H.; Cai, W. Formation of $ZrO_2$ coating on the NiTi alloys for improving their surface properties. *Nucl. Instrum. Methods Phys. Res. Sect. B Beam Interact. Mater. At.* **2006**, *251*, 402–406. [CrossRef]
4. Millán-Ramos, B.; Morquecho-Marín, D.; Silva-Bermudez, P.; Ramírez-Ortega, D.; Depablos-Rivera, O.; García-López, J.; Fernández-Lizárraga, M.; Victoria-Hernández, J.; Letzig, D.; Almaguer-Flores, A.; et al. Biocompatibility and electrochemical evaluation of $ZrO_2$ thin films deposited by reactive magnetron sputtering on MgZnCa alloy. *J. Magnes. Alloy.* **2021**, *9*, 2019–2038. [CrossRef]
5. Bodurov, I.; Vlaeva, I.; Viraneva, A.; Yovcheva, T.; Sainov, S. Modified design of a laser refractometer. *Nanosci. Nanotechnol.* **2016**, *16*, 31–33.
6. Khojier, K.; Savaloni, H.; Jafari, F. Structural, electrical, and decorative properties of sputtered zirconium thin films during post-annealing process. *J. Theor. Appl. Phys.* **2013**, *7*, 55. [CrossRef]
7. Venkataraj, S.; Geurts, J.; Weis, H.; Kappertz, O.; Njoroge, W.K.; Jayavel, R.; Wuttig, M. Structural and optical properties of thin lead oxide films produced by reactive direct current magnetron sputtering. *J. Vac. Sci. Technol. A* **2001**, *19*, 2870–2878. [CrossRef]
8. Horti, N.C.; Kamatagi, M.D.; Nataraj, S.K.; Sannaikar, M.S.; Inamdar, S.R. Photoluminescence properties of zirconium oxide ($ZrO_2$) nanoparticles. *AIP Conf. Proc.* **2020**, *2274*, 020002. [CrossRef]
9. King, A.; Singh, R.; Nayak, B.B. Phase and photoluminescence analysis of dual-color emissive $Eu_{3+}$-doped $ZrO_2$ nanoparticles for advanced security features in anti-counterfeiting. *Colloids Surf. A Physicochem. Eng. Asp.* **2021**, *631*, 127715. [CrossRef]

10. Shajahan, S.; Basu, A. Corrosion, oxidation and wear study of electro-co-deposited $ZrO_2$-$TiO_2$ reinforced Ni-W coatings. *Surf. Coat. Technol.* **2020**, *393*, 125729. [CrossRef]
11. Wang, L.; Hu, X.; Nie, X. Deposition and properties of zirconia coatings on a zirconium alloy produced by pulsed DC plasma electrolytic oxidation. *Surf. Coat. Technol.* **2013**, *221*, 150–157. [CrossRef]
12. Zhao, Y.; Wang, T.; Zhang, D.; Fan, S.; Shao, J.; Fan, Z. Laser conditioning of $ZrO_2$:$Y_2O_3$/$SiO_2$ mirror coatings prepared by E-beam evaporation. *Appl. Surf. Sci.* **2005**, *239*, 171–175. [CrossRef]
13. Zhang, Q.; Li, X.; Shen, J.; Wu, G.; Wang, J.; Chen, L. ZrO2 Thin films and $ZrO_2$/$SiO_2$ optical reflection filters deposited by sol–gel method. *Mater. Lett.* **2000**, *45*, 311–314. [CrossRef]
14. Pirvaram, A.; Talebzadeh, N.; Rostami, M.; Leung, S.N.; O'Brien, P.G. Evaluation of a $ZrO_2$/$ZrO_2$-aerogel one-dimensional photonic crystal as an optical filter for thermophotovoltaic applications. *Therm. Sci. Eng. Prog.* **2021**, *25*, 100968. [CrossRef]
15. Mahmoodi, S.; Moradi, M.; Saeidi, F.S. The nearly perfect optical filter composed of [$SiO_2$/$ZrO_2$] Stacks using one-dimensional photonic crystals. *J. Nanostruct.* **2021**, *11*, 618–627. [CrossRef]
16. Manicone, P.F.; Rossi Iommetti, P.; Raffaelli, L. An Overview of zirconia ceramics: Basic properties and clinical applications. *J. Dent.* **2007**, *35*, 819–826. [CrossRef]
17. Wang, M.; Gao, J. Atomic layer deposition of ZnO thin film on $ZrO_2$ dental implant surface for enhanced antibacterial and bioactive performance. *Mater. Lett.* **2021**, *285*, 128854. [CrossRef]
18. Kusano, E. Homologous substrate-temperature dependence of structure and properties of $TiO_2$, $ZrO_2$, and $HfO_2$ thin films deposited by reactive sputtering. *J. Vac. Sci. Technol. A* **2019**, *37*, 051508. [CrossRef]
19. Patel, U.S.; Patel, K.H.; Chauhan, K.V.; Chawla, A.K.; Rawal, S.K. Investigation of various properties for zirconium oxide films synthesized by sputtering. *Procedia Technol.* **2016**, *23*, 336–343. [CrossRef]
20. Houska, J.; Rezek, J.; Cerstvy, R. Dependence of the $ZrO_2$ growth on the crystal orientation: Growth Simulations and magnetron sputtering. *Appl. Surf. Sci.* **2022**, *572*, 151422. [CrossRef]
21. Beer, S.M.J.; Samelor, D.; Abdel Aal, A.; Etzkorn, J.; Rogalla, D.; Turgambaeva, A.E.; Esvan, J.; Kostka, A.; Vahlas, C.; Devi, A. Direct liquid injection chemical vapor deposition of $ZrO_2$ films from a heteroleptic Zr precursor: Interplay between film characteristics and corrosion protection of stainless steel. *J. Mater. Res. Technol.* **2021**, *13*, 1599–1614. [CrossRef]
22. Espinoza-Pérez, L.J.; López-Honorato, E.; González, L.A. Development of $ZrO_2$ and YSZ coatings deposited by PE-CVD below 800 °C for the protection of Ni alloys. *Ceram. Int.* **2020**, *46*, 15621–15630. [CrossRef]
23. Shin, H.; Jeong, D.-K.; Lee, J.; Sung, M.M.; Kim, J. Formation of $TiO_2$ and $ZrO_2$ nanotubes using atomic layer deposition with ultraprecise control of the wall thickness. *Adv. Mater.* **2004**, *16*, 1197–1200. [CrossRef]
24. Cassir, M.; Goubin, F.; Bernay, C.; Vernoux, P.; Lincot, D. Synthesis of $ZrO_2$ thin films by atomic layer deposition: Growth kinetics, structural and electrical properties. *Appl. Surf. Sci.* **2002**, *193*, 120–128. [CrossRef]
25. Shahmohammadi, M.; Sun, Y.; Yuan, J.C.-C.; Mathew, M.T.; Sukotjo, C.; Takoudis, C.G. In vitro corrosion behavior of coated $Ti_6Al_4V$ with $TiO_2$, $ZrO_2$, and $TiO_2$/$ZrO_2$ mixed nanofilms using atomic layer deposition for dental implants. *Surf. Coat. Technol.* **2022**, *444*, 128606. [CrossRef]
26. Tarafdar, A.; Panda, A.B.; Pramanik, P. Synthesis of $ZrO_2$–$SiO_2$ Mesocomposite with High $ZrO_2$ content via a novel sol–gel method. *Microporous Mesoporous Mater.* **2005**, *84*, 223–228. [CrossRef]
27. Lin, C.; Zhang, C.; Lin, J. Phase transformation and photoluminescence properties of nanocrystalline ZrO2 powders prepared via the pechini-type sol–gel process. *J. Phys. Chem. C* **2007**, *111*, 3300–3307. [CrossRef]
28. Della Giustina, G.; Garoli, D.; Romanato, F.; Brusatin, G. Zirconia based functional sol–gel resist for UV and high resolution lithography. *Microelectron. Eng.* **2013**, *110*, 436–440. [CrossRef]
29. Kintaka, K.; Nishii, J.; Tohge, N. Diffraction gratings of photosensitive $ZrO_2$ gel films fabricated with the two-ultraviolet-beam interference method. *Appl. Opt.* **2000**, *39*, 489–493. [CrossRef]
30. Yamada, I.; Ikeda, Y. Sol-gel zirconia diffraction grating using a soft imprinting process. *Appl. Opt.* **2017**, *56*, 5054–5059. [CrossRef]
31. Ridaoui, H.; Wieder, F.; Ponche, A.; Soppera, O. Direct ArF laser photopatterning of metal oxide nanostructures prepared by the sol-gel route. *Nanotechnology* **2010**, *21*, 065303. [CrossRef]
32. Park, H.-H.; Zhang, X.; Lee, S.-W.; Kim, K.; Choi, D.-G.; Choi, J.-H.; Lee, J.; Lee, E.-S.; Park, H.-H.; Hill, R.H.; et al. Facile nanopatterning of zirconium dioxide films via direct ultraviolet-assisted nanoimprint lithography. *J. Mater. Chem.* **2010**, *21*, 657–662. [CrossRef]
33. Dinachali, S.S.; Saifullah, M.S.M.; Ganesan, R.; Thian, E.S.; He, C. A universal scheme for patterning of oxides via thermal nanoimprint lithography. *Adv. Funct. Mater.* **2013**, *23*, 2201–2211. [CrossRef]
34. Hochedel, M.; Bichotte, M.; Arnould, F.; Celle, F.; Veillas, C.; Pouit, T.; Dubost, L.; Kämpfe, T.; Dellea, O.; Crespo-Monteiro, N.; et al. Microstructuring technology for large and cylindrical receivers for Concentrated Solar Plants (CSP). *Microelectron. Eng.* **2021**, *248*, 111616. [CrossRef]
35. Shavdina, O.; Berthod, L.; Kampfe, T.; Reynaud, S.; Veillas, C.; Verrier, I.; Langlet, M.; Vocanson, F.; Fugier, P.; Jourlin, Y. Large area fabrication of periodic $TiO_2$ nanopillars using microsphere photolithography on a photopatternable sol-gel film. *Langmuir* **2015**, *31*, 7877–7884. [CrossRef]
36. Ai, B.; Yu, Y.; Möhwald, H.; Zhang, G.; Yang, B. Plasmonic films based on colloidal lithography. *Adv. Colloid Interface Sci.* **2014**, *206*, 5–16. [CrossRef]

37. Valour, A.; Usuga Higuita, M.A.; Crespo-Monteiro, N.; Reynaud, S.; Hochedel, M.; Jamon, D.; Donnet, C.; Jourlin, Y. Micro–nanostructured TiN thin film: Synthesis from a photo-patternable $TiO_2$ sol–gel coating and rapid thermal nitridation. *J. Phys. Chem. C* **2020**, *123*, 25480–25488. [CrossRef]
38. Briche, S.; Tebby, Z.; Riassetto, D.; Messaoud, M.; Gamet, E.; Pernot, E.; Roussel, H.; Dellea, O.; Jourlin, Y.; Langlet, M. New insights in photo-patterned sol-gel-derived $TiO_2$ films. *J. Mater. Sci* **2011**, *46*, 1474–1486. [CrossRef]
39. Bardosova, M.; Pemble, M.E.; Povey, I.M.; Tredgold, R.H. The Langmuir-Blodgett approach to making colloidal photonic crystals from silica spheres. *Adv. Mater.* **2010**, *22*, 3104–3124. [CrossRef]
40. Oda, S.; Uchiyama, H.; Kozuka, H. Thermoplasticity of Sol–Gel-derived titanoxanes chemically modified with benzoylacetone. *J. Sol. Gel Sci. Technol.* **2014**, *70*, 441–450. [CrossRef]
41. Wang, Z.; Zhao, G.; Zhang, W.; Feng, Z.; Lin, L.; Zheng, Z. Low-cost micro-lens arrays fabricated by photosensitive sol–gel and multi-beam laser interference. *Photonics Nanostruct. Fundam. Appl.* **2012**, *10*, 667–673. [CrossRef]

Article

# Influence of Addition of Antibiotics on Chemical and Surface Properties of Sol-Gel Coatings

Beatriz Toirac [1,*], Amaya Garcia-Casas [1,2], Miguel A. Monclús [3], John J. Aguilera-Correa [4,5], Jaime Esteban [4,5] and Antonia Jiménez-Morales [1,5,6]

[1] Materials Science and Engineering and Chemical Engineering Department, Carlos III University of Madrid, 28911 Madrid, Spain; amayagarciacasas@gmail.com (A.G.-C.); toni@ing.uc3m.es (A.J.-M.)
[2] CIDETEC, Basque Research and Technology Alliance (BRTA), 20014 Donostia-San Sebastián, Spain
[3] Micro- and Nano-Mechanics Department, Madrid Institutes for Advanced Studies (IMDEA)—Materials, 28906 Madrid, Spain; miguel.monclus@imdea.org
[4] Clinical Microbiology Department, IIS-Fundación Jiménez Díaz, UAM, 28040 Madrid, Spain; john.aguilera@fjd.es (J.J.A.-C.); jesteban@fjd.es (J.E.)
[5] CIBERINFEC, ISCIII—CIBER de Enfermedades Infecciosas, Instituto Carlos III, 28029 Madrid, Spain
[6] Alvaro Alonso Barba Technological Institute of Chemistry and Materials, Carlos III University of Madrid, 28911 Madrid, Spain
* Correspondence: btoirac@ing.uc3m.es

**Abstract:** Infection is one of the most common causes that leads to joint prosthesis failure. In the present work, biodegradable sol-gel coatings were investigated as a promising controlled release of antibiotics for the local prevention of infection in joint prostheses. Accordingly, a sol-gel formulation was designed to be tested as a carrier for 8 different individually loaded antimicrobials. Sols were prepared from a mixture of MAPTMS and TMOS silanes, tris(tri-methylsilyl)phosphite, and the corresponding antimicrobial. In order to study the cross-linking and surface of the coatings, a battery of examinations (Fourier-transform infrared spectroscopy, solid-state $^{29}$Si-NMR spectroscopy, thermogravimetric analysis, SEM, EDS, AFM, and water contact angle, thickness, and roughness measurements) were conducted on the formulations loaded with Cefoxitin and Linezolid. A formulation loaded with both antibiotics was also explored. Results showed that the coatings had a microscale roughness attributed to the accumulation of antibiotics and organophosphites in the surface protrusions and that the existence of chemical bonds between antibiotics and the siloxane network was not evidenced.

**Keywords:** antibiotics-loaded sol-gel coatings; AFM; SEM; solid-state $^{29}$Si-NMR spectroscopy; Fourier-transform infrared spectroscopy

**Citation:** Toirac, B.; Garcia-Casas, A.; Monclús, M.A.; Aguilera-Correa, J.J.; Esteban, J.; Jiménez-Morales, A. Influence of Addition of Antibiotics on Chemical and Surface Properties of Sol-Gel Coatings. *Materials* 2022, 15, 4752. https://doi.org/10.3390/ma15144752

Academic Editor: Aleksej Zarkov

Received: 8 June 2022
Accepted: 5 July 2022
Published: 7 July 2022

**Publisher's Note:** MDPI stays neutral with regard to jurisdictional claims in published maps and institutional affiliations.

**Copyright:** © 2022 by the authors. Licensee MDPI, Basel, Switzerland. This article is an open access article distributed under the terms and conditions of the Creative Commons Attribution (CC BY) license (https://creativecommons.org/licenses/by/4.0/).

## 1. Introduction

Orthopedic implant-associated infections (IAI) are especially challenging for orthopedic trauma services [1]. Infections are caused by bacteria or fungi attached to the implant surface, leading to biofilm formation on the implant surface. Orthopedic IAI can have devastating consequences for patients and represents a significant economic cost in hospital expenses. When IAI occurs, the traditional protocol can include surgery (irrigation and debridement, obliteration of dead space, intravenous administration of antibiotics, and biomaterial removal) associated with a prolonged systemic antibiotic treatment [2]. The recovery chances of limb functionality, even if the infected implant is successfully removed, are quite limited. Sometimes, these procedures usually lead to a bad outcome, such as arthrodesis, amputation, suppressive treatment, and even death [3–5]. Therefore, avoiding the growth of nosocomial pathogens can be more effective than trying to eliminate the biofilm [6].

The orthopedic IAI prevention by systemic administration of antibiotics probably has reached its limit in its effectiveness, and the increasing number of resistant organisms could be a problem in the future. To overcome this problem, possibly will be necessary to use a broader spectrum antibiotics with, in some cases, an increased number of side effects [6]. In addition, after prosthetic implantation, the tissue may be damaged, avascular, or even necrotic. These consequences locally decrease the antibiotic concentration systemically supplied and require the administration of local antibiotics for hours or days [7]. This solution achieves optimal concentrations for prevention and minimizes the adverse effects of systemic treatment. The use of surface coatings has been extensively investigated to prevent/treat orthopedic IAI, using either a non-degradable antibacterial surface or local release of antibiotics as solutions to coat the implant [3,8–11]. Local release of antibiotics from the biodegradable polymer coating requires sustained and controlled release of the antibiotic to inhibit microbial adhesion, colonization, and subsequent biofilm formation. In addition to its biodegradability and desired antibiotic release profile, the polymer system needs to have adequate mechanical strength and matrix formation [12].

Some of the approaches that have been widely used to prevent orthopedic IAI using local antibiotic release from coatings are calcium or silicon bone cements [13,14] and polymer hydrogels [7,8,15,16]. However, they have their limitations. Antibiotic-loaded cements have a burst release and limited release of embedded antibiotics (only 10% of the antibiotics, estimated) because of the diffusion through surface roughness, superficial pores, and surface erosion. However, the release of the antibiotic in the hydrogels is steady and is closely related to the crosslinking structure [3].

Hybrid sol-gel coatings are an example of this last-mentioned alternative. Sol-gel technology is a versatile method used to produce a wide diversity of materials. Among the advantages it offers are the simple sol-gel processing conditions and the possibility of tuning organic-inorganic hybrid materials for specific requirements [17,18]. Some studies have previously been conducted in which antibiotics are introduced into biodegradable sol-gel coatings, but these investigations are very scarce [19,20]. T. Nichol et al. [21] developed a sol-gel coating loaded with gentamicin for cementless hydroxyapatite-coated titanium orthopedic prostheses. They controlled the release of the antibiotic within the desirable time frame of 48 h. S. Radin et al. [19] synthesized sol-gel coatings loaded with vancomycin and studied the effect of the processing parameters on the coating degradation and antibiotic release.

In previous works, we have studied the effect of adding fluconazole and anidulafungin to a sol-gel coating in terms of electrochemical [22] and microbiological characterization [23]. Besides, in vitro studies and an in vivo model have been carried out on a moxifloxacin-loaded sol-gel coating [24]. However, the effect of adding antibiotics on the sol-gel network has not been addressed. The addition of large organic molecules to a sol-gel synthesis can act in detriment to the cross-linking of the network, resulting in poor adhesion of the coating to the substrate. For this reason, this work systematically studies the incorporation of eight antimicrobials of different natures into sol-gel coatings. This first step in knowing if it is possible to successfully incorporate these antibiotics into sol-gel coatings is important to develop a technology with personalized treatments based on each patient's infection.

We also investigated the simultaneous incorporation of two antibiotics into the coating to broaden the antibacterial spectrum against Gram-positive and Gram-negative bacteria. This objective addresses the prevention of both monomicrobial and polymicrobial infections; moreover, introducing antibiotics with different mechanisms of action can lead to a synergistic effect that increases their action against the bacteria to be prevented, reducing the possibility of antibiotic-resistance emergence [5,25,26].

For optimal design and synthesis of the coatings used for the desired application, this research focuses on studying the following two key factors: crosslinking and the surface of the coatings.

Achieving a controlled and constant antibiotic release rate is essential for the targeted applications. In these coatings, the release rate of the antibiotics is conditioned by the coating degradation [18]. Therefore, the design and study of the crosslinking of the formulations are very important to estimate and control the release rate of the antibiotics.

The surface of the coatings is also an important factor and must be studied. The hydrophobicity and roughness of these have an important effect on the initial bacterial attachment as well as the osseointegration capacity [27]. In most previous studies, it has been determined that surface roughness is directly related to the degree of bacterial adhesion. In addition, hydrophobicity influences each microbe differently depending on its nature. For instance, *S. aureus*, the most common bacteria in orthopedic IAI, prefers hydrophobic surfaces to adhere to, according to some research [3].

This research pursues the main objective of studying the influence of the addition of different antimicrobials (especially cefoxitin and linezolid) on the chemical and surface properties of the sol-gel coatings. Achieving the incorporation of these antimicrobials without compromising the sol-gel network will allow us to propose this technology as a versatile processing method to synthesize personalized coatings to prevent local joint prosthesis infections.

## 2. Materials and Methods

### 2.1. Biofunctionalized Coatings Using Antibiotics: Materials and Sample Preparation

Organic–inorganic hybrid coatings were synthesized from methacryloxypropyltrimethoxy silane (MAPTMS, 98%, Acros Organics, Thermo Fisher Scientific, Waltham, MA, USA) and tetramethyl orthosilane (TMOS, 98%, Acros Organics). Sols were prepared from a mixture of MAPTMS and TMOS with a 1:2 molar ratio as described by El Hadad et al. [28]. To enhance cellular proliferation, as demonstrated in a previous work [29], tris(tri-methylsilyl) phosphite (92%, Sigma Aldrich, St. Louis, MI, USA) was added to the sol. The molar ratio of silanes to phosphorus precursor was fixed at 50. Ethanol was added as a solvent to avoid phase separation and water as a reagent to initiate the hydrolysis reaction. Ethanol and water were added in stoichiometric amounts. All species were mixed before addition of water. After the dropwise addition of the aqueous solution, suspensions were stirred for 24 h in a glove box.

Coatings were made by adding different antimicrobials to the formulation. Eight antimicrobials (7 antibiotics and 1 antifungal) were used as biofunctionalizers in the coatings. Antibiotics: gentamicin sulfate salt (GEN, Sigma Aldrich), cefoxitin sodium salt (FOX, Sigma Aldrich), vancomycin hydrochloride hydrate (VAN, Sigma Aldrich), dicloxacillin sodium salt monohydrate (DCX, Sigma Aldrich), clindamycin hydrochloride (CLI, Sigma Aldrich), ampicillin (AMP, Sigma Aldrich), and linezolid (LNZ, 98%, Acros Organics). Antifungal: Amphotericin B (AMB, Sigma Aldrich). The used concentration was correlated with the maximum amount of water solubility of the antibiotic for most incorporated antibiotics. However, AMP, LNZ, and AMB antibiotics are insoluble or poorly soluble in water, so ethanol was used to dilute them, and the used concentration was the highest without producing supersaturation. The concentrations used in the formulations are summarized in Table 1. These antimicrobials were previously dissolved or suspended in water before adding them to the mixture (part of the ethanol volume was used to dissolve them or improve the antimicrobial agent solubility). All reagents were used as received from Sigma-Aldrich and Acros Organics.

**Table 1.** Molar ratio between the concentration of the silanes and the antibiotics used in the sol-gel formulations.

| Antibiotic | Molar Ratio Silanes:Antibiotic |
|---|---|
| GEN | 540:1 |
| FOX | 166:1 |
| VAN | 550:1 |
| DCX | 185:1 |
| CLI | 172:1 |
| AMP | 682:1 |
| LNZ | 165:1 |
| AMB | 902:1 |

Titanium sample pieces of 15 mm diameter × 25 mm thickness, prepared by a conventional powder metallurgy route, described in a previous work [30], were used as substrates (TiPM). The variation in surface roughness affects the coating distribution, so substrates were treated before the film deposition to achieve homogeneous surface conditions. Prior to the application of the sol-gel coating, the substrates were ground with successively finer SiC paper up to 1000 grit size, cleaned ultrasonically with acetone and alcohol, and dried.

A dipping device (KSV instrument-KSV DC) with a controlled withdrawal speed was used for the film deposition. Plates were immersed into the dissolution at a rate of 200 mm/s and were immediately removed at the same rate.

Finally, samples were dried at 60 °C for 60 min inside an oven. An example of the visual appearance of the coated pieces is shown in Figure S1. During the thermal treatment, the condensation of the remaining OH groups was promoted, and evaporation of the solvents occurred, resulting in the final network (xerogel).

The surface morphology and composition of the as-prepared coatings were assessed by scanning electron microscopy (Teneo FEI, W filament, Lincoln, NE, USA). The lateral area of the samples was covered by a Cu layer to increase the conductivity. Images were taken at low vacuum and applying 2 kV and 0.2 nA. Energy dispersive spectrometry (EDS) measurements were performed using an X-ray microanalysis system along with an Octane Plus detector (EDAX, Pleasanton, CA, USA) of 30 mm$^2$ area, which allowed for semi-quantitative analysis of the chemical composition by applying 5 kV and 0.2 nA.

Thickness of films was measured using an ultrasonic thickness NEURTEK instrument (Eibar, Spain).

### 2.2. Biofunctionalized Sol-Gel Coatings with FOX and LNZ: Materials and Sample Preparation

The following two antibiotics were chosen to perform a more detailed characterization: FOX and LNZ. The selection was based on the broadening of the antibacterial spectrum of the coatings for gram-positive and gram-negative bacteria and the pursuit of a synergistic effect with the administration of both antibiotics. Three coatings were prepared for each antibiotic with different antibiotic concentrations. A coating including the maximum concentrations of the two antibiotics was also prepared. A coating without antibiotics was synthesized for comparison. The concentrations of the prepared coatings are listed in Table 2.

**Table 2.** List of the identifying names of the coatings according to the doped antibiotic and its concentration.

| Denotation [1] | Antibiotic Concentration (mmol) |
|---|---|
| Control | Non-Antibiotic |
| lc.FOX | 0.037 mmol FOX |
| mc.FOX | 0.073 mmol FOX |
| hc.FOX | 0.147 mmol FOX |
| lc.LNZ | 0.037 mmol LNZ |
| mc.LNZ | 0.074 mmol LNZ |
| hc.LNZ | 0.148 mmol LNZ |
| hc.FOX-LNZ | 0.147 mmol FOX + 0.148 mmol LNZ |

[1] lc: lowest concentration, mc: medium concentration, hc: highest concentration.

### 2.3. Chemical Characterization

The evolution of the hydrolysis-condensation reaction was monitored by Fourier-transform infrared (FT-IR) spectroscopy. Each sample was prepared by adding a drop of the synthesis to a pressed KBr disc. Spectra were recorded with a Thermo Scientific NICOLET iS50 FT-IR System (Waltham, MA, USA) at room temperature in absorbance mode, covering the mid-infrared range from 500 to 4000 cm$^{-1}$ and with 4 cm$^{-1}$ resolution. For each sol, three measurements were carried out.

Sols were dried at room temperature for 7 days and ground into powder in an agate mortar before the characterization by $^{29}$Si-NMR and TGA [31].

Solid-state $^{29}$Si-NMR spectroscopy was used to determine the Si-O-Si crosslinking densification after curing treatment. The spectra were recorded in a Bruker AVANCE 400 spectrometer (Billerica, MA, USA) equipped with fast Fourier transform unit. The frequency used was 79.48 MHz (9.4 T). Samples were spun at 10 kHz around an axis inclined 54°44′ with respect to the external magnetic field. The used pulse length was 5 µs (90° pulse), the relaxation delay was 10s and 6000 accumulations were acquired. Spectra were referenced to TMS.

For the thermogravimetric analysis (TGA), around 30 mg of the xerogel was placed on alumina crucibles. Spectra were recorded from 30 to 900 °C at a rate of 10 °C/min in air atmosphere (PerkinElmer, STA 6000 instrument, Waltham, MA, USA). Duplicate measurements were made on each sample.

### 2.4. Surface Characterization

The surface morphology, composition, and thickness of the produced coatings were evaluated using SEM, EDS, and ultrasonic transducer respectively as described in Section 2.1.

The wettability of the coatings was determined by measuring the static contact angle of Phosphate-Buffered Saline (PBS) (pH = 7.4) onto sol–gel surfaces using an automatic contact angle meter (DATAPHYSICS OCA 20 Goniometer, DataPhysics Instruments GmbH, Filderstadt, Germany). A sessile drop of 3 µL was deposited on the surfaces at room temperature. The water contact angle was determined by the half-angle method. The value given is the mean of 6 measurements.

For the calculation of thickness and contact angle values, a statistical analysis was performed. Mean and standard deviation values were calculated using the one-way ANOVA statistical technique using as error protection method the Tukey HSD method, which provided a confidence limit of 95%.

The topographical features of the coated samples were inspected by atomic force microscopy (AFM) for scan area sizes of 5 × 5 and 40 × 40 µm$^2$ with 512 × 512 pixels resolution. The AFM instrument used was an XE-150 Park System operated under non-contact mode in ambient conditions at a scan rate of 0.3 Hz. The used silicon cantilever tip (N-type, µmasch, USA) had a nominal radius of 8 nm. XEI software version 4.3.0 (Park System Corp., Suwon si, Korea) was used for surface roughness analysis and Gwyddion software version 2.54 (gwyddion.net (accessed on 30 October 2021)) was used for image treatment. Statistical comparisons were made using the one-way ANOVA test, with $p = 0.05$ as the minimal level of significance and Tukey test was used to identify differences between groups.

## 3. Results

### 3.1. Synthesis and Characterization of Biofunctionalized Coatings

The obtained sol in most of the formulations was transparent, but the sols with GEN, VAN, and FOX showed some turbidity. Besides having an adequate viscosity that facilitates the uniform coverage of the substrate, sols did not evidence the separation of phases.

Dried coatings were simple sight observed obtaining coatings without imperfections such as cracks or macropores. A more thorough inspection was performed using SEM. Figure 1 shows the micrographs of all formulations using the Backscatter Detector (Teneo FEI, Lincoln, NE, USA).

Inspection of the surfaces showed the formation of smooth, uniform, homogeneous, and crack-free coatings on the substrates in most formulations. In coatings containing AMP and AMB, isolated cracks were found throughout their surfaces.

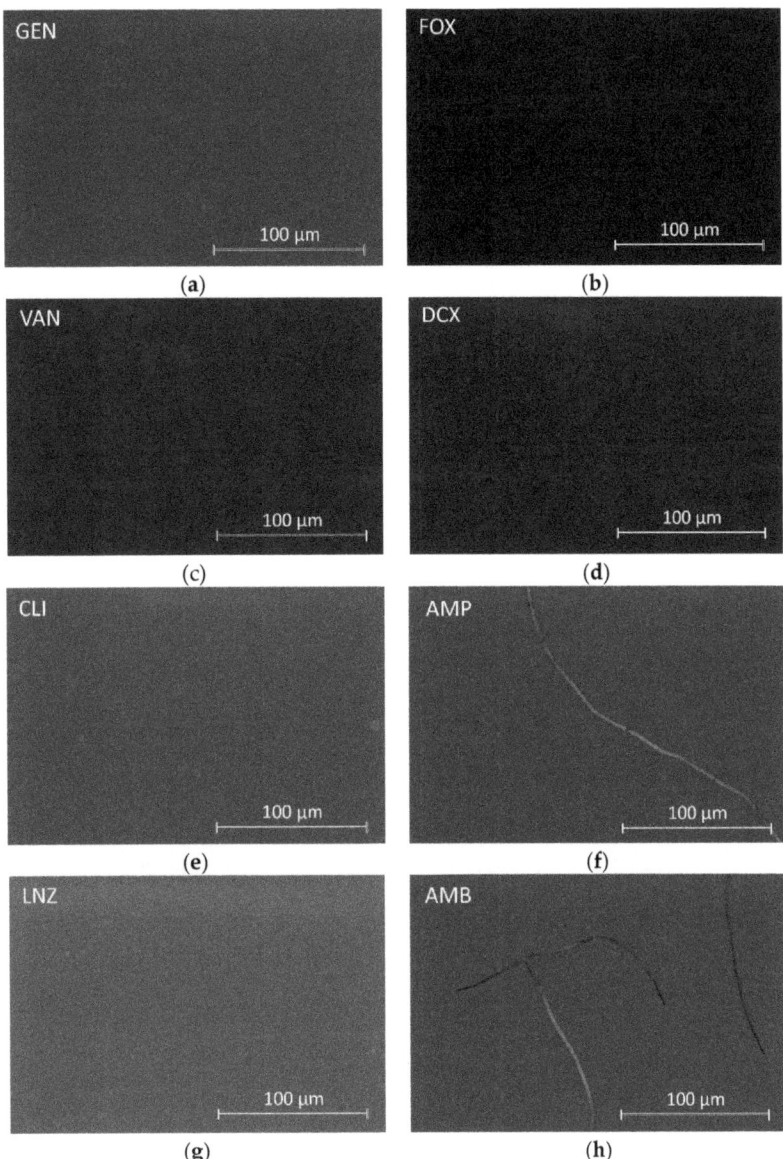

**Figure 1.** Micrographs obtained with CBS detector by SEM of the biofunctionalized coatings with different antibiotics: (**a**) gentamicin, (**b**) cefoxitin, (**c**) vancomycin, (**d**) dicloxacillin, (**e**) clindamycin, (**f**) ampicillin, (**g**) linezolid, and (**h**) amphotericin B at the concentrations described in Table 1.

On some of the surfaces of coatings, bright spots were observed distributed in a darker matrix, suggesting the presence of two well-differentiated phases. These coatings are those containing GEN, FOX, VAN, DCX, CLI, and LNZ. These bright spots change in size and quantity with each coating. The micrograph in Figure 2 is an example of such a surface observation on a FOX-containing coating.

**Figure 2.** Micrograph of the biofunctionalized coating with the antibiotic cefoxitin (FOX) at a silanes:antibiotic molar ratio of 166:1 obtained with CBS detector by SEM.

When EDS analyses were performed in dark areas, the C, Si, and O elements were identified, corresponding to the organic precursors of the synthesis. EDS analyses on bright spots revealed the presence of P, corresponding to the organophosphite compound. In coatings containing FOX and DCX, Na was also identified in the bright spots, while in coatings with FOX, GEN, VAN, CLI, and LNZ, N was also detected in these spots.

Figure 3 shows the thickness of each coating and the molecular weight of the used antibiotics. Thicknesses ranged between 10 and 20 µm, with a median value of 12.5 µm. Significant differences between the obtained thicknesses were observed.

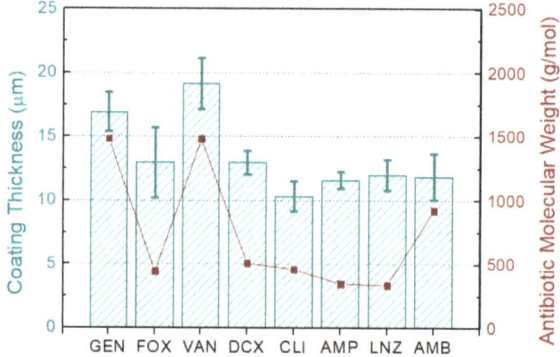

**Figure 3.** Correlation between the incorporated-antibiotics molecular weight in the coatings and their thicknesses.

### 3.2. Chemical Characterization of Coatings Loaded with FOX and LNZ

From this section, the characterization results carried out on the coatings loaded with FOX and LNZ at different concentrations are shown.

The sensitive analytical method of FTIR was used to study the polysiloxane network obtained in sols loaded with FOX and LNZ. Identification of the functional groups present in the formulations is possible with this technique. Furthermore, the detection of possible structural changes in the siloxane network due to the introduction of antibiotics or possible chemical bonds between antibiotics and the siloxane network could be observed. The FTIR spectra of the sols are plotted in Figure 4.

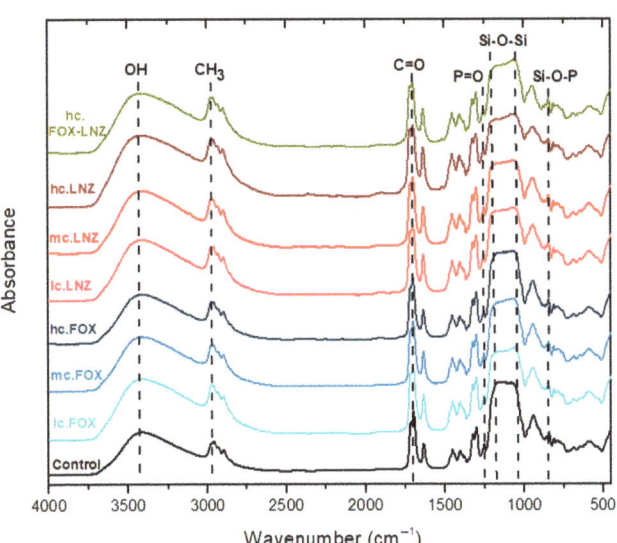

**Figure 4.** Representative FTIR absorption spectra of the studied sols containing FOX (hc., mc., and lc.), LNZ (hc., mc., and lc.), and FOX-LNZ (hc.). As reference the spectrum of the control sol is depicted.

The formation of the silica network was evidenced by identifying the bands associated with the vibrational modes of the Si-O-Si chains, detected at ~815 cm$^{-1}$ (weak band), ~1060 cm$^{-1}$, and ~1160 cm$^{-1}$; the Si-O-Si chains result from the condensation process. A broad band at ~3420 cm$^{-1}$, related to the vibrational modes of OH groups, including those from SiOH formed through hydrolysis, was also observed. These bands proved the existence of hydrolysis and condensation phenomena [28,32,33].

The existence of the bands at 1254 and ~850 cm$^{-1}$, related to the stretching vibration of a P=O bond [29] and the Si-O-P bending [34], respectively, was a clear indication of phosphorus presence and its incorporation into the silica network.

Absorption bands around 2960 and 2900 cm$^{-1}$ were attributed to the presence of C–H bonds [32,35,36]. The bands at ~1720 cm$^{-1}$ and ~1640 cm$^{-1}$ were associated with the stretching vibrations of C=O carbonyl groups and the C=C groups of the methacrylate groups from the MAPTMS precursor, respectively [28]. The band at ~1450 cm$^{-1}$ was attributed to the symmetrical and asymmetrical CH$_3$ deformational modes [36]. The asymmetric and symmetric stretching vibrations of C–O and C–O–C bonds were attributed to bands at ~1320 and 1300 cm$^{-1}$, respectively. Finally, the band at ~950 cm$^{-1}$ was assigned to the C=C vibrations of the C=C–C=O group [28].

Neither differences in the shapes of the absorption bands nor the emergence of new bands were observed in the spectra.

The $^{29}$Si-NMR technique was used to study the siloxane network formed in each of the systems and their condensation degree. The chemical study was extended with this technique, as it offers more detailed information with better resolution thanks to the higher sensitivity to short-range interactions in comparison with the FTIR analysis. The $^{29}$Si-NMR analysis allowed the quantification of the crosslinking degree within the silicate network.

In $^{29}$Si-NMR spectroscopy, there is a well-established nomenclature for identifying each chemical shift of silicon. The T$^n$ and Q$^n$ structures represent the trialkoxysilane and tetraalkoxysilane functionality, respectively, while the superscript n indicates the number of produced siloxane bonds. In the case of the formulations presented in this study, T species were related to MAPTMS precursor and Q species to TMOS [28,36]. Table 3 lists all the possible signals that can be obtained in this case [29]. Figure 5 shows the solid-state $^{29}$Si-NMR spectra of the obtained xerogels.

Table 3. Chemical shift (δ) of TMOS and MAPTMS employing silicon nuclei.

| Signal | Nature of Silicon Unit | Precursor | δ (ppm) |
|---|---|---|---|
| $T^0$ | Unhydrolyzed | MAPTMS | −42 |
| $T^1$ | Once condensed (one siloxane bond) | | −49 |
| $T^2$ | Doubly condensed (two siloxane bonds) | | −58 |
| $T^3$ | Fully condensed (three siloxane bonds) | | −67 |
| $Q^0$ | Unhydrolyzed | TMOS | −82 |
| $Q^1$ | Once condensed (one siloxane bond) | | −86 |
| $Q^2$ | Doubly condensed (two siloxane bonds) | | −92 |
| $Q^3$ | Triple condensed (three siloxane bonds) | | −101 |
| $Q^4$ | Fully condensed (four siloxane bonds) | | −110 |

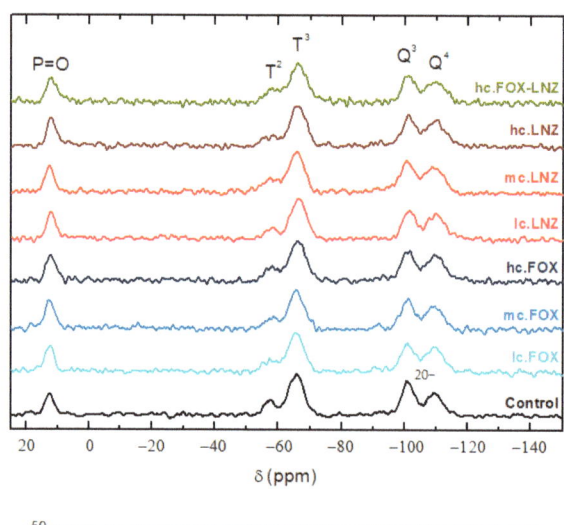

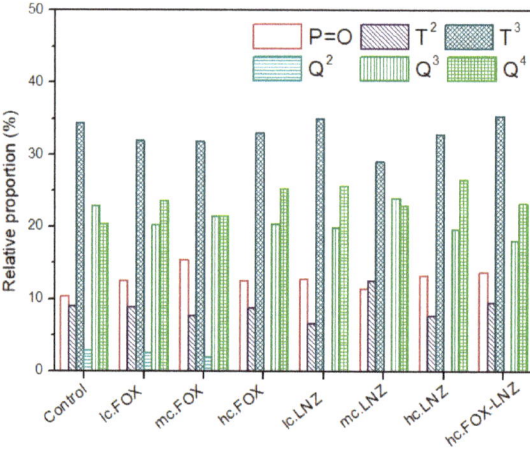

Figure 5. Solid-state $^{29}$Si-NMR spectra of each formulation (up). Relationship of the signals (down).

The signals associated with $T^2$, $T^3$, $Q^3$, and $Q^4$ were observed in all xerogels at −57, −66, −101, and −110 ppm, respectively. In each of the systems, the achievement of a high degree of crosslinking of the siloxane network was observed. This statement was supported by the following observations:

- The absence of non-hydrolyzed species and species with a single siloxane bond in both precursors;
- The $T^3$ predominance over the $T^2$ signal in the MAPTMS precursor and the prevalence of the $Q^3$ and $Q^4$ signals in the TMOS precursor.

At ~12.5 ppm, another signal appeared, denoted as P=O in the spectra, that corresponds to the creation of a -P=O- bond in the entourage of the Si nuclei of the sol-gel network. The integration mechanism of this compound into the network was described elsewhere by Garcia-Casas et al. [29]. Briefly, the organophosphite undergoes a first reaction that leads to the oxidation of trivalent phosphorus to pentavalent phosphorus, allowing the hydrolyzation of the trimethylsilyl chain and its subsequent condensation.

In summary, these xerogels showed the formation of three-dimensional networks, dominated by $T^3$ building blocks, accompanied by small amounts of P=O, $T^2$, $Q^3$, and $Q^4$ units. Table 4 summarizes the proportion of each detected signal in the systems.

Table 4. Relative proportions of T and Q species in the organic-inorganic hybrid materials from the solid-state $^{29}$Si NMR spectra in Figure 5.

| Sample | Proportions [a] (%) | | | | | | Relative [b] Proportions (%) | | Relative [c] Proportions (%) | | | Ratio [d] (%) | | |
|---|---|---|---|---|---|---|---|---|---|---|---|---|---|---|
| | P=O | $T^2$ | $T^3$ | $Q^2$ | $Q^3$ | $Q^4$ | $T^2$ | $T^3$ | $Q^2$ | $Q^3$ | $Q^4$ | P=O | $T^n$ | $Q^n$ |
| Control | 10.37 | 9.00 | 34.44 | 2.89 | 22.89 | 20.41 | 20.72 | 79.28 | 6.26 | 49.56 | 44.18 | 10.37 | 43.43 | 46.20 |
| lc.FOX | 12.55 | 9.00 | 31.92 | 2.58 | 20.25 | 23.70 | 21.99 | 78.01 | 5.54 | 43.53 | 50.93 | 12.55 | 40.91 | 46.54 |
| mc.FOX | 15.45 | 7.73 | 31.88 | 1.90 | 21.51 | 21.53 | 19.53 | 80.47 | 4.22 | 47.86 | 47.92 | 15.45 | 39.61 | 44.94 |
| hc.FOX | 12.50 | 8.78 | 32.99 | - | 20.42 | 25.31 | 21.03 | 78.97 | - | 44.64 | 55.36 | 12.50 | 41.77 | 45.73 |
| lc.LNZ | 12.75 | 6.64 | 35.04 | - | 19.91 | 25.66 | 15.94 | 84.06 | - | 43.69 | 56.31 | 12.74 | 41.69 | 45.57 |
| mc.LNZ | 11.51 | 12.55 | 29.02 | - | 23.98 | 22.94 | 30.19 | 69.81 | - | 51.12 | 48.88 | 11.51 | 41.57 | 46.92 |
| hc.LNZ | 13.26 | 7.73 | 32.78 | - | 19.67 | 26.56 | 19.09 | 80.91 | - | 42.55 | 57.45 | 13.26 | 40.51 | 46.23 |
| hc.FOX-LNZ | 13.78 | 9.56 | 35.36 | - | 18.06 | 23.24 | 21.28 | 78.72 | - | 43.73 | 56.27 | 13.78 | 44.92 | 41.30 |

[a] Peak area % was calculated by the deconvolution technique. Error value assumed is ±5%. [b] (Each T species/total T species) × 100%. [c] (Each Q species/total Q species) × 100%. [d] Si-P = (Si-P signal/total signals) × 100%, $T^n$ = (total T species/total signals) × 100%, $Q^n$ = (total Q species/total signals) × 100%.

The antibiotics' introduction slightly modified the contribution of each precursor to the crosslinking. An increase in the intensity of the P=O signal and attenuation of $T^n$ and $Q^n$ species were observed. Moreover, the $T^n$ species intensity decreased more than the $Q^n$ species signal. The number of fully condensed species increased in Q species.

Thermal characterization using TGA was performed to quantify the inorganic contribution of the network (Si-O-Si bonds). The thermogravimetric analysis (TGA) and the first derivative of the TGA (DTG) plots of the xerogels obtained during each synthesis are shown in Figure 6A,B. Thermogravimetric analyses of the antibiotics used are also shown in Figure 6C,D.

Thermogravimetric analysis of xerogels revealed a sharp inflection above 350 °C corresponding to the partial thermal degradation of organic matter (oligomers and unreacted organopolysiloxanes). This stage of degradation was accompanied by a subsequent stage (between 350 °C and 500 °C) due to the complete thermal degradation of the organic matter and water elimination from further silanol condensation [29]. Small losses after 800 °C in all samples are attributed to further burning of the residual organics [36].

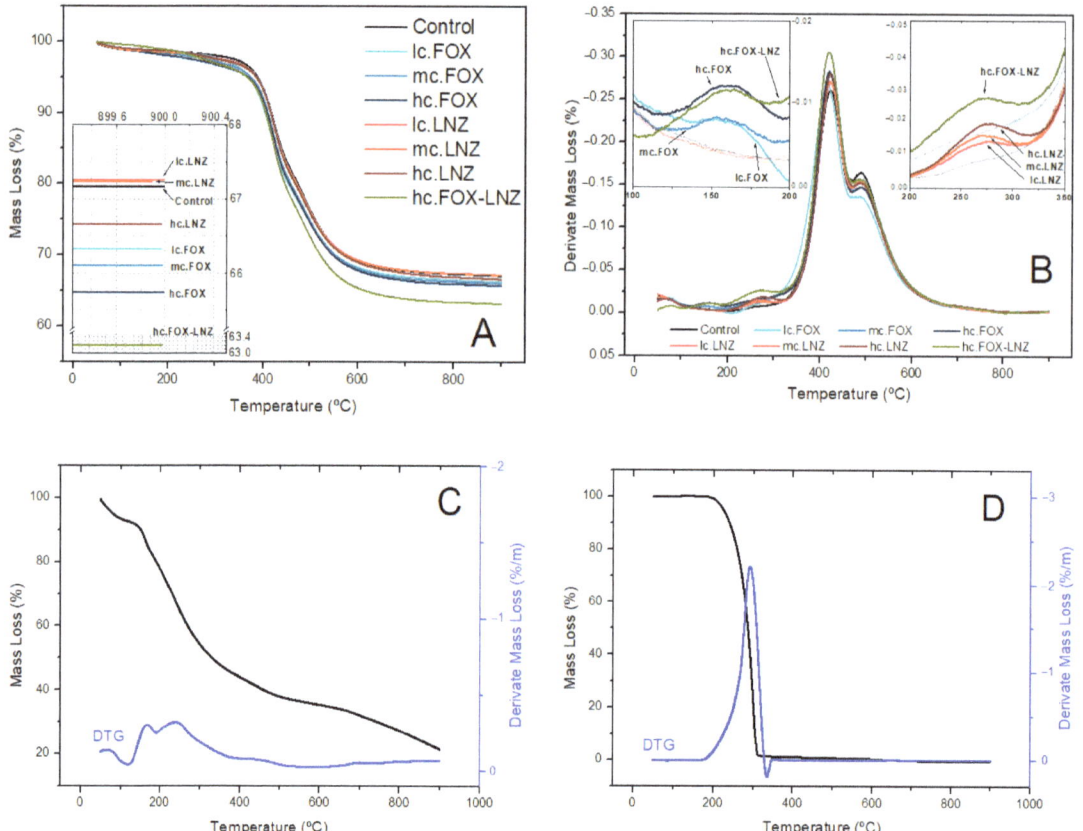

**Figure 6.** TGA (**A**) and DTG (**B**) profiles of the xerogels obtained during the syntheses. TGA and DTG profiles of the cefoxitin sodium Salt (FOX) (**C**) and linezolid (LNZ) (**D**).

A thermal degradation associated with the by-products elimination of the condensation (alcohol and water) and the unreacted reagents (pb$_{TMOS}$ = 121 °C, pb$_{MAPTMS}$ = 190 °C; pborganophosphite = 78–81 °C) occurred at temperatures below 350 °C [29]. The degradation of the antibiotics introduced into the formulations was another contribution present at these temperatures. Xerogels including FOX had a peak in DTG plots around 150 °C and those including LNZ around 275 °C, which coincided with the maximum thermal degradation of each antibiotic (Figure 6C,D).

The mass loss of the xerogels revealed that the hc.FOX-LNZ formulation decreased by almost 5 wt.% the inorganic contribution of Control (67.16% Control vs 63.21% hc.FOX-LNZ), see inset Figure 6A. The difference in mass loss between the rest of the formulations was only 1 wt.%.

### 3.3. Surface Characterization of Coatings Loaded with FOX and LNZ

The surface of the synthetized coatings was inspected by SEM. Figure 7 shows SEM micrographs of the coatings using the Backscatter Detector. Inspection of the surfaces showed the formation of uniform, homogeneous, and crack-free coatings on the substrates.

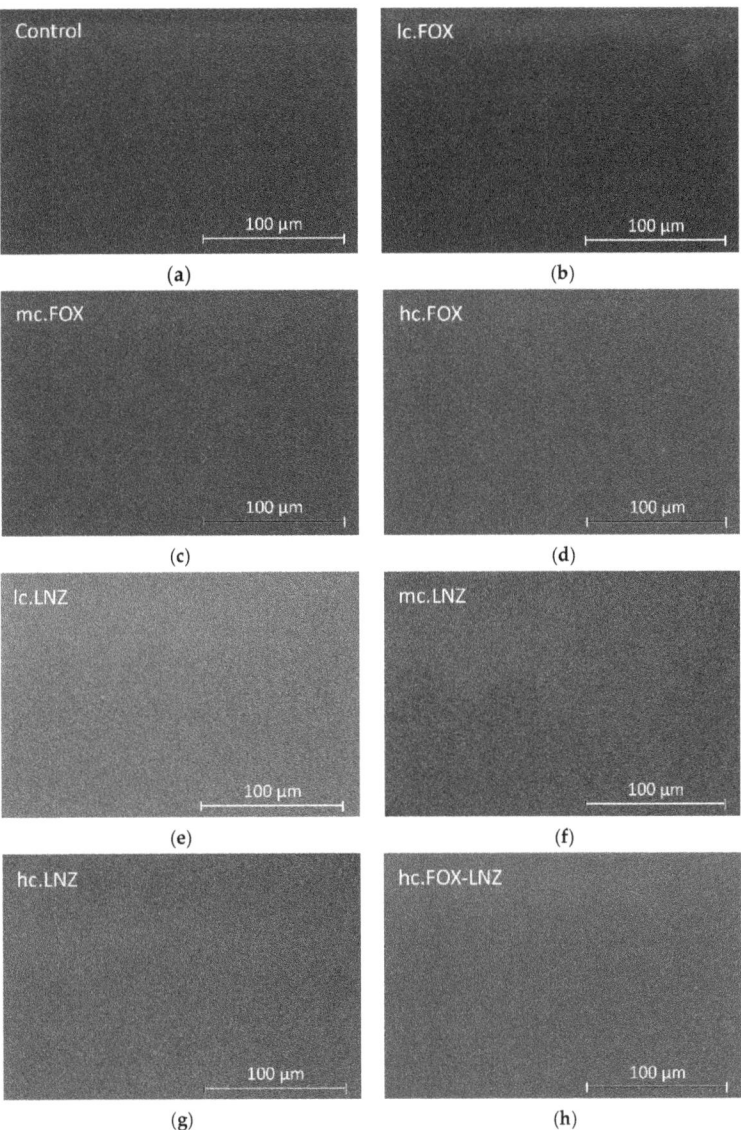

**Figure 7.** SEM micrographs for prepared coatings: (**a**) Control, (**b**) lc.FOX, (**c**) mc.FOX, (**d**) hc.FOX, (**e**) lc.LNZ, (**f**) mc.LNZ, (**g**) hc.LNZ, and (**h**) hc.FOX-LNZ deposited on TiPM substrates.

For a better understanding and a more detailed morphology study of the achieved coatings, AFM was used. Figure 8 compares 5 × 5 µm² AFM images of the surface morphologies of the sol–gel coated samples. In general, topographical images for the antibiotic-loaded coatings displayed a microscale roughness with irregular-shaped and randomly grown granular surfaces, with a significant number of protrusions (hills) exhibiting heights in the range of 0.03–0.16 µm. On the other hand, the size of clusters in the Control coating (Figure 8(a.1–a.3)) was much smaller, with measured heights in the range of 5–25 nm.

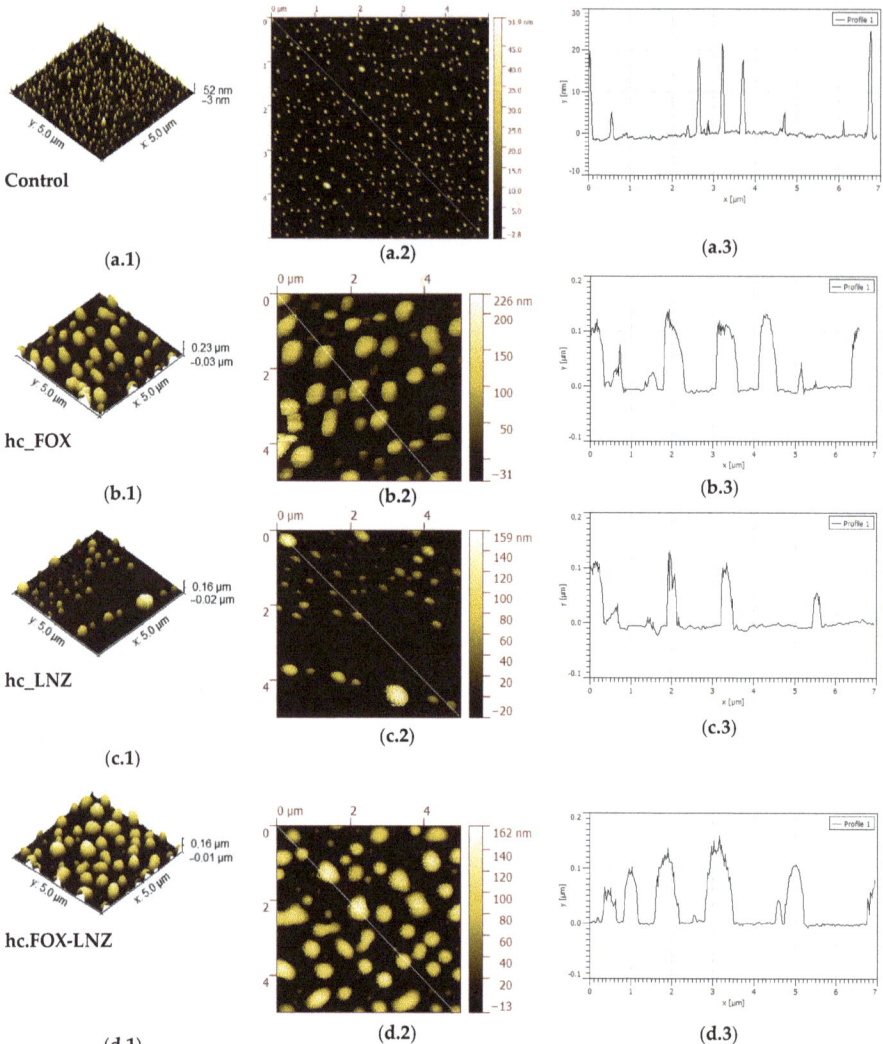

**Figure 8.** Representative non-contact mode AFM 3D topography images (**a.1**), (**b.1**), (**c.1**), and (**d.1**) of the thin films prepared onto TiPM substrates by sol-gel dip coating with (**a.2**), (**b.2**), (**c.2**), and (**d.2**) respective 2D surface morphology and (**a.3**), (**b.3**), (**c.3**), and (**d.3**) section profiles recorded along de white line indicated (scan area = 5 × 5 µm$^2$). From top to bottom: Control, hc.FOX, hc.LNZ, and hc.FOX-LNZ film measurements (Dark colors indicate depressions, light colors protrusions).

A semiquantitative microanalysis was performed using SEM and EDS to confirm the protrusion's identity. A very low vacuum with high magnifications was necessary to obtain these images (Figure 9). According to the obtained compositional analysis for both the hc.FOX and hc.LNZ samples, in the bright areas (corresponding to the protuberances) there was a decrease in the C and Si elements and an increase in N and P elements. In the case of hc.FOX, the Na element was present only in the bright areas.

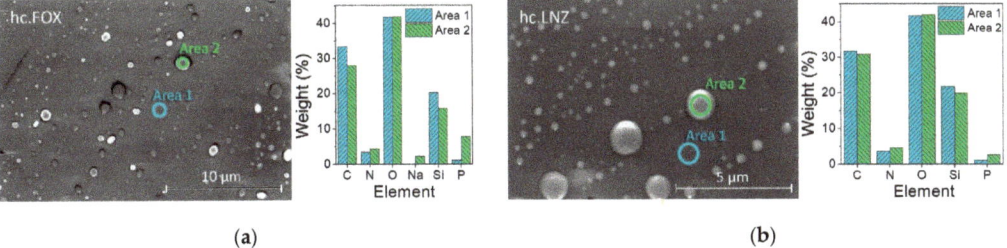

**Figure 9.** SEM/EDS micrographs for prepared coatings (a) hc.FOX (b) and hc.LNZ. (**Left**) SEM micrographs. (**Right**) Graphical results of EDS scan area on the indicated circles (green for the bright area and blue for the dark area).

In previous works where the Control coating was characterized, these formations were discovered and an increase in the size of these protuberances research was evidenced with the increase in the organophosphite concentration [29]. In this, in addition to the emergence of these protuberances associated with the organophosphite introduction, an increase in the concentration of elements associated with the antibiotics (Na or N) in these areas was also found.

The surface texture of the coatings was studied by calculating several important roughness parameters from $40 \times 40$ µm$^2$ scans. These parameters play an important role in the patient's biological response to the prosthesis. Increasing implant-surface roughness is a critical factor for the implant fixation period and fixation strength with body tissues since it increases the values of cell adhesion, proliferation, and differentiation [36]. The numerical values for the most typical roughness parameters of the prepared coatings are summarized in Table 5.

**Table 5.** Surface roughness parameters and their standard deviations obtained by AFM images analysis.

| Surface | $S_a$ (nm) | $S_q$ (nm) | $S_z$ (µm) | $S_{sk}$ (µm) | $S_{ku}$ (µm) |
|---|---|---|---|---|---|
| TiPM | 169.79 ± 42.80 | 227.96 ± 45.05 | 2.111 ± 0.564 | 0.659 ± 1.548 | 6.330 ± 3.872 |
| Control | 9.61 ± 2.43 | 12.43 ± 3.03 | 0.087 ± 0.025 | 0.493 ± 0.674 | 3.809 ± 1.248 |
| lc.FOX | 22.30 ± 5.45 | 29.21 ± 5.85 | 0.322 ± 0.200 | 1.865 ± 2.323 | 17.479 ± 26.406 |
| mc.FOX | 49.58 ± 8.22 | 76.29 ± 9.42 | 0.642 ± 0.116 | 2.393 ± 0.915 | 10.741 ± 3.080 |
| hc.FOX | 51.83 ± 13.53 | 79.30 ± 27.82 | 1.014 ± 0.690 | 2.707 ± 2.541 | 23.482 ± 27.116 |
| lc.LNZ | 63.60 ± 27.16 | 131.92 ± 72.30 | 1.630 ± 0.760 | 4.864 ± 2.014 | 36.828 ± 24.462 |
| mc.LNZ | 43.85 ± 15.36 | 89.18 ± 40.28 | 1.488 ± 0.620 | 5.580 ± 2.802 | 58.149 ± 32.564 |
| hc.LNZ | 38.95 ± 13.84 | 67.85 ± 55.44 | 0.870 ± 0.748 | 2.553 ± 3.185 | 26.995 ± 34.596 |
| hc.FOX-LNZ | 62.04 ± 21.79 | 127.09 ± 92.40 | 1.467 ± 0.959 | 4.429 ± 3.075 | 39.812 ± 34.642 |

$S_a$ = arithmetic mean height; $S_q$ = root mean square height; $S_z$ = maximum height; $S_{sk}$ = skewness and $S_{ku}$ = kurtosis of the surface. The values are average of 8 measurements (maps of $40 \times 40$ µm).

The lower roughness parameters (average roughness $S_a$, RMS roughness $S_q$, and maximum height $S_z$) measured for the coatings proved that the coating application smoothed the surface compared to the bare TiPM substrate. This result supports SEM micrographs where homogeneity and non-existence of uncoated substrate areas in the coatings were apparent.

The values for these same roughness parameters were lower for the Control sample, confirming what was observed in the AFM images where the Control coating presents much smaller protrusions compared to the other coatings.

The skewness ($S_{sk}$) positive values expose that the surface peaks and asperities are predominant over valleys in the studied coatings. In addition, kurtosis ($S_{ku}$) reports the sharpness of profile peaks that in all cases indicate the presence of inordinately high peaks.

Among the LNZ-containing coatings, no significant differences were found in any of the five roughness parameters studied.

Among the coatings containing cefoxitin, lc.FOX was statistically different from mc.FOX and hc.FOX when comparing $S_a$ and $S_q$. For maximum height ($S_z$), mc.FOX was different from hc.FOX.

Figure 10 shows the coating thickness measurements. All the synthesized coatings had values between 10 and 15 µm without statistically significant differences between them.

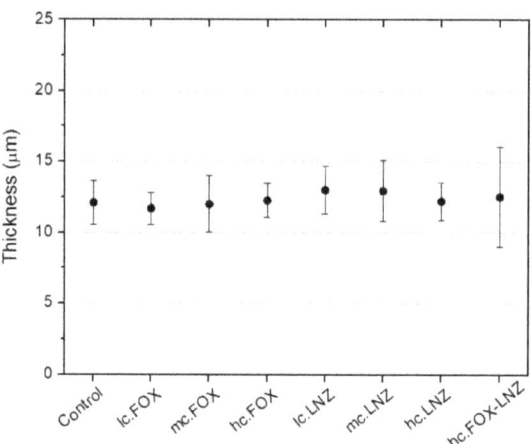

**Figure 10.** Thickness of sol-gel coatings prepared onto TiPM by dip-coating. No statistically significant differences were found between coatings (2-way ANOVA, $p < 0.05$). Bars indicate the standard deviations.

The wettability of the formulations and the substrate were studied by means of contact angle measurements. The results are shown in Figure 11. The contact angle value was higher in all the coatings compared to the substrate as a result, among other factors, of the decrease in surface roughness. Furthermore, between the coatings, there were also significant differences. The most hydrophilic coating was the one with the combination of antibiotics.

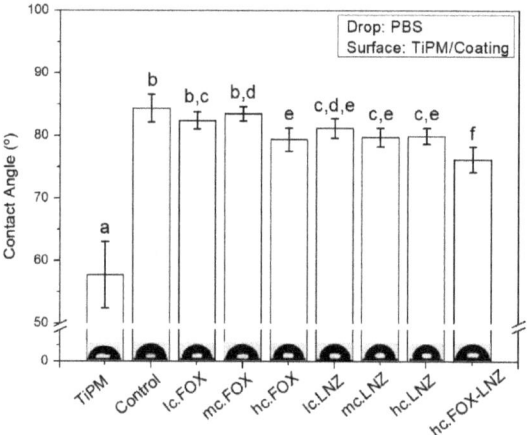

**Figure 11.** Contact angle results for films deposited on TiPM substrates. Bars with different letters denote statistical significance, $p < 0.05$ one-way ANOVA using Tukey HSD method. N = 10–12. All values are means ± SE.

## 4. Discussion

In this study, eight antibiotics (GEN, FOX, VAN, DCX, CLI, AMP, LNZ, AMB) were separately introduced into organic-inorganic sol-gel coatings for TiPM metal prostheses. Furthermore, coatings with different FOX and LNZ concentrations (separately and mixed) were synthesized and examined in terms of morphology and chemical characteristics.

The formation of smooth, uniform, homogeneous, and crack-free coatings was successfully achieved after introducing most of the antibiotics. Isolated cracks were found in coatings loaded with AMP and AMB, possibly related to the poor solubility or insolubility of these antibiotics in water.

The obtained differences in the thickness of the coatings could be related to multiple factors such as molecular weight, chemical nature, solubility, and concentration of the incorporated antibiotic. Although more testing would be needed to determine the contribution of each of these factors to the final thickness, the molecular weight is a highly influential factor; this can be observed in the five coatings (GEN, FOX, VAN, DCX, CLI) with a very similar amount by weight of added antibiotic in the sol-gel and yet markedly different thicknesses, which correlate well with the molecular weight, increasing for higher molecular weight antibiotics (GEN and VAN), as shown in Figure 3.

Coatings had a microscale roughness with irregular-shaped and randomly grown granular surfaces. These granular growths, together with the matrix, corresponded to two well-differentiated micro-phases. The EDS result indicates that the hills (bright areas) were the result of antibiotic and organophosphite accumulations in these areas. NMR analysis supports this claim since the increase in the number of P=O bonds in the network crosslinking occurred in formulations containing antibiotics. With these results, it could be theorized that cluster growth is due to a spontaneous organization in domains or micro-phases ("two-phase systems"), obtaining inorganic segments of bioactive silica-rich regions (matrix) and organic segments of methacryloxypropyl chain and P-O-Si bonds-rich regions (protrusions). The antibiotics would be housed within the organic segment, manifesting in a greater volume in that area. This organization could be possible due to non-covalent interactions (e.g., hydrogen bonds, Van der Waals forces, electrostatic forces, interactions) without external intervention. The factors that could influence this organization are the molecular weight of the introduced antibiotic and, possibly, its hydrophilicity. In the antibiotic-free coating, these microphases were evidenced, with the organophosphite being the contributor to the granular formations, although significant differences were found between the size of the protrusions in this coating and the antibiotic-laden coatings. This result shows that the antibiotics only lodge in these protrusions, being able to favor their formation and increase their size, but the spontaneous organization is not due to their introduction.

In fact, chemical studies suggested the absence of chemical bonds between antibiotics and the siloxane network, despite the slight modification of the network crosslinking caused by the antibiotic introduction. FT-IR results inferred the apparent non-modification of the siloxane network due to the introduction of antibiotics and the absence of chemical bonds between antibiotics and the siloxane network. However, bands related to the introduced antibiotics may be present, but the low concentrations of the antibiotics would only result in very weak bands, if any, being indistinguishable from the main bands of the silicate network.

Solid-state $^{29}$Si-NMR results suggested the favoring of the participation of the phosphorus-based compound in the network crosslinking when an antibiotic is introduced. This phenomenon could be explained because the antibiotics represent a steric hindrance in the network and both, tris(tri-methylsilyl) phosphite and MAPTMS (due to its long organic chain), give the antibiotic room to accommodate.

TGA results, as FT-IR and NMR results, indicated a decrease in network crosslinking in formulations with antimicrobials. Despite all the formulations presenting a greater mass loss compared to Control sample, in most of the cases, it is difficult to elucidate any conclusion on this matter since it was only a 1 wt.% difference. The apparent differences in mass loss could be related to the antibiotic-loaded amounts and their influence on the network crosslinking. The peaks coinciding with the degradation of the antibiotics evidenced that with increasing their concentration, a greater contribution to the degradation occurred (see insets Figure 6B).

Despite these chemical studies, being able to demonstrate the influence or chemical bonds between the antibiotics introduced and the siloxane network is not a straightforward task. The large difference between the number of moles of both parts during the synthesis is the main obstacle.

The relationship between concentration, the introduction of antibiotics, and the size or distribution of protrusions was unclear. From the AFM analysis, no behavior was found that relates to all these parameters. These protuberances behaved quite randomly in coatings loaded with antibiotics. However, all the resulting surface morphologies in the antibiotic coatings were very promising in terms of decreasing microbial adherence. Previous studies proved that these morphologies have an antimicrobial effect only related to their microstructure [37,38].

The wettability differences found between coatings could be attributed to their roughness. In contact angle measurements where the deposited drop has a low viscosity, it is reported that the increase in roughness decreases the contact angle because of more surface area [32]. These results can be contrasted with the obtained roughness parameters, observing an opposite trend between both parameters. Wettability is a key factor affecting not only protein adsorption and cell attachment but also the degradation kinetics of coatings when in contact with a physiological medium. Hydrophobic surfaces or surfaces with an intermediate wettability (60–90°) are the suggestion by most literature to achieve good biocompatibility through protein adsorption and to prevent bacterial adhesion [39–41].

In coatings loaded with FOX and LNZ (separately), no conclusions could be reached about the relationship between concentration, type of introduced antibiotics, and size or distribution of protrusions. Neither were there great differences found in network crosslinking, roughness, thickness, or wettability. However, the coating loaded with the two antibiotics simultaneously exhibited weaker network crosslinking along with a more hydrophilic surface than the rest. This result indicates that the concentration of antibiotics can influence the final characteristics of the coating, although it is necessary to vary the concentration considerably to obtain significant differences.

The roughness and hydrophilicity results make the coating loaded with the two antibiotics a good candidate for combined therapy to fight different bacteria, including *S. aureus*. However, in the next step, biological assessments will be needed to shed more light on the antibacterial effectiveness of the coating.

## 5. Conclusions

The introduction of different antibiotics into biodegradable sol-gel coatings for metal joint prostheses was successfully carried out. Sol-gel formulations were designed as carriers for different antibiotics separately (GEN, FOX, VAN, DCX, CLI, AMP, LNZ, AMB) and for two antibiotics simultaneously (FOX and LNZ). Synthesis containing either AMP or AMB must be optimized to obtain crack-free coatings.

The surface of coatings contains a "two-phase system" with an inorganic silica-rich matrix and organic-rich protrusions. The granular surfaces shown in the coatings resulted in microscale roughness and were attributed to the accumulation of antibiotics and organophosphites in the surface protrusions. The existence of chemical bonds between the introduced antibiotics and the siloxane network could not be demonstrated, although the presence of the antibiotics indirectly modified the network crosslinking. The influence of antibiotic introduction on network crosslinking, roughness, thickness, and wettability is evidenced only for significant variations of introduced concentrations.

The coating loaded with two antibiotics showed somehow a weaker crosslinking. However, introducing two antibiotics broadens the antibacterial spectrum of the coating, resulting in a good strategy to prevent local joint prosthesis infections. Thus, further studies are required to verify the antibacterial activity of those coatings.

**Supplementary Materials:** The following supporting information can be downloaded at: https://www.mdpi.com/article/10.3390/ma15144752/s1, Figure S1: Visual appearance of the powder metallurgical titanium pieces coated with the sol-gel formulations.

**Author Contributions:** Conceptualization, A.J.-M., A.G.-C, and J.E.; methodology, B.T., A.G.-C., M.A.M, and J.J.A.-C.; validation, A.J.-M., A.G.-C, and J.E.; formal analysis, B.T. and A.G.-C.; investigation, B.T., A.G.-C., M.A.M, and J.J.A.-C.; resources, A.J.-M. and J.E.; data curation, B.T., A.G.-C., M.A.M, and J.J.A.-C.; writing—original draft preparation, B.T. and A.G.-C.; writing—review and editing, B.T., A.G.-C., M.A.M., J.J.A.-C., A.J.-M, and J.E.; visualization, B.T. and A.G.-C.; supervision, A.J.-M. and J.E.; project administration, A.J.-M. and J.E.; funding acquisition, A.J.-M. and J.E. All authors have read and agreed to the published version of the manuscript.

**Funding:** This research was funded by the Regional Government of Madrid through the program ADITIMAT (ref. P2018/NMT-4411) and CIBERINFEC-CIBER of Infectious Diseases (CB21/13/00043). B.T. would like to thank the Spanish Ministry of Education, Culture, and Sports for the support through FPU grant (FPU17/05977).

**Institutional Review Board Statement:** Not applicable.

**Informed Consent Statement:** Not applicable.

**Data Availability Statement:** Not applicable.

**Acknowledgments:** Authors also wish to express their gratitude to Juan Pedro Fernández for allowing us to use IMDEA Materials facilities, to the solid $^{29}$Si MASNMR service of the Materials Science Institute of Madrid for the measures, as well as C. Moral for help during SEM inspection.

**Conflicts of Interest:** The authors declare no conflict of interest.

# References

1. Drago, L.; Clerici, P.; Morelli, I.; Ashok, J.; Benzakour, T.; Bozhkova, S.; Alizadeh, C.; Del Sel, H.; Sharma, H.K.; Peel, T.; et al. The world association against infection in orthopaedics and trauma (WAIOT) procedures for microbiological sampling and processing for periprosthetic joint infections (PJIs) and other implant-related infections. *J. Clin. Med.* **2019**, *8*, 933. [CrossRef] [PubMed]
2. Slullitel, P.A.; Oñativia, J.I.; Buttaro, M.A.; Sánchez, M.L.; Comba, F.; Zanotti, G.; Piccaluga, F. State-of-the-art diagnosis and surgical treatment of acute peri-prosthetic joint infection following primary total hip arthroplasty. *EFORT Open Rev.* **2018**, *3*, 434–441. [CrossRef] [PubMed]
3. Pan, C.; Zhou, Z.; Yu, X. Coatings as the useful drug delivery system for the prevention of implant-related infections. *J. Orthop. Surg. Res.* **2018**, *13*, 220. [CrossRef]
4. Hooshmand, B.; Youssef, D.; Riederer, K.M.; Szpunar, S.M.; Bhargava, A. 381. Clinical Outcome of Polymicrobial Prosthetic Joint Infection Managed with Debridement, Antibiotics, and Implant Retention (DAIR). *Open Forum Infect. Dis.* **2019**, *6*, S198. [CrossRef]
5. Flurin, L.; Greenwood-quaintance, K.E.; Patel, R. Microbiology of polymicrobial prosthetic joint infection. *Diagn. Microbiol. Infect. Dis.* **2019**, *94*, 255–259. [CrossRef] [PubMed]
6. Lisoń, J.; Taratuta, A.; Paszenda, Z.; Szindler, M.; Basiaga, M. Perspectives in Prevention of Biofilm for Medical Applications. *Coatings* **2022**, *12*, 197. [CrossRef]
7. Esteban, J.; Vallet-Regí, M.; Aguilera-Correa, J.J. Antibiotics-and heavy metals-based titanium alloy surface modifications for local prosthetic joint infections. *Antibiotics* **2021**, *10*, 1270. [CrossRef]
8. Tobin, E.J. Recent coating developments for combination devices in orthopedic and dental applications: A literature review. *Adv. Drug Deliv. Rev.* **2017**, *112*, 88–100. [CrossRef]
9. Chouirfa, H.; Bouloussa, H.; Migonney, V.; Falentin-Daudré, C. Review of titanium surface modification techniques and coatings for antibacterial applications. *Acta Biomater.* **2019**, *83*, 37–54. [CrossRef]
10. Raphel, J.; Holodniy, M.; Goodman, S.B.; Heilshorn, S.C. Multifunctional coatings to simultaneously promote osseointegration and prevent infection of orthopaedic implants. *Biomaterials* **2016**, *84*, 301–314. [CrossRef]
11. Shahid, A.; Aslam, B.; Muzammil, S.; Aslam, N.; Shahid, M.; Almatroudi, A.; Allemailem, K.S.; Saqalein, M.; Nisar, M.A.; Rasool, M.H.; et al. The prospects of antimicrobial coated medical implants. *J. Appl. Biomater. Funct. Mater.* **2021**, *19*, 22808000211040304. [CrossRef] [PubMed]
12. Blanco, I. Polysiloxanes in theranostics and drug delivery: A review. *Polymers* **2018**, *10*, 755. [CrossRef] [PubMed]
13. Gupta, S.; Majumdar, S.; Krishnamurthy, S. Bioactive glass: A multifunctional delivery system. *J. Control. Release* **2021**, *335*, 481–497. [CrossRef] [PubMed]
14. Baino, F.; Fiorilli, S.; Vitale-Brovarone, C. Composite biomaterials based on sol-gel mesoporous silicate glasses: A review. *Bioengineering* **2017**, *4*, 15. [CrossRef] [PubMed]

15. Asri, R.I.M.; Harun, W.S.W.; Hassan, M.A.; Ghani, S.A.C.; Buyong, Z. A review of hydroxyapatite-based coating techniques: Sol-gel and electrochemical depositions on biocompatible metals. *J. Mech. Behav. Biomed. Mater.* **2016**, *57*, 95–108. [CrossRef]
16. Nasr Azadani, R.; Sabbagh, M.; Salehi, H.; Cheshmi, A.; Beena Kumari, A.R.; Erabi, G. Sol-gel: Uncomplicated, routine and affordable synthesis procedure for utilization of composites in drug delivery: Review. *J. Compos. Compd.* **2021**, *2*, 57–70. [CrossRef]
17. Figueira, R.; Fontinha, I.; Silva, C.; Pereira, E. Hybrid Sol-Gel Coatings: Smart and Green Materials for Corrosion Mitigation. *Coatings* **2016**, *6*, 12. [CrossRef]
18. Catauro, M.; Ciprioti, S.V. Characterization of hybrid materials prepared by sol-gel method for biomedical implementations. A critical review. *Materials* **2021**, *14*, 1788. [CrossRef]
19. Radin, S.; Ducheyne, P. Controlled release of vancomycin from thin sol-gel films on titanium alloy fracture plate material. *Biomaterials* **2007**, *28*, 1721–1729. [CrossRef]
20. Qu, H.; Knabe, C.; Radin, S.; Garino, J.; Ducheyne, P. Percutaneous external fixator pins with bactericidal micron-thin sol-gel films for the prevention of pin tract infection. *Biomaterials* **2015**, *62*, 95–105. [CrossRef]
21. Nichol, T.; Callaghan, J.; Townsend, R.; Stockley, I.; Hatton, P.V.; Le Maitre, C.; Smith, T.J.; Akid, R. The antimicrobial activity and biocompatibility of a controlled gentamicin-releasing single-layer sol-gel coating on hydroxyapatite-coated titanium. *Bone Jt. J.* **2021**, *103-B*, 522–529. [CrossRef] [PubMed]
22. Toirac, B.; Garcia-Casas, A.; Cifuentes, S.C.; Aguilera-Correa, J.J.; Esteban, J.; Mediero, A.; Jiménez-Morales, A. Electrochemical characterization of coatings for local prevention of Candida infections on titanium-based biomaterials. *Prog. Org. Coat.* **2020**, *146*, 105681. [CrossRef]
23. Romera, D.; Toirac, B.; Aguilera-Correa, J.J.; García-Casas, A.; Mediero, A.; Jiménez-Morales, A.; Esteban, J. A biodegradable antifungal-loaded sol-gel coating for the prevention and local treatment of yeast prosthetic-joint infections. *Materials* **2020**, *13*, 3144. [CrossRef] [PubMed]
24. Aguilera-Correa, J.J.; Garcia-Casas, A.; Mediero, A.; Romera, D.; Mulero, F.; Cuevas-López, I.; Jiménez-Morales, A.; Esteban, J. A New Antibiotic-Loaded Sol-Gel Can Prevent Bacterial Prosthetic Joint Infection: From in vitro Studies to an in vivo Model. *Front. Microbiol.* **2020**, *10*, 2935. [CrossRef] [PubMed]
25. Chen, H.; Du, Y.; Xia, Q.; Li, Y.; Song, S.; Huang, X. Role of linezolid combination therapy for serious infections: Review of the current evidence. *Eur. J. Clin. Microbiol. Infect. Dis.* **2020**, *39*, 1043–1052. [CrossRef] [PubMed]
26. Beitdaghar, M.; Ahmadrajabi, R.; Karmostaji, A.; Saffari, F. In vitro activity of linezolid alone and combined with other antibiotics against clinical enterococcal isolates. *Wien. Med. Wochenschr.* **2019**, *169*, 215–221. [CrossRef]
27. Wang, M.; Tang, T. Surface treatment strategies to combat implant-related infection from the beginning. *J. Orthop. Transl.* **2019**, *17*, 42–54. [CrossRef]
28. El Hadad, A.A.; Carbonell, D.; Barranco, V.; Jiménez-Morales, A.; Casal, B.; Galván, J.C. Preparation of sol-gel hybrid materials from γ-methacryloxypropyltrimethoxysilane and tetramethyl orthosilicate: Study of the hydrolysis and condensation reactions. *Colloid Polym. Sci.* **2011**, *289*, 1875–1883. [CrossRef]
29. Garcia-Casas, A.; Aguilera-Correa, J.J.; Mediero, A.; Esteban, J.; Jimenez-Morales, A. Functionalization of sol-gel coatings with organophosphorus compounds for prosthetic devices. *Colloids Surf. B Biointerfaces* **2019**, *181*, 973–980. [CrossRef]
30. Bolzoni, L.; Ruiz-Navas, E.M.; Gordo, E. Powder metallurgy CP-Ti performances: Hydride-dehydride vs. sponge. *Mater. Des.* **2014**, *60*, 226–232. [CrossRef]
31. Alcantara-Garcia, A.; Garcia-Casas, A.; Jimenez-Morales, A. Electrochemical study of the synergic effect of phosphorus and cerium additions on a sol-gel coating for Titanium manufactured by powder metallurgy. *Prog. Org. Coat.* **2018**, *124*, 267–274. [CrossRef]
32. Ramezanzadeh, M.; Mahdavian, M.; Alibakhshi, E.; Akbarian, M.; Ramezanzadeh, B. Evaluation of the corrosion protection performance of mild steel coated with hybrid sol-gel silane coating in 3.5 wt.% NaCl solution. *Prog. Org. Coat.* **2018**, *123*, 190–200. [CrossRef]
33. Juan-Díaz, M.J.; Martínez-Ibáñez, M.; Hernández-Escolano, M.; Cabedo, L.; Izquierdo, R.; Suay, J.; Gurruchaga, M.; Goñi, I. Study of the degradation of hybrid sol–gel coatings in aqueous medium. *Prog. Org. Coat.* **2014**, *77*, 1799–1806. [CrossRef]
34. El Hadad, A.A.; Barranco, V.; Jiménez-Morales, A.; Hickman, G.J.; Galván, J.C.; Perry, C.C. Triethylphosphite as a network forming agent enhances in vitro biocompatibility and corrosion protection of hybrid organic-inorganic sol-gel coatings for Ti6Al4V alloys. *J. Mater. Chem. B* **2014**, *2*, 7955–7963. [CrossRef]
35. Romero-Gavilán, F.; Barros-Silva, S.; Garcia-Cañadas, J.; Palla, B.; Izquierdo, R.; Gurruchaga, M.; Goñi, I.; Suay, J. Control of the degradation of silica sol-gel hybrid coatings for metal implants prepared by the triple combination of alkoxysilanes. *J. Non-Cryst. Solids* **2016**, *453*, 66–73. [CrossRef]
36. El Hadad, A.A.; Barranco, V.; Jiménez-Morales, A.; Peón, E.; Hickman, G.J.; Perry, C.C.; Galván, J.C. Enhancing in vitro biocompatibility and corrosion protection of organic-inorganic hybrid sol-gel films with nanocrystalline hydroxyapatite. *J. Mater. Chem. B* **2014**, *2*, 3886–3896. [CrossRef]
37. Wu, S.; Zuber, F.; Maniura-Weber, K.; Brugger, J.; Ren, Q. Nanostructured surface topographies have an effect on bactericidal activity. *J. Nanobiotechnol.* **2018**, *16*, 20. [CrossRef]

38. Elbourne, A.; Crawford, R.J.; Ivanova, E.P. Nano-structured antimicrobial surfaces: From nature to synthetic analogues. *J. Colloid Interface Sci.* **2017**, *508*, 603–616. [CrossRef] [PubMed]
39. Stallard, C.P.; McDonnell, K.A.; Onayemi, O.D.; O'Gara, J.P.; Dowling, D.P. Evaluation of protein adsorption on atmospheric plasma deposited coatings exhibiting superhydrophilic to superhydrophobic properties. *Biointerphases* **2012**, *7*, 31. [CrossRef]
40. Dou, X.Q.; Zhang, D.; Feng, C.; Jiang, L. Bioinspired Hierarchical Surface Structures with Tunable Wettability for Regulating Bacteria Adhesion. *ACS Nano* **2015**, *9*, 10664–10672. [CrossRef]
41. Yuan, Y.; Hays, M.P.; Hardwidge, P.R.; Kim, J. Surface characteristics influencing bacterial adhesion to polymeric substrates. *RSC Adv.* **2017**, *7*, 14254–14261. [CrossRef]

Article

# Sol-Gel Synthesis and Characterization of the Cu-Mg-O System for Chemical Looping Application

Timofey M. Karnaukhov [1,2], Grigory B. Veselov [1], Svetlana V. Cherepanova [1,3] and Aleksey A. Vedyagin [1,*]

1. Department of Materials Science and Functional Materials, Boreskov Institute of Catalysis SB RAS, 630090 Novosibirsk, Russia; karnaukhovtm@catalysis.ru (T.M.K.); g.veselov@catalysis.ru (G.B.V.); svch@catalysis.ru (S.V.C.)
2. Faculty of Natural Sciences, Novosibirsk State University, 630090 Novosibirsk, Russia
3. Physical Faculty, Novosibirsk State University, 630090 Novosibirsk, Russia
* Correspondence: vedyagin@catalysis.ru

**Abstract:** A sol-gel technique was applied to prepare the two-component oxide system Cu-Mg-O, where MgO plays the role of oxide matrix, and CuO is an active chemical looping component. The prepared samples were characterized by scanning electron microscopy, low-temperature nitrogen adsorption, and X-ray diffraction analysis. The reduction behavior of the Cu-Mg-O system was examined in nine consecutive reduction/oxidation cycles. The presence of the MgO matrix was shown to affect the ability of CuO towards reduction and re-oxidation significantly. During the first reduction/oxidation cycle, the main characteristics of the oxide system (particle size, crystallization degree, etc.) undergo noticeable changes. Starting from the third cycle, the system exhibits a stable operation, providing the uptake of similar hydrogen amounts within the same temperature range. Based on the obtained results, the two-component Cu-Mg-O system can be considered as a prospective chemical looping agent.

**Keywords:** sol-gel synthesis; magnesium oxide matrix; copper oxide; chemical looping; reduction/oxidation cycling

**Citation:** Karnaukhov, T.M.; Veselov, G.B.; Cherepanova, S.V.; Vedyagin, A.A. Sol-Gel Synthesis and Characterization of the Cu-Mg-O System for Chemical Looping Application. *Materials* **2022**, *15*, 2021. https://doi.org/10.3390/ma15062021

Academic Editor: Aleksej Zarkov

Received: 2 February 2022
Accepted: 7 March 2022
Published: 9 March 2022

**Publisher's Note:** MDPI stays neutral with regard to jurisdictional claims in published maps and institutional affiliations.

**Copyright:** © 2022 by the authors. Licensee MDPI, Basel, Switzerland. This article is an open access article distributed under the terms and conditions of the Creative Commons Attribution (CC BY) license (https://creativecommons.org/licenses/by/4.0/).

## 1. Introduction

The catalysts based on copper and copper oxide are traditionally attractive for researchers and industrialists mainly due to their high activity in a number of industrially important processes at relatively low temperatures (about 100–300 °C). For instance, copper-containing systems were intensively studied in the water gas shift reaction [1–8] and steam reforming of methanol and methane [9–12]. In these cases, the main advantages of using copper as an active component are its low cost if compared with precious metals, high dispersity of copper species, and their strong interaction with the support that allows tuning the catalytic properties. Besides the mentioned applications, copper-based catalysts are highly efficient in the processes of total and partial (selective) oxidation [13–25], hydrogenation [26–29], dehydrogenation of alcohols [30–33], reduction of nitrogen oxides [34,35], etc.

Such a high activity of the copper-containing systems in the oxidation reactions is stipulated, in many ways, by the high reactive capacity of copper oxide CuO. As a rule, the reactions with its involvement are highly exothermic, while the kinetic parameters of the copper oxide reduction are quite advantageous. All this causes the increased attention to CuO as a prospective oxygen carrier for the developed chemical looping technologies, among which the chemical looping combustion processes [36–38] should be mentioned specially.

At present, it is not a secret that the size of the active component significantly affects its reactive capacity and, therefore, catalytic activity. In most cases, the higher the dispersion of the particles, the higher its activity. The Cu/CuO-containing systems are not an exception [39–41]. On the other hand, one of the main problems connected with these systems is their instability during the utilization at elevated temperatures. The copper nanoparticles have an addiction to their sintering and agglomeration [42]. This trend is more crucial in the case of nanocrystalline copper-based materials. Thus, the analysis of the melting behavior of copper gives the following predictions. If the bulk copper undergoes melting at the temperature of 1083 °C, then the copper particles of 20 nm in size should start to melt at ~1000 °C. Further decrease in particle size to 10 and 5 nm is expected to result in the melting temperature values of ~750 and ~500 °C. Therefore, an approaching of the process temperature to the expected melting temperature should lead to a noticeable increase in the lability of copper species and their agglomeration into the thermodynamically favorable large particles. In order to solve the mentioned problem, copper and its oxide are deposited on various supports, and the metal-support interactions play the key roles here, providing the long-term stability of the supported copper species during the catalyst exploitation [1,9,30,32–34,43–45].

To keep the high dispersity of the active component during the catalytic process, the small particles of this component should be uniformly distributed in the matrix of the support [46]. Aluminum and zirconium oxides are known to provide good enough thermal stability of the copper catalysts [34,47]. Moreover, alumina can form a number of joint phases with copper, including spinel materials exhibiting high activity in the redox reaction [48,49]. Similar effects are reported for the systems with iron oxide used as support [1]. In addition, the copper nanoparticles can be stabilized using zeolites. The latter facilitates the adsorption of the reagents, thus improving the catalytic activity [50]. Titanium oxide is less thermally stable; however, it provides the chemical stability of the copper catalysts towards poisoning [51–53]. Oppositely, carbonaceous supports possess high thermal stability and attractive mechanical and catalytic properties [30,32,54–56].

The stability of the copper-containing systems is principally defined by the preparation method. The sol-gel techniques give a number of advantages, including the mentioned small particle size and uniform distribution of one component within the matrix of another component serving as the support. Recently, such a sol-gel approach was successfully applied to prepare the two- and three-component oxide systems based on the MgO matrix [57–60]. Magnesium oxide is a unique material possessing attractive textural characteristics such as developed specific surface area and porosity. Its melting temperature is 2802 °C. Therefore, MgO is resistant to sintering even at significantly elevated process temperatures. All the mentioned properties provide the high thermal stability of the active components distributed within the MgO matrix. In addition, magnesium oxide can also form joint phases with the oxides of the majority of transition metals, thus, strengthening the metal-support interaction. This feature gives a broad spectrum of possibilities for controllable tuning of the catalytic and redox properties [61].

For the oxide systems obtained via the sol-gel approach, the preparation conditions significantly affect the characteristics of the final materials and, therefore, define their area of application. For instance, the amount of used alcohol influences the porosity and the agglomeration degree of the primary particles [62]. The next set of effects is connected to the applied alkaline agent, structure-directing surfactant, and calcination temperature [63–66]. These factors determine the textural and morphological features of the oxides. Among the advantages of sol-gel synthesis of two-component MgO-based systems, a uniform distribution of the second phase within the MgO matrix should be mentioned [65,66]. Such a distribution provides appropriate dispersity of the second phase. Thus, Barad et al. reported the suppression of the grain growth process by confining yttrium oxide nanocrystallites within a polycrystalline magnesium oxide [67]. On the other hand, the second component can interact with the matrix with the formation of new joint phases [65,68,69]. Both the size of the distributed particles and the presence of the joint phases govern the catalytic and optical properties [65,68–71].

In the present work, the oxide system Cu-Mg-O containing 15 wt% of CuO was synthesized by the sol-gel method. Recently, the effect of the CuO concentration on the textural properties was studied; the loading of 15 wt% was found to be an optimal value [72]. The copper salt-precursor was added at the stage of the gel formation. The performed investigation of the oxide system by low-temperature nitrogen adsorption, scanning electron microscopy, X-ray diffraction analysis, and temperature-programmed reduction/oxidation has revealed that the copper species are stabilized within the MgO matrix and keep their dispersity during the consecutive reduction/oxidation cycles.

## 2. Materials and Methods

### 2.1. Preparation of the Samples

#### 2.1.1. Sol-Gel Synthesis of Cu-Mg-OH and Cu-Mg-O Systems

The two-component xerogel (Cu-Mg-OH) and oxide (Cu-Mg-O) systems were obtained via a sol-gel technique. The piece of magnesium ribbon (1 g, purity of 99.9%, Sigma-Aldrich, St. Louis, MO, USA) was dissolved in 43 mL of methanol (Avantor Performance Materials, Gliwice, Poland). In order to stabilize the gel, toluene (Component-Reaktiv, Moscow, Russia) was added into the formed solution of magnesium methoxide. The methanol-to-toluene ratio was 1:1. Then, the solution was dropwise mixed with an aqueous solution of copper (II) nitrate (Baltic Enterprise, Saint-Petersburg, Russia). The obtained gel was dried at room temperature for 2 h. After an additional drying at 200 °C for 2 h, the xerogel sample was denoted as Cu-Mg-OH. Finally, the sample was calcined in the air at slow heating to 500 °C for 6 h. The obtained oxide was labeled as Cu-Mg-O. The resulting copper loading was 15 wt%, with respect to CuO.

#### 2.1.2. Preparation of Bulk CuO

The reference sample of bulk CuO was prepared by the thermal decomposition of copper (II) nitrate (Baltic Enterprise, Saint-Petersburg, Russia) in a furnace at slow heating to 500 °C for 6 h.

### 2.2. Characterization and Testing of the Prepared Samples

#### 2.2.1. Low-Temperature Nitrogen Adsorption

The specific surface area (SSA), pore volume ($V_{pore}$), and average pore diameter ($D_{av}$) of the samples were calculated from the low-temperature nitrogen adsorption data. The pore size distributions were obtained from the isotherms of nitrogen adsorption at 77 K using an ASAP-2400 (Micromeritics, Norcross, GA, USA) instrument. The measurement uncertainty of this method was ±3%.

#### 2.2.2. Differential Thermal Analysis (DTA)

The thermogravimetric (TG), differential thermogravimetric (DTG), and differential scanning calorimetry (DSC) profiles were registered using a Netzsch STA 409 PC/PG simultaneous thermal analyzer (NETZSCH-Gerätebau GmbH, Selb, Germany). The sample was heated in an inert atmosphere (nitrogen) within a temperature range of 25–700 °C with the ramping rate of 5 °C/min. The measurement uncertainty of this method was ±1%.

#### 2.2.3. Scanning Electron Microscopy (SEM)

The microscopic studies of the Cu-Mg-OH and Cu-Mg-O samples were performed using a JSM-6460 (JEOL Ltd., Tokyo, Japan) scanning electron microscope.

#### 2.2.4. Temperature-Programmed Reduction ($H_2$-TPR)

The temperature-programmed reduction of the samples was carried out in a hydrogen flow. The gas mixture containing 10 vol% $H_2$ in Ar was passed through the reactor with the sample with a flow rate of 40 mL/min. The temperature was increased from 20 to 900 °C with a ramping rate of 10 °C/min. Prior to the experiments, each sample was kept in an argon flow at 150 °C to remove the adsorbed water. The hydrogen concentration at the reactor outlet was measured using the standard thermal conductivity detector working at 70 mA. The instrument was calibrated by a direct method, by which the hydrogen-containing flow was controllably diluted with argon. The measurement uncertainty of this method was ±3%.

### 2.2.5. Temperature-Programmed Reduction/Oxidation Cycling

The temperature-programmed reduction/oxidation cycling experiments were performed in a flow-through reactor system, which allows for regulating the gas flows and switching the reductive and oxidative gas mixtures in an automatic mode. The sample (200 mg) was loaded inside the quartz reactor. The reactor was purged with a nitrogen flow of 40 mL/min for 30 min and then fed with the reductive gas mixture containing 10 vol% $H_2$ in $N_2$ (total flow rate of 45.7 mL/min). The reactor was heated from 30 to 700 °C with a ramping rate of 10 °C/min. The hydrogen concentration in the outlet gas mixture was monitored using a hydrogen gas analyzer GAMMA-100 (FSUE "SPA "Analitpribor", Smolensk, Russia). When the temperature reached 700 °C, the reactor was maintained at this temperature for 15 min and was cooled down to 30 °C in a nitrogen flow (40 mL/min). Then, the reactor inlet gas mixture was switched to an air flow (10 mL/min), and the reactor was heated to 500 °C with a ramping rate of 20 °C/min. After remaining at the final point for 30 min, the reactor was cooled down to 30 °C in an air flow, and the reactor inlet gas mixture was switched back to the reductive mixture. The described reduction/oxidation cycles were repeated nine times. The measurement uncertainty of this method was ±5%.

### 2.2.6. In Situ X-ray Diffraction Analysis at the Temperature-Programmed Reduction/Oxidation Conditions

The X-ray diffraction (XRD) analysis of the Cu-Mg-O system was performed in an in-situ regime using a D8 diffractometer (Bruker, Karlsruhe, Germany). The reduction and oxidation procedures were carried out directly in the reactor chamber of the diffractometer. Initially, the sample was heated from 25 to 700 °C in hydrogen, cooled down in hydrogen, passivated in helium at 25 °C, and re-oxidized in a gas mixture containing 5 vol% $O_2$ in helium at heating from 25 to 700 °C. Finally, the sample was cooled down to 25 °C in a helium flow. The temperature ramping rate was 10 °C/min. The gas flow rates were 20 mL/min. The registration of the XRD patterns was made within the 2θ range of 15–85° with a step of 0.05° and an accumulation time of 3 s at each temperature point (25, 300, 500, and 700 °C). The lattice parameters were defined from the patterns recorded at room temperature only (without thermal expansion). All the calculations were made using a TOPAS (Bruker, Karlsruhe, Germany) software based on the Rietveld method [73].

## 3. Results and Discussion

First, the as-prepared samples after drying at 200 °C (Cu-Mg-OH, xerogel) were studied by the DTA technique. Figure 1 presents the corresponding TG, DTG, and DSC profiles. It should be noted that the decomposition of bulk $Cu(NO_3)_2$ finishes at around 250 °C. In the case of dispersed particles, for example, when copper nitrate is supported on alumina, the complete decomposition can occur at even lower temperatures [74]. However, in the case of sol-gel-prepared systems, the residual nitrate species still can exist within the MgO matrix even after drying at 200 °C for 2 h. The first weight loss of about 3.8 wt% is observed below 200 °C and is attributed to the removal of the physically adsorbed water molecules. This process is accompanied by a large endothermal effect in the DSC curve. The main weight loss of 30.8 wt% is registered within a range of 240–360 °C. The presence of two endothermic effects in the DSC profile indicates that two processes take place consecutively. The decomposition of the residual species of copper nitrate is followed by the degradation of magnesium hydroxide. As a result, the second process occurs at lower temperatures if compared with the magnesium hydroxide systems reported in the literature [75,76]. As recently found [61], the intercalation of $Cu^{2+}$ ions into the interlayer space of $Mg(OH)_2$ simplifies the dehydration process. The last weight loss of 1.9 wt% is connected to the elimination of the residual species of organic molecules (toluene and methanol) [77].

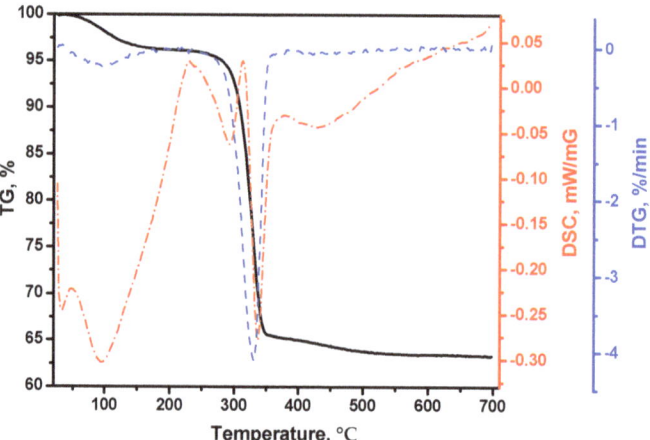

**Figure 1.** TG, DTG, and DSC profiles of the Cu-Mg-OH xerogel sample.

The xerogel Cu-Mg-OH samples and the samples calcined at 500 °C (Cu-Mg-O, oxide) were examined by scanning electron microscopy. The obtained SEM images are shown in Figure 2. Both the samples are represented by poorly crystallized layered agglomerates consisting of nanosized primary particles. At the same time, the xerogel sample (Figure 2a,b) looks fluffy and lacy, while the oxide system (Figure 2c,d) seems to be more dense and compacted. It is natural to suppose that the initial xerogel with a developed structure undergoes collapsing under the calcination conditions.

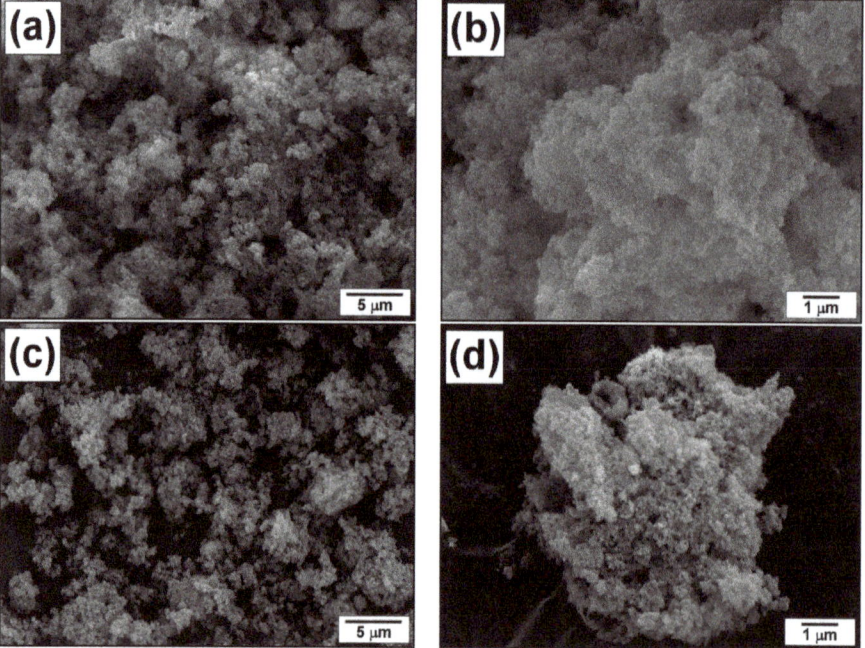

**Figure 2.** SEM images of the prepared samples: Cu-Mg-OH xerogel (**a**,**b**); Cu-Mg-O oxide (**c**,**d**).

The low-temperature nitrogen adsorption studies have revealed the changes in the porous structure that occurred during the calcination procedure. The nitrogen adsorption/desorption isotherms for the xerogel and oxide samples are presented in Figure 3a. Despite both the isotherms being characterized by the presence of a hysteresis loop, their shapes are significantly different. The quantitative parameters calculated from these results are summarized in Table 1. As seen, the SSA value drops down from 410 to 120 m$^2$/g, i.e., more than three times, while the decrease in pore volume is not so crucial. An increase in the average pore size from 12 to 33 nm is connected with the shift of the maximum in pore size distribution towards larger sizes and the disappearance of the small pores (Figure 3b). It should be mentioned that the introduction of copper species inside the MgO matrix significantly worsens the textural properties of the latter. Thus, for pure MgO prepared via the same procedures with hydrolysis by distilled water instead of an aqueous solution of salt, the SSA value was as high as 243 m$^2$/g [59].

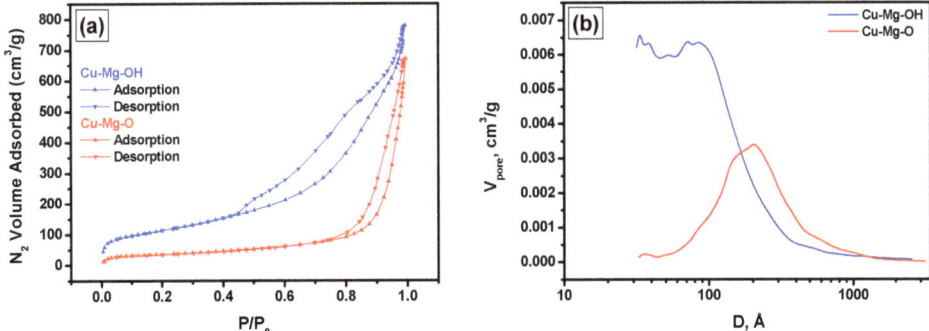

**Figure 3.** Low-temperature nitrogen adsorption/desorption isotherms (a) and pore size distribution (b) for the Cu-Mg-OH xerogel and Cu-Mg-O oxide sample.

**Table 1.** Low-temperature nitrogen adsorption data for the Cu-Mg-OH xerogel and Cu-Mg-O oxide samples.

| Sample | SSA, m$^2$/g | V$_{pores}$, cm$^3$/g | D$_{av}$, nm |
|---|---|---|---|
| Cu-Mg-OH | 410 ± 12 | 1.23 ± 1.1 | 12 ± 2 |
| Cu-Mg-O | 120 ± 4 | 1.04 ± 1.0 | 33 ± 4 |

In order to investigate the effect of the MgO matrix on the reduction/oxidation behavior of copper, the prepared Cu-Mg-O sample was compared with the bulk CuO oxide by the H$_2$-TPR method. Since the Cu-Mg-O system contains just 15 wt% of CuO, the hydrogen uptake intensities were normalized with respect to 1 g of CuO. Figure 4 demonstrates the resulting H$_2$-TPR profiles. As evident, the profiles differ from each other. In the case of bulk CuO, its reduction takes place in a temperature range of 100–300 °C with a maximum at ~240 °C. The H$_2$-TPR profile for the Cu-Mg-O sample has two hydrogen uptake peaks. The main peak appears at the lower temperature of 225 °C, thus, indicating the higher dispersity of the CuO particles if compared with the bulk copper oxide. The second uptake peak has a maximum at ~340 °C and gives a shoulder on the cumulative profile. This peak corresponds to the reduction in copper species strongly interacting with the matrix. In the case of the bulk Cu-Mg-O system, the whole reduction process occurs in a range of 160–400 °C.

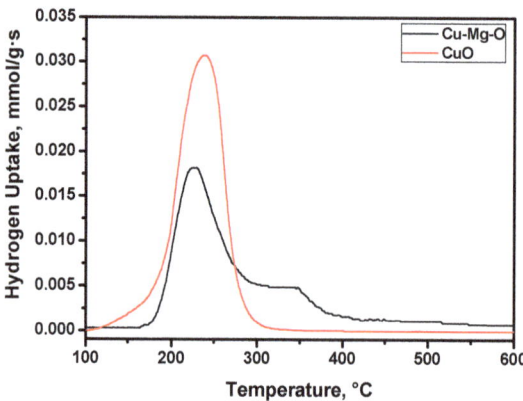

**Figure 4.** H$_2$-TPR profiles for the bulk CuO oxide and prepared Cu-Mg-O system.

It should be remembered that, in the chemical looping concept, a cyclic regime of exploitation is considered. During these cycles, the properties of the oxygen carriers can be noticeably changed; the appropriate system should demonstrate a stable and reproducible behavior. In order to examine the efficiency and prospectivity of the prepared Cu-Mg-O system, it was tested in nine consecutive reduction/oxidation cycles. The obtained TPR profiles are shown in Figure 5. As was already discussed, the first reduction profile of the Cu-Mg-O system consists of two peaks, and the main uptake peak is shifted to lower temperatures if compared with the reference sample of bulk CuO (Figure 4). The second and subsequent reduction profiles of the two-component system are even more shifted to the left side (Figure 5). Moreover, the second reduction cycle is also represented by two uptake peaks, but the first of them has a maximum at ~75 °C. It can be supposed that these easily reducible copper particles are formed from the dispersed copper species that strongly interacted with MgO due to their reduction during the first reduction cycle. Already in the third reduction cycle, this low-temperature peak is disappeared. Note that the total hydrogen uptake is almost constant in all the cycles. The profiles from the third to ninth cycles show no remarkable difference. At the same time, the position of the main uptake peak slightly shifts towards high temperatures from 194 °C for the first cycle to 204 °C for the ninth cycle. It can be concluded here that the MgO matrix provides the thermal stabilization of the dispersed copper species. Just a small sintering effect is observed under the redox cycling conditions.

At the final stage of the research, the Cu-Mg-O system was studied by an in-situ XRD technique. Figure 6 illustrates the corresponding XRD patterns recorded at 25, 300, 500, and 700 °C under reductive (Figure 6a) and oxidative (Figure 6b) conditions. The quantitative parameters obtained from the XRD data are collected in Table 2.

**Table 2.** Quantitative characteristics (average size of crystallites <D>; the lengths a, b, and c of the three cell edges meeting at a vertex, and the angle β between edges a and c) obtained from the XRD data.

| Phase (wt%) | Initial | | After Reduction | | After Oxidation | |
|---|---|---|---|---|---|---|
| | a, Å | <D>, nm | a, Å | <D>, nm | Lattice Parameter | <D>, nm |
| MgO | 4.222(1) | 8 | 4.220(1) | 12 | a = 4.216(1) Å | 16 |
| Cu (13%) | - | - | 3.621(1) | 10 | | |
| CuO (6%) | - | - | - | - | a = 4.691(2) Å<br>b = 3.423(1) Å<br>c = 5.137(2) Å<br>β = 99.42(3) ° | 25 |

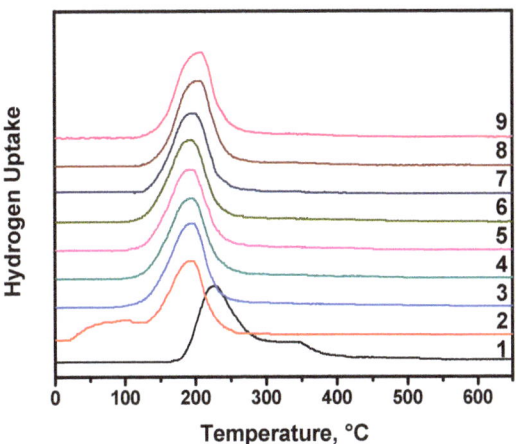

**Figure 5.** TPR profiles for the Cu-Mg-O system registered in nine consecutive reduction/oxidation cycles.

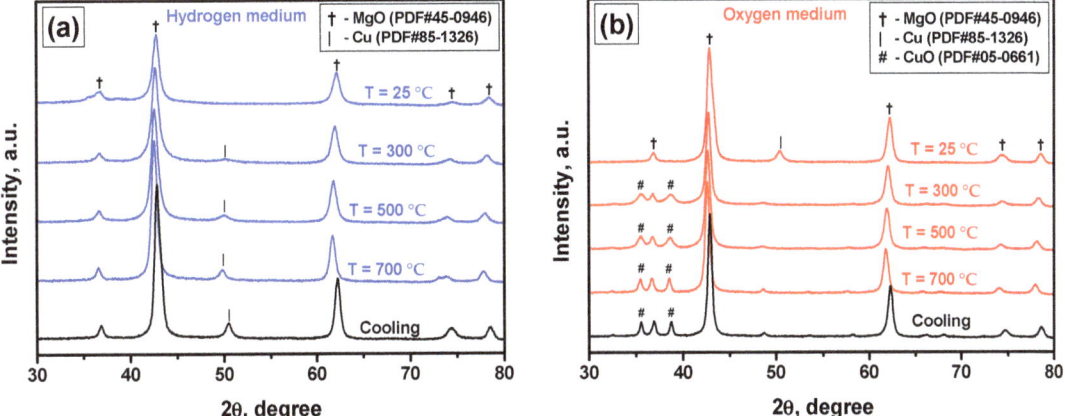

**Figure 6.** XRD patterns of the Cu-Mg-O system recorded in an in-situ regime at hydrogen (a) and oxygen (b) atmospheres.

According to the presented patterns, for the initial sample, only the reflections at ~36, 43, 62, 74, and 78 degrees corresponding to the magnesium oxide phase are detected (Figure 6a). Supposedly, all the copper species are in a roentgen-amorphous state. These species undergo reduction with the temperature rise. The corresponding reflection at ~50 degrees appears already at 300 °C, and its intensity grows with the further temperature increase, indicating the formation of metallic copper nanocrystallites. The average size of the formed metallic copper particles is estimated to be 10 nm (Table 2). The MgO phase also undergoes crystallization, and the reflections become narrow and more intensive. Therefore, the initially observed poorly crystalline structure (see Figure 2) transforms into a nanocrystalline one. The ordering of the oxide structure is accompanied by the enlargement of the primary particles.

The oxidative part of the redox cycle oppositely shows the disappearance of the Cu(0) phase and the formation of the CuO phase (Figure 6b). The corresponding reflections of the oxide phase are seen at ~35 and 38 degrees starting from 300 °C. After 700 °C, they seem to be well-crystallized, with an average size of ~25 nm. These results confirm the previously made assumption concerning the subsequent formation of Cu(0) and CuO phases from the roentgen-amorphous well-dispersed copper species that strongly interacted with the MgO matrix. As follows from the data presented in Table 2, the lattice parameter of the initial sample is enlarged if compared with the conventional magnesium oxide (a = 4.211 Å). After the reduction/oxidation cycle, this parameter approximates the standard value that testifies to the exit of the copper species from the MgO matrix.

## 4. Conclusions

In modern industry, the oxides of transition metals are widely applied in chemical looping processes. However, being used in the bulk form, they undergo rapid agglomeration and sintering that significantly diminishes their efficiency in the redox cycles. Therefore, the use of an inert oxide matrix that preserves the high dispersity of the active component is an actual task. In the present work, the two-component Cu-Mg-O system was synthesized by the sol-gel method. The obtained material possesses a layered mesoporous structure with the developed surface area. The redox behavior of the two-component oxide system differs from that for the bulk CuO reference sample. In the initial state, the copper species are roentgen-amorphous and partly exhibit a strong interaction with the MgO matrix. During the first reduction/oxidation cycle, the final formation of the phase composition of the system takes place. The formed CuO nanoparticles undergo reproducible reduction to metallic Cu(0) nanoparticles with a maximum of hydrogen uptake at near 200 °C. Due to the presence of the MgO matrix, the dispersity of the CuO/Cu(0)-active species remains the same during the redox cycling. Therefore, this system can be considered as a prospective chemical looping material.

**Author Contributions:** Conceptualization, A.A.V.; methodology, A.A.V. and S.V.C.; investigation, T.M.K., G.B.V. and S.V.C.; writing—original draft preparation, T.M.K.; writing—review and editing, A.A.V.; funding acquisition, A.A.V. All authors have read and agreed to the published version of the manuscript.

**Funding:** This work was supported by the Ministry of Science and Higher Education of the Russian Federation [project No. AAAA-A21-121011390054-1].

**Institutional Review Board Statement:** Not applicable.

**Informed Consent Statement:** Not applicable.

**Data Availability Statement:** The data presented in this study are available on request from the corresponding author.

**Acknowledgments:** Characterization of the samples was performed using the equipment of the Center of Collective Use "National Center of Catalysts Research".

**Conflicts of Interest:** The authors declare no conflict of interest. The funders had no role in the design of the study; in the collection, analyses, or interpretation of data; in the writing of the manuscript, or in the decision to publish the results.

## References

1. Lin, X.; Zhang, Y.; Yin, L.; Chen, C.; Zhan, Y.; Li, D. Characterization and catalytic performance of copper-based WGS catalysts derived from copper ferrite. *Int. J. Hydrogen Energy* **2014**, *39*, 6424–6432. [CrossRef]
2. Shishido, T.; Yamamoto, M.; Atake, I.; Li, D.; Tian, Y.; Morioka, H.; Honda, M.; Sano, T.; Takehira, K. Cu/Zn-based catalysts improved by adding magnesium for water–gas shift reaction. *J. Mol. Catal. A-Chem.* **2006**, *253*, 270–278. [CrossRef]
3. Atake, I.; Nishida, K.; Li, D.; Shishido, T.; Oumi, Y.; Sano, T.; Takehira, K. Catalytic behavior of ternary Cu/ZnO/Al$_2$O$_3$ systems prepared by homogeneous precipitation in water-gas shift reaction. *J. Mol. Catal. A-Chem.* **2007**, *275*, 130–138. [CrossRef]
4. Nishida, K.; Atake, I.; Li, D.; Shishido, T.; Oumi, Y.; Sano, T.; Takehira, K. Effects of noble metal-doping on Cu/ZnO/Al$_2$O$_3$ catalysts for water–gas shift reaction. *Appl. Catal. A-Gen.* **2008**, *337*, 48–57. [CrossRef]
5. Sagata, K.; Imazu, N.; Yahiro, H. Study on factors controlling catalytic activity for low-temperature water–gas-shift reaction on Cu-based catalysts. *Catal. Today* **2013**, *201*, 145–150. [CrossRef]
6. Shishido, T.; Nishimura, S.; Yoshinaga, Y.; Ebitani, K.; Teramura, K.; Tanaka, T. High sustainability of Cu–Al–Ox catalysts against daily start-up and shut-down (DSS)-like operation in the water–gas shift reaction. *Catal. Commun.* **2009**, *10*, 1057–1061. [CrossRef]

7. Li, L.; Song, L.; Wang, H.; Chen, C.; She, Y.; Zhan, Y.; Lin, X.; Zheng, Q. Water-gas shift reaction over CuO/CeO$_2$ catalysts: Effect of CeO$_2$ supports previously prepared by precipitation with different precipitants. *Int. J. Hydrogen Energy* **2011**, *36*, 8839–8849. [CrossRef]
8. Tanaka, Y. Water gas shift reaction for the reformed fuels over Cu/MnO catalysts prepared via spinel-type oxide. *J. Catal.* **2003**, *215*, 271–278. [CrossRef]
9. Yousefi Amiri, T.; Moghaddas, J. Cogeled copper–silica aerogel as a catalyst in hydrogen production from methanol steam reforming. *Int. J. Hydrogen Energy* **2015**, *40*, 1472–1480. [CrossRef]
10. Turco, M.; Cammarano, C.; Bagnasco, G.; Moretti, E.; Storaro, L.; Talon, A.; Lenarda, M. Oxidative methanol steam reforming on a highly dispersed CuO/CeO$_2$/Al$_2$O$_3$ catalyst prepared by a single-step method. *Appl. Catal. B-Environ.* **2009**, *91*, 101–107. [CrossRef]
11. Clancy, P.; Breen, J.P.; Ross, J.R.H. The preparation and properties of coprecipitated Cu–Zr–Y and Cu–Zr–La catalysts used for the steam reforming of methanol. *Catal. Today* **2007**, *127*, 291–294. [CrossRef]
12. Díez-Martín, L.; Grasa, G.; Murillo, R.; Martini, M.; Gallucci, F.; van Sint Annaland, M. Determination of the oxidation kinetics of high loaded CuO-based materials under suitable conditions for the Ca/Cu H$_2$ production process. *Fuel* **2018**, *219*, 76–87. [CrossRef]
13. Song, W.; Perez Ferrandez, D.M.; van Haandel, L.; Liu, P.; Nijhuis, T.A.; Hensen, E.J.M. Selective Propylene Oxidation to Acrolein by Gold Dispersed on MgCuCr$_2$O$_4$ Spinel. *ACS Catal.* **2015**, *5*, 1100–1111. [CrossRef]
14. Li, G.; Vassilev, P.; Sanchez-Sanchez, M.; Lercher, J.A.; Hensen, E.J.M.; Pidko, E.A. Stability and reactivity of copper oxo-clusters in ZSM-5 zeolite for selective methane oxidation to methanol. *J. Catal.* **2016**, *338*, 305–312. [CrossRef]
15. Gonçalves, R.V.; Wojcieszak, R.; Wender, H.; Dias, C.S.B.; Vono, L.L.R.; Eberhardt, D.; Teixeira, S.R.; Rossi, L.M. Easy Access to Metallic Copper Nanoparticles with High Activity and Stability for CO Oxidation. *ACS Appl. Mater. Interf.* **2015**, *7*, 7987–7994. [CrossRef]
16. Acharyya, S.S.; Ghosh, S.; Adak, S.; Tripathi, D.; Bal, R. Fabrication of CuCr$_2$O$_4$ spinel nanoparticles: A potential catalyst for the selective oxidation of cycloalkanes via activation of Csp3–H bond. *Catal. Commun.* **2015**, *59*, 145–150. [CrossRef]
17. Liu, C.-H.; Lai, N.-C.; Lee, J.-F.; Chen, C.-S.; Yang, C.-M. SBA-15-supported highly dispersed copper catalysts: Vacuum–thermal preparation and catalytic studies in propylene partial oxidation to acrolein. *J. Catal.* **2014**, *316*, 231–239. [CrossRef]
18. Senanayake, S.D.; Stacchiola, D.; Rodriguez, J.A. Unique Properties of Ceria Nanoparticles Supported on Metals: Novel Inverse Ceria/Copper Catalysts for CO Oxidation and the Water-Gas Shift Reaction. *Acc. Chem. Res.* **2013**, *46*, 1702–1711. [CrossRef]
19. Belin, S.; Bracey, C.L.; Briois, V.; Ellis, P.R.; Hutchings, G.J.; Hyde, T.I.; Sankar, G. CuAu/SiO$_2$ catalysts for the selective oxidation of propene to acrolein: The impact of catalyst preparation variables on material structure and catalytic performance. *Catal. Sci. Technol.* **2013**, *3*, 2944–2957. [CrossRef]
20. Duzenli, D.; Seker, E.; Senkan, S.; Onal, I. Epoxidation of Propene by High-Throughput Screening Method Over Combinatorially Prepared Cu Catalysts Supported on High and Low Surface Area Silica. *Catal. Lett.* **2012**, *142*, 1234–1243. [CrossRef]
21. Bracey, C.L.; Carley, A.F.; Edwards, J.K.; Ellis, P.R.; Hutchings, G.J. Understanding the effect of thermal treatments on the structure of CuAu/SiO$_2$ catalysts and their performance in propene oxidation. *Catal. Sci. Technol.* **2011**, *1*, 76–85. [CrossRef]
22. Tüysüz, H.; Galilea, J.L.; Schüth, F. Highly Diluted Copper in a Silica Matrix as Active Catalyst for Propylene Oxidation to Acrolein. *Catal. Lett.* **2009**, *131*, 49–53. [CrossRef]
23. Su, W.; Wang, S.; Ying, P.; Feng, Z.; Li, C. A molecular insight into propylene epoxidation on Cu/SiO$_2$ catalysts using O$_2$ as oxidant. *J. Catal.* **2009**, *268*, 165–174. [CrossRef]
24. Desyatykh, I.V.; Vedyagin, A.A.; Kotolevich, Y.S.; Tsyrul'nikov, P.G. Preparation of CuO-CeO$_2$ catalysts deposited on glass cloth by surface self-propagating thermal synthesis. *Combust. Explos. Shock Waves* **2011**, *47*, 677–682. [CrossRef]
25. Desyatykh, I.V.; Vedyagin, A.A.; Mishakov, I.V.; Shubin, Y.V. CO oxidation over fiberglasses with doped Cu-Ce-O catalytic layer prepared by surface combustion synthesis. *Appl. Surf. Sci.* **2015**, *349*, 21–26. [CrossRef]
26. Hu, Q.; Fan, G.; Yang, L.; Li, F. Aluminum-Doped Zirconia-Supported Copper Nanocatalysts: Surface Synergistic Catalytic Effects in the Gas-Phase Hydrogenation of Esters. *ChemCatChem* **2014**, *6*, 3501–3510. [CrossRef]
27. Dhakshinamoorthy, A.; Navalon, S.; Sempere, D.; Alvaro, M.; Garcia, H. Reduction of alkenes catalyzed by copper nanoparticles supported on diamond nanoparticles. *Chem. Commun.* **2013**, *49*, 2359–2361. [CrossRef]
28. Ungureanu, A.; Dragoi, B.; Chirieac, A.; Royer, S.; Duprez, D.; Dumitriu, E. Synthesis of highly thermostable copper-nickel nanoparticles confined in the channels of ordered mesoporous SBA-15 silica. *J. Mater. Chem.* **2011**, *21*, 12529–12541. [CrossRef]
29. Munnik, P.; Wolters, M.; Gabrielsson, A.; Pollington, S.D.; Headdock, G.; Bitter, J.H.; de Jongh, P.E.; de Jong, K.P. Copper Nitrate Redispersion To Arrive at Highly Active Silica-Supported Copper Catalysts. *J. Phys. Chem. C* **2011**, *115*, 14698–14706. [CrossRef]
30. Ponomareva, E.A.; Krasnikova, I.V.; Egorova, E.V.; Mishakov, I.V.; Vedyagin, A.A. Ethanol dehydrogenation over copper supported on carbon macrofibers. *Mendeleev Commun.* **2017**, *27*, 210–212. [CrossRef]
31. Shelepova, E.V.; Ilina, L.Y.; Vedyagin, A.A. Theoretical predictions on dehydrogenation of methanol over copper-silica catalyst in a membrane reactor. *Catal. Today* **2019**, *331*, 35–42. [CrossRef]
32. Shelepova, E.V.; Vedyagin, A.A.; Ilina, L.Y.; Nizovskii, A.I.; Tsyrulnikov, P.G. Synthesis of carbon-supported copper catalyst and its catalytic performance in methanol dehydrogenation. *Appl. Surf. Sci.* **2017**, *409*, 291–295. [CrossRef]
33. Vedyagin, A.; Kotolevich, Y.; Tsyrulikov, P.; Khramov, E.; Nizovskii, A. Methanol dehydrogenation over Cu/SiO$_2$ catalysts. *Int. J. Nanotechnol.* **2016**, *13*, 185–199. [CrossRef]

34. Patel, A.; Shukla, P.; Pan, G.T.; Chong, S.; Rudolph, V.; Zhu, Z. Influence of copper loading on mesoporous alumina for catalytic NO reduction in the presence of CO. *J. Environ. Chem. Eng.* **2017**, *5*, 2350–2361. [CrossRef]
35. Torreabreu, C.; Ribeiro, M.; Henriques, C.; Delahay, G. NO TPD and H2-TPR studies for characterisation of CuMOR catalysts The role of Si/Al ratio, copper content and cocation. *Appl. Catal. B-Environ.* **1997**, *14*, 261–272. [CrossRef]
36. Popescu, I.; Tanchoux, N.; Tichit, D.; Marcu, I.-C. Total oxidation of methane over supported CuO: Influence of the MgxAlyO support. *Appl. Catal. A-Gen.* **2017**, *538*, 81–90. [CrossRef]
37. Adanez, J.; Abad, A.; Garcia-Labiano, F.; Gayan, P.; de Diego, L.F. Progress in Chemical-Looping Combustion and Reforming technologies. *Prog. Energy Combust. Sci.* **2012**, *38*, 215–282. [CrossRef]
38. Wang, P.; Means, N.; Shekhawat, D.; Berry, D.; Massoudi, M. Chemical-Looping Combustion and Gasification of Coals and Oxygen Carrier Development: A Brief Review. *Energies* **2015**, *8*, 10605–10635. [CrossRef]
39. Pease, R.N.; Taylor, H.S. The Reduction of Copper Oxide by Hydrogen. *J. Am. Chem. Soc.* **2002**, *43*, 2179–2188. [CrossRef]
40. Hierl, R. Surface properties and reduction behavior of calcined $CuO/Al_2O_3$ and $CuO$-$NiO/Al_2O_3$ catalysts. *J. Catal.* **1981**, *69*, 475–486. [CrossRef]
41. Dumas, J.M.; Geron, C.; Kribii, A.; Barbier, J. Preparation of supported copper catalysts. *Appl. Catal.* **1989**, *47*, L9–L15. [CrossRef]
42. Zhan, L.; Zhu, X.; Qin, X.; Wu, M.; Li, X. Sintering mechanism of copper nanoparticle sphere-plate of crystal misalignment: A study by molecular dynamics simulations. *J. Mater. Res. Technol.* **2021**, *12*, 668–678. [CrossRef]
43. Kikugawa, M.; Yamazaki, K.; Shinjoh, H. Characterization and catalytic activity of $CuO/TiO_2$-$ZrO_2$ for low temperature CO oxidation. *Appl. Catal. A-Gen.* **2017**, *547*, 199–204. [CrossRef]
44. Barnes, P.A.; Tiernan, M.J.; Parkes, G.M.B. Sample Controlled Thermal Analysis Temperature Programmed Reduction of Bulk and Supported Copper Oxide. *J. Therm. Anal. Calorim.* **1999**, *56*, 733–737. [CrossRef]
45. Vedyagin, A.A.; Nizovskii, A.I.; Golohvast, K.S.; Tsyrulnikov, P.G. Nanocomposites on the basis of layered silicates as the catalysts for the dehydrogenation of methanol. *Nanotechnol. Russ.* **2014**, *9*, 693–699. [CrossRef]
46. Din, I.U.; Shaharun, M.S.; Naeem, A.; Tasleem, S.; Rafie Johan, M. Carbon nanofibers based copper/zirconia catalysts for carbon dioxide hydrogenation to methanol: Effect of copper concentration. *Chem. Eng. J.* **2018**, *334*, 619–629. [CrossRef]
47. Sun, C.; Zhu, J.; Lv, Y.; Qi, L.; Liu, B.; Gao, F.; Sun, K.; Dong, L.; Chen, Y. Dispersion, reduction and catalytic performance of CuO supported on $ZrO_2$-doped $TiO_2$ for NO removal by CO. *Appl. Catal. B-Environ.* **2011**, *103*, 206–220. [CrossRef]
48. Pepe, F. Catalytic behavior and surface chemistry of copper/alumina catalysts for isopropanol decomposition. *J. Catal.* **1985**, *91*, 69–77. [CrossRef]
49. Friedman, R. Characterization of $Cu/Al_2O_3$ catalysts. *J. Catal.* **1978**, *55*, 10–28. [CrossRef]
50. Lai, N.-C.; Tsai, M.-C.; Liu, C.-H.; Chen, C.-S.; Yang, C.-M. Efficient selective oxidation of propylene by dioxygen on mesoporous-silica-nanoparticle-supported nanosized copper. *J. Catal.* **2018**, *365*, 411–419. [CrossRef]
51. Takahashi, N.; Suda, A.; Hachisuka, I.; Sugiura, M.; Sobukawa, H.; Shinjoh, H. Sulfur durability of NOx storage and reduction catalyst with supports of $TiO_2$, $ZrO_2$ and $ZrO_2$-$TiO_2$ mixed oxides. *Appl. Catal. B-Environ.* **2007**, *72*, 187–195. [CrossRef]
52. Ito, K.; Kakino, S.; Ikeue, K.; Machida, M. NO adsorption/desorption property of $TiO_2$–$ZrO_2$ having tolerance to $SO_2$ poisoning. *Appl. Catal. B-Environ.* **2007**, *74*, 137–143. [CrossRef]
53. Venezia, A.M.; Di Carlo, G.; Pantaleo, G.; Liotta, L.F.; Melaet, G.; Kruse, N. Oxidation of $CH_4$ over Pd supported on $TiO_2$-doped $SiO_2$: Effect of Ti(IV) loading and influence of $SO_2$. *Appl. Catal. B-Environ.* **2009**, *88*, 430–437. [CrossRef]
54. Deerattrakul, V.; Dittanet, P.; Sawangphruk, M.; Kongkachuichay, P. $CO_2$ hydrogenation to methanol using Cu-Zn catalyst supported on reduced graphene oxide nanosheets. *J. CO2 Util.* **2016**, *16*, 104–113. [CrossRef]
55. Sun, Y.; Chen, L.; Bao, Y.; Wang, G.; Zhang, Y.; Fu, M.; Wu, J.; Ye, D. Roles of nitrogen species on nitrogen-doped CNTs supported Cu-$ZrO_2$ system for carbon dioxide hydrogenation to methanol. *Catal. Today* **2018**, *307*, 212–223. [CrossRef]
56. Kang, I.; Heung, Y.Y.; Kim, J.H.; Lee, J.W.; Gollapudi, R.; Subramaniam, S.; Narasimhadevara, S.; Hurd, D.; Kirikera, G.R.; Shanov, V.; et al. Introduction to carbon nanotube and nanofiber smart materials. *Compos. Part B-Eng.* **2006**, *37*, 382–394. [CrossRef]
57. Karnaukhov, T.M.; Vedyagin, A.A.; Cherepanova, S.V.; Rogov, V.A.; Stoyanovskii, V.O.; Mishakov, I.V. Study on reduction behavior of two-component Fe Mg O oxide system prepared via a sol-gel technique. *Int. J. Hydrogen Energy* **2017**, *42*, 30543–30549. [CrossRef]
58. Vedyagin, A.A.; Karnaukhov, T.M.; Cherepanova, S.V.; Stoyanovskii, V.O.; Rogov, V.A.; Mishakov, I.V. Synthesis of binary Co–Mg–O oxide system and study of its behavior in reduction/oxidation cycling. *Int. J. Hydrogen Energy* **2019**, *44*, 20690–20699. [CrossRef]
59. Karnaukhov, T.M.; Vedyagin, A.A.; Cherepanova, S.V.; Rogov, V.A.; Mishakov, I.V. Sol–gel synthesis and characterization of the binary Ni–Mg–O oxide system. *J. Sol-Gel Sci. Technol.* **2019**, *92*, 208–214. [CrossRef]
60. Veselov, G.B.; Karnaukhov, T.M.; Bauman, Y.I.; Mishakov, I.V.; Vedyagin, A.A. Sol-Gel-Prepared Ni-Mo-Mg-O System for Catalytic Transformation of Chlorinated Organic Wastes into Nanostructured Carbon. *Materials* **2020**, *13*, 4404. [CrossRef]
61. Vedyagin, A.A.; Mishakov, I.V.; Karnaukhov, T.M.; Krivoshapkina, E.F.; Ilyina, E.V.; Maksimova, T.A.; Cherepanova, S.V.; Krivoshapkin, P.V. Sol–gel synthesis and characterization of two-component systems based on MgO. *J. Sol-Gel Sci. Technol.* **2017**, *82*, 611–619. [CrossRef]
62. Nakonieczny, D.S.; Antonowicz, M.; Paszenda, Z.K.; Radko, T.; Drewniak, S.; Bogacz, W.; Krawczyk, C. Experimental investigation of particle size distribution and morphology of alumina-yttria-ceria-zirconia powders obtained via sol–gel route. *Biocyber. Biomed. Eng.* **2018**, *38*, 535–543. [CrossRef]

63. Mashayekh-Salehi, A.; Moussavi, G.; Yaghmaeian, K. Preparation, characterization and catalytic activity of a novel mesoporous nanocrystalline MgO nanoparticle for ozonation of acetaminophen as an emerging water contaminant. *Chem. Eng. J.* **2017**, *310*, 157–169. [CrossRef]
64. Ouraipryvan, P.; Sreethawong, T.; Chavadej, S. Synthesis of crystalline MgO nanoparticle with mesoporous-assembled structure via a surfactant-modified sol–gel process. *Mater. Lett.* **2009**, *63*, 1862–1865. [CrossRef]
65. Possato, L.G.; Gonçalves, R.G.L.; Santos, R.M.M.; Chaves, T.F.; Briois, V.; Pulcinelli, S.H.; Martins, L.; Santilli, C.V. Sol-gel synthesis of nanocrystalline MgO and its application as support in Ni/MgO catalysts for ethanol steam reforming. *Appl. Surf. Sci.* **2021**, *542*, 148744. [CrossRef] [PubMed]
66. Alhaji, A.; Razavi, R.S.; Ghasemi, A.; Loghman-Estarki, M.R. Modification of Pechini sol–gel process for the synthesis of MgO-$Y_2O_3$ composite nanopowder using sucrose-mediated technique. *Ceram. Int.* **2017**, *43*, 2541–2548. [CrossRef]
67. Barad, C.; Kimmel, G.; Shamir, D.; Hirshberg, K.; Gelbstein, Y. Lattice variations in nanocrystalline $Y_2O_3$ confined in magnesia (MgO) matrix. *J. Alloys Compd.* **2019**, *801*, 375–380. [CrossRef]
68. Bayal, N.; Jeevanandam, P. Synthesis of $TiO_2$–MgO mixed metal oxide nanoparticles via a sol–gel method and studies on their optical properties. *Ceram. Int.* **2014**, *40*, 15463–15477. [CrossRef]
69. Todan, L.; Dascalescu, T.; Preda, S.; Andronescu, C.; Munteanu, C.; Culita, D.C.; Rusu, A.; State, R.; Zaharescu, M. Porous nanosized oxide powders in the MgO-$TiO_2$ binary system obtained by sol-gel method. *Ceram. Int.* **2014**, *40*, 15693–15701. [CrossRef]
70. Ahmed, K.; Rabah, M.; Khaled, M.; Mohamed, B.; Mokhtar, M. Optical and structural properties of Mn doped MgO powders synthesized by Sol-gel process. *Optik* **2016**, *127*, 8253–8258. [CrossRef]
71. Xu, S.; Li, J.; Kou, H.; Shi, Y.; Pan, Y.; Guo, J. Spark plasma sintering of $Y_2O_3$–MgO composite nanopowder synthesized by the esterification sol–gel route. *Ceram. Int.* **2015**, *41*, 3312–3317. [CrossRef]
72. Karnaukhov, T.; Vedyagin, A.; Mishakov, I.; Bedilo, A.; Volodin, A. Synthesis and Characterization of Nanocrystalline M-Mg-O and Carbon-Coated MgO Systems. *Mater. Sci. Forum* **2018**, *917*, 157–161. [CrossRef]
73. Tsubota, M.; Kitagawa, J. A necessary criterion for obtaining accurate lattice parameters by Rietveld method. *Sci. Rep.* **2017**, *7*, 15381. [CrossRef] [PubMed]
74. Małecka, B.; Łącz, A.; Drożdż, E.; Małecki, A. Thermal decomposition of d-metal nitrates supported on alumina. *J. Therm. Anal. Calorim.* **2014**, *119*, 1053–1061. [CrossRef]
75. Dhaouadi, H.; Chaabane, H.; Touati, F. $Mg(OH)_2$ Nanorods Synthesized by A Facile Hydrothermal Method in the Presence of CTAB. *Nano-Micro Lett.* **2011**, *3*, 153–159. [CrossRef]
76. Chen, Y.; Zhou, T.; Fang, H.; Li, S.; Yao, Y.; He, Y. A Novel Preparation of Nano-sized Hexagonal $Mg(OH)_2$. *Proc. Eng.* **2015**, *102*, 388–394. [CrossRef]
77. Ilyina, E.V.; Mishakov, I.V.; Vedyagin, A.A.; Cherepanova, S.V.; Nadeev, A.N.; Bedilo, A.F.; Klabunde, K.J. Synthesis and characterization of mesoporous $VO_x$/MgO aerogels with high surface area. *Microporous Mesoporous Mater.* **2012**, *160*, 32–40. [CrossRef]

MDPI
St. Alban-Anlage 66
4052 Basel
Switzerland
www.mdpi.com

*Materials* Editorial Office
E-mail: materials@mdpi.com
www.mdpi.com/journal/materials

Disclaimer/Publisher's Note: The statements, opinions and data contained in all publications are solely those of the individual author(s) and contributor(s) and not of MDPI and/or the editor(s). MDPI and/or the editor(s) disclaim responsibility for any injury to people or property resulting from any ideas, methods, instructions or products referred to in the content.

www.ingramcontent.com/pod-product-compliance
Lightning Source LLC
LaVergne TN
LVHW070635100526
838202LV00012B/815